GEOMETRY OF CONVEX SETS

GEOMETRY OF CONVEX SETS

I. E. LEONARD
Department of Mathematical and Statistical Sciences
University of Alberta, Edmonton, AB, Canada

J. E. LEWIS
Department of Mathematical and Statistical Sciences
University of Alberta, Edmonton, AB, Canada

Published by John Wiley & Sons, Inc., Hoboken, New Jersey
Published simultaneously in Canada

For general information on our other products and services or for technical support, please contact our Customer Care Department within the United States at (800) 762-2974, outside the United States at (317) 572-3993 or fax (317) 572-4002.

Wiley also publishes its books in a variety of electronic formats. Some content that appears in print may not be available in electronic formats. For more information about Wiley products, visit our web site at www.wiley.com.

Library of Congress Cataloging-in-Publication Data:

Leonard, I. Ed., 1938-
  Geometry of convex sets / I.E. Leonard, J.E. Lewis.
    pages cm
  Includes bibliographical references and index.
  ISBN 978-1-119-02266-4 (cloth)
  1. Convex sets. 2. Geometry. I. Lewis, J. E. (James Edward) II. Title.
  QA640.L46 2016
  516′.08–dc23
                          2015021566

Cover image courtesy of I. E. Leonard

Set in 10/12pt TeXGyreTermes by SPi Global, Chennai, India

Printed in the United States of America

10 9 8 7 6 5 4 3 2 1

1   2016

Contents

Preface

This book is designed for a one-semester course studying the geometry of convex sets in n-dimensional space. It is meant for students in education, arts, science, and engineering and is usually given in the third or fourth year of university. Prerequisites are courses in elementary geometry, linear algebra, and some familiarity with coordinate geometry.

The geometry of convex sets in n-dimensional space usually starts with the basic definitions of the linear concepts of addition and scalar multiplication of vectors and then defines the notion of convexity for subsets of the n-dimensional space. Many properties of convex sets can be discovered using just the linear structure. However, in order to prove more interesting results, we need the notion of a metric space, that is, the notion of distance, so that we can talk about open and closed sets, bounded sets, compact sets, etc. It is the interplay between the linear and topological concepts that makes the notion of convexity so interesting. Convexity can be defined using only the linear structure, but it is the marriage of the linear structure and the topological structure that allows us to see the really beautiful results.

In the first chapter, we introduce the linear or vector space notions of addition and scalar multiplication, linear subspaces, linear functionals, and hyperplanes. We also get a start on the convexity and topology by defining different distances in n-space and studying the geometric properties of subsets, subspaces, and hyperplanes.

The second chapter discusses the notion of topology in the setting of metrics derived from a norm on the n-dimensional space. First, we define the basic topological notions of open sets and interior points, closed sets and accumulation points, boundary points, the interior of a set, the closure of a set, and the boundary of a set. Next we discuss the basic properties of compact sets and give some geometric examples. We also give examples to show that in the setting of an n-dimensional linear space which is a metric space, all these notions can be given in terms of convergence of sequences and subsequences.

In the third chapter, we introduce the notion of convexity and discuss the basic properties of convex sets. Here we define the convex hull of a set as well as the interior and closure of convex sets. We prove Carathéodory's theorem and show that the convex hull of a compact set is compact. We discuss flats or affine subspaces and their properties and go on to discuss the various separation theorems for convex sets. Finally, we discuss the notions of extreme points and exposed points of convex sets, and polyhedral sets and polytopes, and prove the existence of extreme points for compact convex sets. We prove the Krein-Milman theorem and Birkhoff's theorem on the extreme points of the set of $n \times n$ doubly stochastic matrices.

In the fourth and final chapter, we prove Helly's theorem and give some applications involving transversals of families of pairwise disjoint compact convex subsets of the plane. We include a proof of Vicensini's theorem and Klee's reduction of the Helly number and give the results of Santalo and Hadwiger on finite families. We also discuss covering problems, piercing or stabbing problems, and construction of sets of constant width in the plane. Finally, we give a proof of Borsuk's problem in the plane using Pál's theorem and also Melzak's proof of Borsuk's problem for smooth sets of constant width in $\mathbb{R}^n$.

The book is based on classroom notes for a course on the geometry of convex sets, a course in a sequence of four undergraduate geometry courses given at the University of Alberta for the past 30 years. The choice of topics is based on years of experience in teaching these courses and covers typical material covered in most introductory geometry courses on convex sets.

Both the authors were influenced by the texts by Hadwiger, Debrunner, and Klee [55], Eggleston [36], and Valentine [117] and later by the graduate texts of Holmes [61] and Wilansky [120]. We do not cover all of the material in this book in a one-semester undergraduate course; however, the topics that we have selected have all been included over the years.

ED AND TED

Edmonton, Alberta, Canada
August, 2015

1 INTRODUCTION TO *N*-DIMENSIONAL GEOMETRY

1.1 FIGURES IN *N*-DIMENSIONS

There is much to be said for a hands-on approach in geometry. We enhance our understanding in two dimensions by drawing plane figures or by experimenting with polygons cut out of paper. In three dimensions, we can construct polyhedrons out of cardboard, wire, or plastic straws. In some cases, we may use a computer program to provide a visual representation. Although everyone knows that a picture does not constitute a "proof," there is no doubt that a decent diagram can be utterly convincing.

However, we cannot make a full-dimensional model of an object that has more than three dimensions. We cannot visualize a cube of four or five dimensions, at least not in the usual sense. Nevertheless, if you can imagine that "hypercubes" of four or five dimensions exist, you might suspect that they are a lot like three-dimensional cubes. You would be correct. Geometry in four or higher dimensions is quite similar to geometry in two and three dimensions, and although we cannot visualize space of any more than three dimensions, we *can* build up a fairly reliable intuition about what such a space is like.

In this section, we examine some n-dimensional objects and learn how to work with them. We begin with the higher dimensional analogs of points, lines, planes, and spheres. We find out that these can be described in a *dimension-free* way and that their geometric properties are as they should be.

We also examine how geometric notions such as perpendicular and parallel lines extend to higher dimensions. The emphasis will be on how n-dimensional space is like two- or three-dimensional space, but we also examine some of the ways in which they differ.

Geometry of Convex Sets, First Edition. I. E. Leonard and J. E. Lewis.
© 2016 John Wiley & Sons, Inc. Published 2016 by John Wiley & Sons, Inc.

Geometry at this level necessarily involves proofs. Even if you have already had a geometry course, you will find that some of the techniques are new. In this first section, we spend some time explaining how a proof proceeds before actually carrying it out. We apologize to anyone who is already familiar with the methods.

1.2 POINTS, VECTORS, AND PARALLEL LINES

1.2.1 Points and Vectors

We denote the usual two-dimensional coordinate space by $\mathbb{R}^2$, three-dimensional coordinate space by $\mathbb{R}^3$, four-dimensional coordinate space by $\mathbb{R}^4$, and so on. In general, $\mathbb{R}^n$ denotes n-dimensional space (the n is not a variable—it stands for some fixed nonnegative integer) and is the set of all n-tuples of real numbers, that is,

$$\mathbb{R}^n = \left\{\, (x_1, x_2, \ldots, x_n) : x_i \in \mathbb{R} \text{ for } i = 1, 2, \ldots, n \,\right\}.$$

Points in $\mathbb{R}^n$ will be denoted by italic letters such as x, y, and z. Numbers are usually (but not always) denoted with lowercase Greek letters, α, β, γ, and so on.*

A *point* in $\mathbb{R}^n$ is just an n-tuple and can also be described by giving its coordinates. In $\mathbb{R}^4$, for example, the ordered 4-tuple $(\alpha_1, \alpha_2, \alpha_3, \alpha_4)$ denotes the point whose ith coordinate is α_i. The *origin* $(0, 0, 0, 0)$ is denoted by $\overline{0}$ (the bar is used to avoid confusion with the real number 0). The symbol $\overline{0}$ is also used to denote the origin in any space $\mathbb{R}^n$.

Points from the same space can be added together, subtracted from each other, and multiplied by scalars (that is, real numbers), and these operations are performed coordinatewise.

---

* The Greek alphabet is as follows:

A	α	Alpha	I	ι	Iota	P	ρ, ϱ	Rho
B	β	Beta	K	κ	Kappa	Σ	σ, ς	Sigma
Γ	γ	Gamma	Λ	λ	Lambda	T	τ	Tau
Δ	δ	Delta	M	μ	Mu	Υ	υ	Upsilon
E	ϵ, ε	Epsilon	N	ν	Nu	Φ	ϕ, φ	Phi
Z	ζ	Zeta	Ξ	ξ	Xi	X	χ	Chi
H	η	Eta	O	o	Omicron	Ψ	ψ	Psi
Θ	θ	Theta	Π	π, ϖ	Pi	Ω	ω	Omega

The alternate pi (ϖ), sigma (ς), and upsilon (υ) are very seldom used.

If $x = (\alpha_1, \alpha_2, \ldots, \alpha_n)$ and $y = (\beta_1, \beta_2, \ldots, \beta_n)$ are two points in $\mathbb{R}^n$ and if ρ is any number, then we define

- $x = y$ if and only if $\alpha_i = \beta_i$ for $i = 1, 2, \ldots, n$ (*equality*)
- $x + y = (\alpha_1 + \beta_1, \alpha_2 + \beta_2, \ldots, \alpha_n + \beta_n)$ (*addition*)
- $\rho x = (\rho\alpha_1, \rho\alpha_2, \ldots, \rho\alpha_n)$ (*scalar multiplication*).

With these definitions of addition and scalar multiplication, $\mathbb{R}^n$ becomes an n-dimensional vector space.*

In other words, **points** behave algebraically as though they were **vectors**. As a consequence, a notation such as $(1, 2)$ can be interpreted as a vector as well as a point. The two interpretations may be tied together geometrically by thinking of the vector as an arrow whose tail is at the origin and whose tip is at the point $(1, 2)$, as in the figure below.

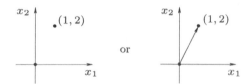

A more abstract notion of a vector is as a class of arrows (an equivalence class, to be more exact), each arrow in the class having the same length and pointing in the same direction. Any particular arrow from the class of a given vector is called a **representative** of the vector. The arrow, in the example above, whose tail is at the point $(0, 0)$ and whose tip is at the point $(1, 2)$, is but one of infinitely many representatives of the vector, as in the figure below.

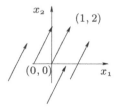

Note. Representatives of a given vector are called **free vectors**, since apart from their length and direction, there is no restriction as to their position.

Although a given n-tuple $(\alpha_1, \alpha_2, \ldots, \alpha_n)$ has infinitely many free vectors associated with it, the n-tuple always represents *one and only one point* in $\mathbb{R}^n$. When we refer to *the* point associated with a given vector, we mean that point.

---

* Unlike the situation in synthetic geometry, the notions of point and line are no longer primitive or undefined terms but are defined in terms of n-tuples (coordinatized space).

1.2.2 Lines

If we were to multiply the point $v = (1, 2)$ by the numbers

$$-1, \qquad \tfrac{1}{2}, \qquad \text{and} \qquad 3,$$

we would get the points

$$(-1, -2), \qquad (\tfrac{1}{2}, 1), \qquad \text{and} \qquad (3, 6),$$

respectively. The three points all lie on the straight line through $\bar{0}$ and v, as in the figure below.

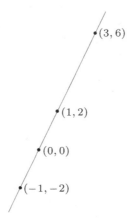

If μ is *any* number, then the point μv also lies on the straight line through $\bar{0}$ and v. This is not just a property of $\mathbb{R}^2$: if v is any point other than $\bar{0}$ in $\mathbb{R}^1$, $\mathbb{R}^2$, or $\mathbb{R}^3$, then the point μv also lies on the straight line through $\bar{0}$ and v.

In fact, we can describe the line L through $\bar{0}$ and v algebraically as the set of all multiples of the point v:

$$L = \{\, \mu v \: : \: -\infty < \mu < \infty \,\}.$$

The line L passes through the origin.

To obtain a line M parallel to L, but passing through a given point p, we simply add p to every point of L:

$$M = \{\, p + \mu v \: : \: -\infty < \mu < \infty \,\},$$

since p is on M (take $\mu = 0$) and the vector v is parallel to L.

The above equations for L and M are also used to describe lines in spaces of dimension $n > 3$. These equations, in fact, define what is meant by straight lines in higher dimensions.

Often a line is described by a ***vector equation***. For example, a point x is on M if and only if

$$x = p + \mu v,$$

for some real number μ. In this vector equation, p and v are fixed, and x is a variable point; thus, the **line passing through the point p in the direction of v** can be written as

$$M = \{\, x \in \mathbb{R}^n \,:\, x = p + \mu v,\ -\infty < \mu < \infty \,\}.$$

To give a concrete example, suppose that $p = (1, -2, 4, 1)$ and $v = (7, 8, -6, 5)$ are points in $\mathbb{R}^4$. Denoting x by (x_1, x_2, x_3, x_4), we can describe the vector equation for the line through p parallel to the vector v as the set of all points $x \in \mathbb{R}^4$ such that

$$(x_1, x_2, x_3, x_4) = (1, -2, 4, 1) + \mu(7, 8, -6, 5),\ -\infty < \mu < \infty,$$

which is the same as

$$(x_1, x_2, x_3, x_4) = (1 + 7\mu, -2 + 8\mu, 4 - 6\mu, 1 + 5\mu),\ -\infty < \mu < \infty.$$

This is sometimes written in ***parametric form***:

$$\begin{cases} x_1 = & 1 + 7\mu \\ x_2 = & -2 + 8\mu \\ x_3 = & 4 - 6\mu \\ x_4 = & 1 + 5\mu, \end{cases}$$

where $-\infty < \mu < \infty$.

Perhaps some words of caution are worthwhile at this point. Giving a name to something does not endow it with any special properties. Simply because we have called an object in $\mathbb{R}^n$ by the same name as something from $\mathbb{R}^2$ or $\mathbb{R}^3$, it does not follow that the n-dimensional object has the same properties as its namesake in $\mathbb{R}^2$ or $\mathbb{R}^3$. We cannot assume that what we are calling straight lines in $\mathbb{R}^n$ automatically have the same properties as do straight lines in two or three dimensions. This has to be proved, which is what the next few theorems do.

Theorem 1.2.1. *In the definition of the line*

$$M = \{\, x \in \mathbb{R}^n \,:\, x = p + \mu v,\ -\infty < \mu < \infty \,\},$$

the point p may be replaced by any point on M.

Proof. It is often worthwhile to rephrase a statement to make sure that we understand what it means. This theorem is saying that if p' is any point on the line M and if we replace p with p' in the equation of M, then we will get exactly the same line.

In a situation similar to this, it helps to give different names to the potentially different lines. We will use M' to denote the line that we get when we replace p by p':

$$M' = \{\, x \in \mathbb{R}^n \,:\, x = p' + \mu v, \ -\infty < \mu < \infty \,\}.$$

Our task is to show that M and M' must be the same line.

How do we show that two lines are the same? Well, lines are just special types of sets, and we will use the standard strategy for showing that two sets are equal—namely, we will show that every point in the first set also belongs to the second set and vice versa.

(i) Note that a typical point x on M' can be written as

$$x = p' + \mu v$$

for some number μ. We claim that x is also on M. To show that this is the case, we have to show that there is some number λ such that x can be written as $p + \lambda v$, which is a point on M.
Now, since p' is on M, there must be some number β such that

$$p' = p + \beta v,$$

and therefore,

$$
\begin{aligned}
x &= p' + \mu v \\
&= (p + \beta v) + \mu v \\
&= p + (\beta + \mu)v.
\end{aligned}
$$

Setting $\lambda = \beta + \mu$ shows that x is on M, and since x was an arbitrary point of M', then $M' \subseteq M$.

(ii) Conversely, we will show that every point on M is also on M'. If we let y be a typical point on M, then

$$y = p + \mu v$$

for some number μ. Using the fact that p' is on M, we have

$$p' = p + \beta v$$

for some number β, so that $p = p' - \beta v$. Therefore,

$$
\begin{aligned}
y &= p + \mu v \\
&= (p' - \beta v) + \mu v \\
&= p' + (-\beta + \mu)v,
\end{aligned}
$$

which shows that y is on M', and since y was an arbitrary point of M, then $M \subseteq M'$.

Thus, $M = M'$, which completes the proof.

$\square$

The proof of the next theorem is somewhat similar and is left as an exercise.

Theorem 1.2.2. *In the definition of the line*

$$M = \{ x \in \mathbb{R}^n : x = p + \mu v, \ -\infty < \mu < \infty \},$$

the vector v may be replaced by any nonzero multiple of v.

Two nonzero vectors are said to be **parallel** if one is a multiple of the other. With this terminology, the previous theorems can be combined as follows.

Corollary 1.2.3. *In the definition of the line*

$$M = \{ x \in \mathbb{R}^n : x = p + \mu v, \ -\infty < \mu < \infty \},$$

the point p may be replaced by any point on the line and the vector v may be replaced by any vector parallel to v.

In the next theorem, we would like to show that if one line is a subset of another line, then the two lines must be the same. Before proving this, we should convince ourselves that we actually have something to prove, since it looks like this might be just another way of stating that a point and vector determine a unique straight line. Perhaps the two statements

(i) "A point and a vector determine a unique straight line"
(ii) "If one line is contained in another, then the two lines must coincide"

are logically equivalent in the sense that one follows from the other without really invoking any geometry.

To see that this is not the case, try replacing the word "line" by the words "solid ball." It is true that a point and a vector determine a unique solid ball, namely, the ball with the point as its center and with a radius equal to the length of the vector. It is clearly *not true* that two solid balls must coincide if one is a subset of the other.

Having talked ourselves into believing that there is something to prove, the next problem is to find a way to carry it out. This is how we will do it:

We will first suppose that M_1 passes through the point p_1 and is parallel to the nonzero vector v_1 and that M_2 passes through the point p_2 and is parallel to the nonzero vector v_2. The assumption is that every point on M_1 is contained in M_2. We will show that this means that v_1 and v_2 are parallel, which is what our intuition suggests should be the case. We will then use the previous theorems to finish the proof.

Theorem 1.2.4. *If M_1 and M_2 are straight lines in $\mathbb{R}^n$, and if $M_1 \subset M_2$, then $M_1 = M_2$.*

Proof. We may suppose that M_1 and M_2 are the lines

$$M_1 = \{\, x \in \mathbb{R}^n \,:\, x = p_1 + \mu v_1, \ -\infty < \mu < \infty \,\}$$
$$M_2 = \{\, x \in \mathbb{R}^n \,:\, x = p_2 + \mu v_2, \ -\infty < \mu < \infty \,\}.$$

Since M_1 is a subset of M_2, then $p_1 \in M_1 \subset M_2$, and we can replace p_2 in M_2 with the point p_1. We can then write M_2 as

$$M_2 = \{\, x \in \mathbb{R}^n \,:\, x = p_1 + \mu v_2, \ -\infty < \mu < \infty \,\}.$$

Again, using the fact that $M_1 \subset M_2$, the point $p_1 + v_1$ of M_1 must also belong to M_2, and from the previous equation, there must be a number μ_0 such that

$$p_1 + v_1 = p_1 + \mu_0 v_2,$$

so that $v_1 = \mu_0 v_2$. But this means that in the last equation for M_2, we can replace the vector v_2 with the parallel vector v_1, that is,

$$M_2 = \{\, x \in \mathbb{R}^n \,:\, x = p_1 + \mu v_1, \ -\infty < \mu < \infty \,\},$$

and the right-hand side of this equation is precisely M_1.

$\square$

Theorem 1.2.5. *Let M be the straight line given by*

$$M = \{\, x \in \mathbb{R}^n \,:\, x = p + \mu v, \ -\infty < \mu < \infty \,\}.$$

If x_1 and x_2 are distinct points on M, then the vector $x_1 - x_2$ is parallel to v.

Proof. To prove this, one only needs to write down what x_1 and x_2 are in terms of p and v and then perform the subtraction. Since each of the two points is on M, there must be numbers μ_1 and μ_2 such that

$$x_1 = p + \mu_1 v \qquad \text{and} \qquad x_2 = p + \mu_2 v,$$

and, therefore,

$$x_1 - x_2 = (p + \mu_1 v) - (p + \mu_2 v)$$
$$= (\mu_1 - \mu_2)v.$$

Since x_1 and x_2 are distinct, $(\mu_1 - \mu_2)v \neq \overline{0}$, and if $\mu_1 = \mu_2$, then

$$(\mu_1 - \mu_2)v = 0 \cdot v = \overline{0}.$$

Therefore, $\mu_1 \neq \mu_2$, so that $x_1 - x_2$ is a nonzero multiple of v, that is, $x_1 - x_2$ is parallel to v.

$\square$

Another way to describe a line is to specify points on it.

Theorem 1.2.6. *A straight line in $\mathbb{R}^n$ is completely determined by any two distinct points on the line.*

Proof. This is proved using Theorem 1.2.5 and Corollary 1.2.3.

If p and q are distinct points on M, then M must be parallel to the vector $q - p$ by Theorem 1.2.5. Since p is on M, Corollary 1.2.3 now implies that M must be the line

$$\{\, x \in \mathbb{R}^n \, : \, x = p + \mu(q - p), \ -\infty < \mu < \infty \,\}.$$

In other words, p and q completely determine M.

$\square$

In the proof of the previous theorem, we showed that if p and q are two distinct points in $\mathbb{R}^n$, then the line determined by p and q will have the equation

$$x = p + \mu(q - p), \ -\infty < \mu < \infty.$$

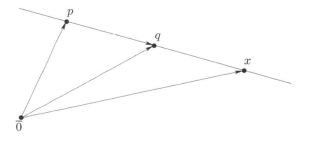

This equation may be rearranged to get the more usual form

$$x = (1 - \mu)p + \mu q, \quad -\infty < \mu < \infty.$$

Note that if $\mu = 0$, then $x = p$, while if $\mu = 1$, then $x = q$. Since p and q are on the line, then the vector $q - p$ must be parallel to the line.

Other equations that also describe the same line are

$$x = \eta p + \mu q, \quad \eta + \mu = 1$$

and

$$x = \mu p + (1 - \mu)q, \quad -\infty < \mu < \infty.$$

Example 1.2.7. *If $p = (-1, -1)$ and $q = (1, 1)$, find the equation of the line passing through p and q.*

Solution. The equation of the line M passing through p and q can be written as

$$\begin{aligned}
(x_1, x_2) &= (1 - \mu)(-1, -1) + \mu(1, 1) \\
&= (-1 + 2\mu, -1 + 2\mu)
\end{aligned}$$

or parametrically as

$$\begin{aligned}
x_1 &= -1 + \lambda \\
x_2 &= -1 + \lambda
\end{aligned}$$

for $-\infty < \lambda < \infty$.

$\square$

As in $\mathbb{R}^2$ or $\mathbb{R}^3$, two distinct lines in $\mathbb{R}^n$ either do not intersect or they meet in exactly one point.

Theorem 1.2.8. *Two distinct lines in $\mathbb{R}^n$ meet in at most one point.*

Proof. Suppose that

$$M_1 = \{p + \mu v : -\infty < \mu < \infty\}$$

and

$$M_2 = \{q + \mu w : -\infty < \mu < \infty\}$$

are distinct lines in $\mathbb{R}^n$ and $M_1 \cap M_2 \neq \emptyset$.

We may assume that the vectors v and w are not parallel. If they are parallel and $M_1 \cap M_2 \neq \emptyset$, then by Theorem 1.2.2, M_1 and M_2 coincide, which contradicts the fact that they are distinct.

Suppose that x_0 and x_1 are in $M_1 \cap M_2$, then

$$x_0 - p = \lambda_0 v$$
$$x_0 - q = \mu_0 w$$

for some real numbers λ_0 and μ_0. Also,

$$x_1 - p = \lambda_1 v$$
$$x_1 - q = \mu_1 w$$

for some numbers λ_1 and μ_1.

Therefore,

$$p - q = -\lambda_0 v + \mu_0 w = \mu_1 w - \lambda_1 v,$$

so that

$$(\mu_1 - \mu_0)w = (\lambda_1 - \lambda_0)v.$$

If $\mu_0 \neq \mu_1$, then

$$w = \left(\frac{\lambda_1 - \lambda_0}{\mu_1 - \mu_0} \right) v,$$

which is a contradiction. Thus, $\mu_0 = \mu_1$.

Similarly, if $\lambda_0 \neq \lambda_1$, then

$$v = \left(\frac{\mu_1 - \mu_0}{\lambda_1 - \lambda_0} \right) w,$$

which is a contradiction. Thus, $\lambda_0 = \lambda_1$.

Therefore,

$$x_1 = p + \lambda_1 v = p + \lambda_0 v = x_0.$$

$\square$

1.2.3 Segments

The part of a straight line between two distinct points p and q in $\mathbb{R}^n$ is called a *straight line segment*, or simply a *line segment*.

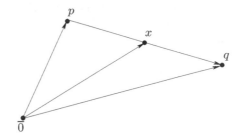

- $[p, q]$ denotes the ***closed line segment*** joining p and q

$$[p, q] = \{\, x \in \mathbb{R}^n \ : \ x = (1 - \mu)p + \mu q \ : \ 0 \le \mu \le 1 \,\}.$$

- (p, q) denotes the ***open line segment*** joining p and q

$$(p, q) = \{\, x \in \mathbb{R}^n \ : \ x = (1 - \mu)p + \mu q \ : \ 0 < \mu < 1 \,\}.$$

- $[p, q)$ and $(p, q]$ denote the ***half open line segments*** joining p and q

$$[p, q) = \{\, x \in \mathbb{R}^n \ : \ x = (1 - \mu)p + \mu q \ : \ 0 \le \mu < 1 \,\},$$
$$(p, q] = \{\, x \in \mathbb{R}^n \ : \ x = (1 - \mu)p + \mu q \ : \ 0 < \mu \le 1 \,\}.$$

Note that as the scalar increases from $\mu = 0$ to $\mu = 1$, the point x moves along the line segment from $x = p$ to $x = q$.

1.2.4 Examples

Example 1.2.9. *Show that the points*

$$a = (-2, -2), \qquad b = (-1, 1), \qquad c = (1, 7)$$

are collinear.

Solution. Note that

$$b - a = (-1, 1) - (-2, -2) = (1, 3)$$
$$c - a = (1, 7) - (-2, -2) = (3, 9) = 3(1, 3)$$

so that $c - a = 3(b - a)$ and $b - a$ and $c - a$ are parallel.

Therefore,

$$x = a + \lambda(b - a), \ -\infty < \lambda < \infty$$

and

$$x = a + \lambda(c - a), \quad -\infty < \lambda < \infty$$

are the same line.

For $\lambda = 0$, $x = a$ is on the line from the first equation, while for $\lambda = 1$, $x = b$ and $x = c$ are on the line from the first and second equation, respectively. Therefore, a, b, and c are collinear.

□

Example 1.2.10. *Let a and b be distinct points in $\mathbb{R}^n$ and let μ and ν be scalars such that $\mu + \nu = 1$. Show that the point $c = \mu a + \nu b$ is on the line through a and b.*

Solution. Note that

$$\begin{aligned} c - a &= \mu a + \nu b - a \\ &= \nu b - (1 - \mu)a \\ &= \nu b - \nu a \\ &= \nu(b - a). \end{aligned}$$

Therefore, $c - a$ is parallel to $b - a$, and the points a, b, and c are collinear.

□

Example 1.2.11. *Given distinct point a_1 and a_2 in $\mathbb{R}^n$, the midpoint of the segment $[a_1, a_2]$ is given by*

$$\tfrac{1}{2}a_1 + \tfrac{1}{2}a_2.$$

Solution. Since $\frac{1}{2} + \frac{1}{2} = 1$, then the point

$$\tfrac{1}{2}a_1 + \tfrac{1}{2}a_2$$

is on the line joining a_1 and a_2.

Also,

$$\begin{aligned} \tfrac{1}{2}a_1 + \tfrac{1}{2}a_2 - a_1 &= \tfrac{1}{2}a_1 - \tfrac{1}{2}a_2 \\ &= \tfrac{1}{2}(a_2 - a_1) \end{aligned}$$

so that

$$\tfrac{1}{2}a_1 + \tfrac{1}{2}a_2 = a_1 + \tfrac{1}{2}(a_2 - a_1).$$

Also,

$$a_2 - \left(\tfrac{1}{2}a_1 + \tfrac{1}{2}a_2\right) = \tfrac{1}{2}a_2 - \tfrac{1}{2}a_1$$
$$= \tfrac{1}{2}(a_2 - a_1).$$

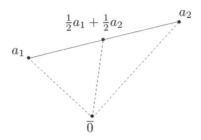

Therefore,

$$\tfrac{1}{2}a_1 + \tfrac{1}{2}a_2$$

is the midpoint of the segment $[a_1, a_2]$.

□

Exercise 1.2.12. *Given three noncollinear points a, b, and c in $\mathbb{R}^n$, show that the medians of the triangle with vertices a, b, and c intersect at a point G, the familiar centroid from synthetic geometry, and that*

$$G = \tfrac{1}{3}a + \tfrac{1}{3}b + \tfrac{1}{3}c.$$

In general, if $a_1, a_2, \ldots, a_k$ are k points in $\mathbb{R}^n$, where $n > 2$, we may define the point

$$\tfrac{1}{k}a_1 + \tfrac{1}{k}a_2 + \cdots + \tfrac{1}{k}a_k$$

to be the **centroid** of the set $\{a_1, a_2, \ldots, a_k\}$. It is then obvious from the previous examples that this definition agrees with the notion of the centroid of a segment or a triangle, that is, for $n = 1$ or $n = 2$.

Example 1.2.13. *Show that the point of intersection of the lines joining the midpoints of the opposite sides of a plane quadrilateral is the centroid of the vertices of the quadrilateral.*

Solution. Let a, b, c, and d be the vertices of the quadrilateral.

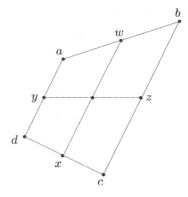

The centroid of the vertices is
$$\tfrac{1}{4}(a+b+c+d).$$

The midpoints of the sides, in opposite pairs, are

$$w = \tfrac{1}{2}a + \tfrac{1}{2}b, \qquad x = \tfrac{1}{2}c + \tfrac{1}{2}d$$
$$y = \tfrac{1}{2}a + \tfrac{1}{2}d, \qquad z = \tfrac{1}{2}b + \tfrac{1}{2}c.$$

The midpoint of the segment $[w, x]$ is

$$\tfrac{1}{2}w + \tfrac{1}{2}x = \tfrac{1}{4}(a+b+c+d),$$

and the midpoint of the segment $[y, z]$ is

$$\tfrac{1}{2}y + \tfrac{1}{2}z = \tfrac{1}{4}(a+b+c+d).$$

Thus, the segments $[w, x]$ and $[y, z]$ intersect at $\tfrac{1}{4}(a+b+c+d)$, the centroid of the vertices of the quadrilateral.

$\square$

Example 1.2.14. *Given a triangle with vertices $\overline{0}$, a, b, as shown in the figure below,*

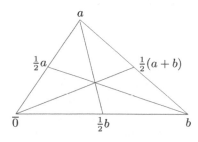

show that there exists a triangle whose sides are equal in length and parallel to the medians of the triangle.

Solution. The medians of the triangle are the following line segments:

$$\text{from } \overline{0} \text{ to } \tfrac{1}{2}(a+b), \qquad \text{from } b \text{ to } \tfrac{1}{2}a, \qquad \text{from } a \text{ to } \tfrac{1}{2}b$$

and the (free) vectors corresponding to these sides are

$$u = \tfrac{1}{2}(a+b), \qquad v = \tfrac{1}{2}a - b, \qquad w = \tfrac{1}{2}b - a$$

and we need only show that these vectors can be positioned to form a triangle.

Let one vertex of the triangle be $\overline{0}$, let a second vertex be the point u, and let the third vertex be the point $u + v$.

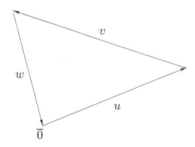

- Clearly, the edge from $\overline{0}$ to u is parallel to and equal in length to u, the first median.
- The second edge from u to $u + v$ is parallel to and equal in length to v, the second median.
- The third edge from $u + v$ to $\overline{0}$ is parallel to and equal in length to the vector $u + v$, but

$$u + v = \tfrac{1}{2}(a+b) + \tfrac{1}{2}a - b = a - \tfrac{1}{2}b = -w,$$

so that the third side of the triangle is parallel to and equal in length to w, the third median.

$\square$

Example 1.2.15. *Given the quadrilateral* $[a, b, c, d]$ *in the plane, the sides* $[a, b]$ *and* $[c, d]$, *when extended, meet at the point* p. *The sides* $[b, c]$ *and* $[a, d]$, *when extended, meet at the point* q. *On the rays from* p *through* b *and* c *are points* u *and* v *so that* $[p, u]$ *and* $[p, v]$ *are of the same lengths as* $[a, b]$ *and* $[c, d]$, *respectively. On the rays from* q *through* a *and* b *are points* x *and* y *so that* $[q, x]$ *and* $[q, y]$ *are of the same lengths as* $[a, d]$ *and* $[b, c]$, *respectively.*

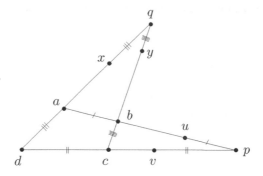

Show that $[u, v]$ is parallel to $[x, y]$.

Solution. In the figure, we have

$$u = p + (a - b)$$
$$v = p + (d - c)$$
$$x = q + (a - d)$$
$$y = q + (b - c),$$

so that

$$u - v = (a - b) - (d - c) = a + c - (b + d)$$
$$x - y = (a - d) - (b - c) = a + c - (b + d),$$

and $[u, v]$ is parallel to $[x, y]$.

$\square$

Example 1.2.16. *Let a, b, c, and d be the vertices of a tetrahedron in $\mathbb{R}^3$.*

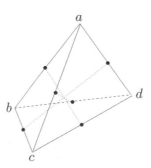

Show that the three lines through the midpoints of the opposite sides are concurrent.

Solution. Note that in each case, the centroid of the tetrahedron

$$\tfrac{1}{4}(a + b + c + d)$$

is the midpoint of the segment.

For example, consider the opposite edges $[a, b]$ and $[c, d]$, their midpoints are $\tfrac{1}{2}(a+b)$ and $\tfrac{1}{2}(c+d)$, and the midpoint of the segment $\left[\tfrac{1}{2}(a + b), \tfrac{1}{2}(c + d)\right]$ is

$$\tfrac{1}{2}\left(\tfrac{1}{2}(a + b) + \tfrac{1}{2}(c + d)\right) = \tfrac{1}{4}(a + b + c + d).$$

$\square$

1.2.5 Problems

A remark about the exercises is necessary. Certain questions are phrased as statements to avoid the incessant use of "prove that." See Problem 1, for example. Such statements are supposed to be proved. Other questions have a "true–false" or "yes–no" quality. The point of such questions is not to guess, but to justify your answer. Questions marked with $*$ are considered to be more challenging. Hints are given for some problems. Of course, a hint may contain statements that must be proved.

1. Let S be a nonempty set in $\mathbb{R}^n$. If every three points of S are collinear, then S is collinear.
2. In $\mathbb{R}^2$, there are two different types of equations that describe a straight line:

   (a) A vector equation: $(x_1, x_2) = (\alpha_1, \alpha_2) + \mu(\beta_1, \beta_2)$.
   (b) A linear equation: $\mu_1 x_1 + \mu_2 x_2 = \delta$.

   Given that the line L has the vector equation

   $$(x_1, x_2) = (4, 5) + \mu(-3, 2),$$

   find a linear equation for L.
3. Given that the line L has the linear equation

   $$\mu_1 x_1 + \mu_2 x_2 = \delta,$$

   show that the point

   $$\left(\frac{\mu_1 \delta}{\mu_1^2 + \mu_2^2}, \frac{\mu_2 \delta}{\mu_1^2 + \mu_2^2}\right)$$

   is on the line and that the vector $(-\mu_2, \mu_1)$ is parallel to the line.
   Hint. If p is on the line and if $p + v$ is also on the line, then v must be parallel to the line.

4. Prove Theorem 1.2.2. In the definition of the line

$$M = \{\, x \in \mathbb{R}^n \,:\, x = p + \mu v, \ -\infty < \mu < \infty \,\}$$

the vector v may be replaced by any nonzero multiple of v.

5. The centroid of three noncollinear points a, b, and c in $\mathbb{R}^n$ is defined to be

$$G = \tfrac{1}{3}(a + b + c).$$

Show that this definition of the centroid yields the synthetic definition of the centroid of the triangle with vertices a, b, and c, namely, the point at which the three medians of the triangle intersect. Prove also that the medians do indeed intersect at a common point.

1.3 DISTANCE IN *N*-SPACE

1.3.1 Metrics

In $\mathbb{R}^n$, several distance functions are used, the most common being the Euclidean distance.

If $x = (\alpha_1, \alpha_2, \ldots, \alpha_n)$ and $y = (\beta_1, \beta_2, \ldots, \beta_n)$, then the ***Euclidean distance*** between x and y is given by

$$d(x, y) - \sqrt{(\alpha_1 - \beta_1)^2 + (\alpha_2 - \beta_2)^2 + \cdots + (\alpha_n - \beta_n)^2}.$$

When $\mathbb{R}^n$ is equipped with the Euclidean distance, it is called ***Euclidean n-space***.

The word "metric" is synonymous with "distance function." A metric has the three basic properties that one expects a distance to have, namely, it is never negative, it is symmetric (the distance from Calgary to Edmonton is the same as the distance from Edmonton to Calgary), and the triangle inequality holds (the total length of two sides of a triangle is never smaller than the length of the third side).

Formally, a ***metric*** on a set X is any mapping $d(\,\cdot\,,\,\cdot\,)$ from $X \times X$ into $\mathbb{R}$ that has the following properties:

(M_1) $d(x, y) \geq 0$, with equality if and only if $x = y$ ***(nonnegativity)***
(M_2) $d(x, y) = d(y, x)$ ***(symmetry)***
(M_3) $d(x, z) \leq d(x, y) + d(y, z)$ ***(triangle inequality)***

for all $x, y, z \in X$. The pair (X, d) is called a ***metric space***.

Besides the ℓ_2 or *Euclidean metric* defined above, two other metrics that are occasionally used in $\mathbb{R}^n$ are:

- the ℓ_1 or "Manhattan" metric given by

$$d(x,y) = |\alpha_1 - \beta_1| + |\alpha_2 - \beta_2| + \cdots + |\alpha_n - \beta_n|,$$

- the ℓ_∞ or "sup" or "supremum" metric given by

$$d(x,y) = \max\{\,|\alpha_1 - \beta_1|, |\alpha_2 - \beta_2|, \ldots, |\alpha_n - \beta_n|\,\}.$$

The metrics that are used on the linear space $\mathbb{R}^n$ are almost always derived from a norm.

1.3.2 Norms

A *norm* on a linear space X is any mapping $\|\cdot\|$ from X to $\mathbb{R}$ that has the following properties:

(N_1) $\|x\| \geq 0$, with equality if and only if $x = 0$ *(nonnegativity)*
(N_2) $\|\lambda x\| = |\lambda| \cdot \|x\|$ *(positive homogeneity)*
(N_3) $\|x + y\| \leq \|x\| + \|y\|$ *(triangle inequality)*

for all $x, y \in X$ and $\lambda \in \mathbb{R}$. The pair $(X, \|\cdot\|)$ is called a *normed linear space*. The norm of a vector is always thought of as being the length of the vector.

If $x = (\alpha_1, \alpha_2, \ldots, \alpha_n) \in \mathbb{R}^n$, the norms of x corresponding to the three metrics that we defined above are:

- $\|x\|_2 = \sqrt{\alpha_1^2 + \alpha_2^2 + \cdots + \alpha_n^2}$ *Euclidean norm* or ℓ_2 *norm*

- $\|x\|_1 = |\alpha_1| + |\alpha_2| + \cdots + |\alpha_n|$ ℓ_1 *norm*

- $\|x\|_\infty = \max\{\,|\alpha_1|, |\alpha_2|, \ldots, |\alpha_n|\,\}$ ℓ_∞ *norm*.

Note. Every norm $\|\cdot\|$ on a linear space always has a corresponding metric (although the converse is *not* true). The metric is derived from the norm by defining

$$d(x,y) = \|x - y\|.$$

In addition to the three properties of nonnegativity, symmetry, and the triangle inequality, a metric that is derived from a norm satisfies the following:

(M_1) $d(x, y) \geq 0$, with equality if and only if $x = y$ *(nonnegativity)*
(M_2) $d(x, y) = d(y, x)$ *(symmetry)*
(M_3) $d(x, z) \leq d(x, y) + d(y, z)$ *(triangle inequality)*
(M_4) $d(x + v, y + v) = d(x, y)$ *(translation invariance)*
(M_5) $d(\lambda x, \lambda y) = |\lambda| \cdot d(x, y)$ *(positive homogeneity)*

for all $x, y, z, v \in \mathbb{R}^n$ and $\lambda \in \mathbb{R}$.

The property of translation invariance means that a "ruler" does not stretch or shrink when it is moved parallel to itself from one part of the space to another.*

Example 1.3.1. *If p and q are points in $\mathbb{R}^n$ with $\|p - q\| = \delta$, and if*

$$x = (1 - \lambda)p + \lambda q,$$

find the distance between x and p and the distance between x and q.

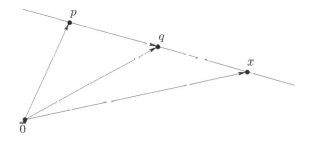

Solution. Note that we do not have to specify which distance function we are using, since the points p, q, and x are collinear.

The distance from x to p is

$$\|x - p\| = \|(1 - \lambda)p + \lambda q - p\| = \|p - \lambda p + \lambda q - p\|$$
$$= \|\lambda(q - p)\| = |\lambda| \cdot \|q - p\|$$
$$= |\lambda| \cdot \delta.$$

---

* However, you might wonder what happens to a ruler when it is rotated, especially when a non-Euclidean norm is being used.

The distance from x to q is

$$\begin{aligned}
\|x - q\| &= \|(1 - \lambda)p + \lambda q - q\| = \|p - \lambda p + \lambda q - q\| \\
&= \|(1 - \lambda)p + (-1 + \lambda)q\| = \|(1 - \lambda)p - (1 - \lambda)q\| \\
&= \|(1 - \lambda)(p - q)\| = |(1 - \lambda)| \cdot \| - p + q\| \\
&= |1 - \lambda| \cdot \|q - p\| = |1 - \lambda| \cdot \delta.
\end{aligned}$$

Note that if $0 \leq \lambda \leq 1$, then the point x is on the line segment between p and q; if $\lambda < 0$, then the point x is on the line joining p and q beyond p; while if $\lambda > 1$, the point x is on the line joining p and q beyond q.

□

Example 1.3.2. *Find all points on the line through p and q that are twice as far from p as they are from q.*

Solution. A point x on the line through p and q can be expressed as

$$x = (1 - \lambda)p + \lambda q$$

for some scalar λ.

From the previous example,

$$\|x - p\| = |\lambda| \cdot \|q - p\| \qquad \text{and} \qquad \|x - q\| = |1 - \lambda| \cdot \|q - p\|,$$

so we must have

$$|\lambda| \cdot \|q - p\| = 2|1 - \lambda| \cdot \|q - p\|.$$

It follows that

$$|\lambda| = 2|1 - \lambda|,$$

which implies that

$$\lambda^2 = (2(1 - \lambda))^2.$$

Expanding and rearranging, we have

$$3\lambda^2 - 8\lambda + 4 = 0$$

or

$$(3\lambda - 2)(\lambda - 2) = 0.$$

The solutions

$$\lambda = \tfrac{2}{3} \qquad \text{and} \qquad \lambda = 2,$$

give us two points that are twice as far from p as they are from q:

$$x_1 = \tfrac{1}{3}p + \tfrac{2}{3}q \qquad \text{and} \qquad x_2 = -p + 2q.$$

□

Remark. As we noted earlier, in the previous two examples, the answers did not depend on the actual distance function but only on the general properties of the norm. When working with distances and norms, one should use the general properties whenever possible. Of course, the actual numerical value of $\|x\|$ depends on the particular norm that is involved, as illustrated by the next example.

Example 1.3.3. *Find the distance between the points* $p = (1,1)$ *and* $q = (-1, 2)$ *using the three different metrics described earlier.*

Solution. This is just a matter of straightforward computation.

For the ℓ_2 or Euclidean norm, we have

$$\|p - q\|_2 = \sqrt{(1 - (-1))^2 + (1 - 2)^2} = \sqrt{5}.$$

For the ℓ_1 norm, we have

$$\|p - q\|_1 = |1 - (-1)| + |1 - 2| = 3.$$

For the ℓ_∞ or supremum norm, we have

$$\|p - q\|_\infty = \max\{\, |1 - (-1)|, |1 - 2| \,\} = 2.$$

$\square$

1.3.3   Balls and Spheres

In $\mathbb{R}^n$ with the ℓ_2 norm, the ***closed ball*** centered at x with radius ρ is the set

$$\overline{B}(x, \rho) = \{\, y \in \mathbb{R}^n \; : \; \|x - y\|_2 \leq \rho \,\}.$$

If we omit the boundary, the ***open ball*** centered at x with radius ρ is the set

$$B(x, \rho) = \{\, y \in \mathbb{R}^n \; : \; \|x - y\|_2 < \rho \,\}.$$

The boundary itself is called a **sphere** (or a **circle** in $\mathbb{R}^2$) centered at x with radius ρ and is the set

$$S(x, \rho) = \{\, y \in \mathbb{R}^n \ : \ \|x - y\|_2 = \rho \,\}.$$

For the particular case when $x = \bar{0}$ and $\rho = 1$, the sets are called the **closed unit ball**, the **open unit ball**, and the **unit sphere**, respectively.

The shape of a ball will depend on the norm being used. The previous three figures show a closed ball, an open ball, and a sphere in $\mathbb{R}^2$ using the Euclidean norm.

The figure below shows the corresponding balls and sphere in $\mathbb{R}^2$ centered at $\bar{0}$ with radius ρ in the ℓ_1 norm and the ℓ_∞ norm.

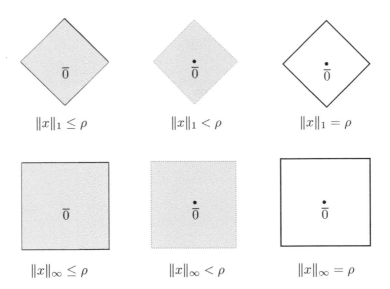

It should be mentioned that the closed unit ball in the ℓ_∞ norm is often called the **unit cube**, that is, the unit cube is the set

$$S(\bar{0}, 1) = \{\, (\alpha_1, \alpha_2, \ldots, \alpha_n) \ : \ |\alpha_i| \leq 1, \ i = 1, 2, \ldots, n \,\}.$$

Example 1.3.4. *Find the points where the line through the origin parallel to the vector $v = (2, 0, -3, 6)$ intersects the unit sphere.*

Note. Whenever a specific distance is to be calculated but no distance function is specified, we always assume that the Euclidean distance is to be used.

Solution. The line has the equation

$$x = \mu v, \quad -\infty < \mu < \infty,$$

and for $x = (x_1, x_2, x_3, x_4)$, this can be written as

$$(x_1, x_2, x_3, x_4) = \mu(2, 0, -3, 6), \quad -\infty < \mu < \infty.$$

This line intersects the unit sphere at the points where

$$\|(x_1, x_2, x_3, x_4)\| = \|\mu(2, 0, -3, 6)\| = 1,$$

that is, where

$$|\mu| \cdot \|(2, 0, -3, 6)\| = |\mu|\sqrt{4 + 0 + 9 + 36} = 7|\mu| = 1,$$

so we must have $\mu = \frac{1}{7}, -\frac{1}{7}$. Therefore, the line intersects the unit sphere at

$$x_1 = (\tfrac{2}{7}, 0, -\tfrac{3}{7}, \tfrac{6}{7}) \quad \text{and} \quad x_2 = (-\tfrac{2}{7}, 0, \tfrac{3}{7}, -\tfrac{6}{7}).$$

$\square$

If A is a subset of $\mathbb{R}^n$ and if v is a vector from $\mathbb{R}^n$, the set $A + v$ defined by

$$A + v = \{x + v \in \mathbb{R}^n : x \in A\}$$

is called a ***translate*** of A.

Note that the set $A + v$ is obtained from the set A by adding the vector v to every point in A. The key to working with translates is to use the fact that

$$x \in A + v \quad \text{if and only if} \quad x - v \in A.$$

Example 1.3.5. *Show that the translate of a closed ball is a closed ball.*

Note that in this case, no specific distance has to be calculated, so we will attempt to solve the problem using only the general properties of the norm.

Solution. Let

$$A = \overline{B}(q, \rho) = \{x \in \mathbb{R}^n : \|x - q\| \le \rho\}.$$

If $A + v$ is indeed a closed ball, then we should expect that it has the same radius as A and that its center is at $q + v$. Thus, we will try to show that

$$A + v = \overline{B}(q + v, \rho) = \{x \in \mathbb{R}^n : \|x - (q + v)\| \le \rho\}.$$

As usual, we will show that the two sets are equal by showing that each is a subset of the other.

(i) We will show first that $\overline{B}(q+v,\rho) \subseteq A+v$, so we let $x \in \overline{B}(q+v,\rho)$, then

$$\|x - (q + v)\| \leq \rho,$$

so that

$$\|(x - v) - q\| \leq \rho,$$

that is,

$$z = x - v \in A = \overline{B}(q, \rho).$$

Therefore,

$$x = z + v \in A + v,$$

and since $x \in \overline{B}(q + v, \rho)$ is arbitrary, then $\overline{B}(q + v, \rho) \subseteq A + v$.

(ii) Conversely, if $x \in A + v = \overline{B}(q, \rho) + v$, then $x = z + v$, where $z \in \overline{B}(q, \rho)$, so that

$$\|z - q\| \leq \rho,$$

that is,

$$\|(x - v) - q\| \leq \rho,$$

so that

$$\|x - (q + v)\| \leq \rho,$$

and $x \in \overline{B}(q + v, \rho)$. Since $x \in A + v$ is arbitrary, then $A + v \subseteq \overline{B}(q + v, \rho)$.

Therefore,

$$A + v = \overline{B}(q + v, \rho) = \{\, x \in \mathbb{R}^n \,:\, \|x - (q + v)\| \leq \rho \,\}.$$

$\square$

Exercise 1.3.6. *Show that*

$$B(q, \rho) + v = B(q + v, \rho)$$

and

$$S(q, \rho) + v = S(q + v, \rho).$$

If A is a subset of $\mathbb{R}^n$ and λ is a real number with $\lambda > 0$, the set

$$\lambda A = \{\, z \in \mathbb{R}^n \,:\, z = \lambda x, \text{ where } x \in A \,\}$$

is called a ***positive homothet*** of A.

Note that the set λA is obtained from the set A by multiplying each vector in A by the positive scalar λ. In fact, the definition of λA applies to all scalars $\lambda \in \mathbb{R}$, not just $\lambda > 0$.

Theorem 1.3.7. *The positive homothet of a closed ball is a closed ball.*

The proof of this theorem is left as an exercise. What about a positive homothet of an open ball? A sphere?

The ℓ_1, ℓ_2, and ℓ_∞ norms are not the only useful norms on $\mathbb{R}^n$. One sometimes encounters the ℓ_p norm.

Given any real number p with $1 < p < \infty$, the ℓ_p **norm** on $\mathbb{R}^n$ is defined to be

$$\|x\|_p = \left(\sum_{k=1}^{n} |x_k|^p \right)^{\frac{1}{p}}$$

for $x = (x_1, x_2, \dots, x_n) \in \mathbb{R}^n$.

Even though the proof that the ℓ_p norms are indeed norms is beyond the scope of this text, we can still determine what the balls and spheres look like in $\mathbb{R}^2$.

Example 1.3.8. *Sketch the closed ball* $\overline{B}(\overline{0}, a)$ *in* $\mathbb{R}^2$ *when it is equipped with the* ℓ_p *norm, where* $1 < p < \infty$.

Solution. The closed ball is the set

$$\overline{B}(\overline{0}, a) = \{ (x_1, x_2) \subset \mathbb{R}^2 : (|x_1|^p + |x_2|^p)^{\frac{1}{p}} \leq a \},$$

and since this is symmetric about the x_1 and x_2 axes, it is sufficient to sketch the curve

$$x_1^p + x_2^p = a^p$$

for $0 \leq x_1 \leq a$ and $0 \leq x_2 \leq a$.

Differentiating implicitly, it is easy to see that this curve has a horizontal tangent at the point $(0, a)$ and a vertical tangent at the point $(a, 0)$ and is concave down on the interval $0 < x_1 < a$.

Using the symmetry of the curve, we get the figure below.

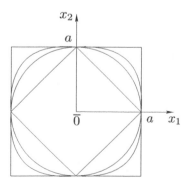

The figure shows the ℓ_1, ℓ_2, ℓ_p, and ℓ_∞ balls. In the figure, $2 < p < \infty$. For the case $1 < p < 2$, the ℓ_p ball would be contained in the ℓ_2 ball and would contain the ℓ_1 ball.

As p increases without bound, it would appear that the ℓ_p balls expand to approach the ℓ_∞ ball, and this is indeed the case, as the next example shows.

Example 1.3.9. *Show that*

$$\lim_{p \to \infty} \|x\|_p = \|x\|_\infty$$

for each $x \in \mathbb{R}^n$.

Solution. If we let $x = (x_1, x_2, \ldots, x_n)$ be any point in $\mathbb{R}^n$, then there is an index n_0 such that

$$|x_{n_0}| = \max\{ |x_k| : 1 \le k \le n \} = \|x\|_\infty,$$

and, therefore,

$$\|x\|_\infty = |x_{n_0}| = (|x_{n_0}|^p)^{\frac{1}{p}} \le \left(\sum_{k=1}^{n} |x_k|^p \right)^{\frac{1}{p}} = \|x\|_p,$$

that is,

$$\|x\|_\infty \le \|x\|_p$$

for all $p \ge 1$.

Also,

$$\|x\|_p = \left(\sum_{k=1}^{n} |x_k|^p \right)^{\frac{1}{p}} = \left(\sum_{k=1}^{n} \left| \frac{x_k}{\|x\|_\infty} \right|^p \right)^{\frac{1}{p}} \cdot \|x\|_\infty,$$

and since $|x_k| \le \|x\|_\infty$ for $1 \le k \le n$, then

$$\|x\|_p \le \|x\|_\infty \cdot n^{\frac{1}{p}}$$

for all $p \ge 1$.

Combining these two inequalities, we obtain

$$\|x\|_\infty \le \|x\|_p \le \|x\|_\infty \cdot n^{\frac{1}{p}}$$

for all $p \ge 1$.

Letting $p \to \infty$, since

$$\lim_{p \to \infty} n^{\frac{1}{p}} = \lim_{p \to \infty} e^{\frac{1}{p} \log n} = e^0 = 1.$$

Since $\|x\|_p$ is stuck between two quantities approaching $\|x\|_\infty$, the limit exists and

$$\lim_{p\to\infty} \|x\|_p = \|x\|_\infty$$

for all $x \in \mathbb{R}^n$.

$\square$

1.4 INNER PRODUCT AND ORTHOGONALITY

An **_inner product_**[*] on $\mathbb{R}^n$ is a mapping $\langle \cdot\, , \cdot \rangle : \mathbb{R}^n \times \mathbb{R}^n \longrightarrow \mathbb{R}$ such that for all x, y, z in $\mathbb{R}^n$ and all λ in $\mathbb{R}$,

   (i) $\langle x, x \rangle \geq 0$ and $\langle x, x \rangle = 0$ if and only if $x = \bar{0}$,
  (ii) $\langle x, y \rangle = \langle y, x \rangle$,
 (iii) $\langle x, y + z \rangle = \langle x, y \rangle + \langle x, z \rangle$ and $\langle x + y, z \rangle = \langle x, z \rangle + \langle y, z \rangle$,
 (iv) $\langle \lambda x, y \rangle = \langle x, \lambda y \rangle = \lambda \langle x, y \rangle$.

The **_standard_** or **_Euclidean inner product_** on $\mathbb{R}^n$ is given by

$$\langle x, y \rangle = x_1 y_1 + x_2 y_2 + \cdots + x_n y_n$$

for $x = (x_1, x_2, \ldots, x_n)$ and $y = (y_1, y_2, \ldots, y_n)$ in $\mathbb{R}^n$.

Exercise 1.4.1. _Show that the Euclidean inner product is an actual inner product as defined above._

The standard inner product is intimately related to the Euclidean norm, since it is immediately apparent from the definition of the inner product that

$$\|x\|_2^2 = \langle x, x \rangle,$$

that is,

$$\|x\|_2 = \sqrt{\langle x, x \rangle} = \left(x_1^2 + x_2^2 + \cdots + x_n^2 \right)^{\frac{1}{2}}.$$

This is just the **_Euclidean norm_** on $\mathbb{R}^n$ and is the usual Euclidean distance from the origin $\bar{0}$ to the point x, that is, it is the Euclidean length of the vector x.

---

[*] Also called a **_scalar product_** or a **_dot product_** on $\mathbb{R}^n$, and from properties (i), (ii), (iii), and (iv), it is also called a **_nondegenerate, positive definite, symmetric, bilinear mapping_**.

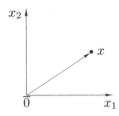

Note that in the definition of the inner product, property (i) says that $\|x\| = 0$ if and only if $x = \bar{0}$.

It is very common to approach the notion of distance in $\mathbb{R}^n$ by first discussing the inner product. However, we have elected to discuss the notion of distance first because there are many different distance functions for $\mathbb{R}^n$ that cannot be derived from an inner product.

In two or three dimensions, orthogonality is synonymous with perpendicularity. In fact, in the Euclidean plane, we have the following notion of orthogonality or perpendicularity.

Theorem 1.4.2. *If x and y are vectors in $\mathbb{R}^2$, then $x \perp y$ if and only if $\langle x, y \rangle = 0$.*

Proof.

If x and y are perpendicular vectors in the plane, then from the Pythagorean theorem and its converse, we have $x \perp y$ if and only if

$$\|x - y\|_2^2 = \|x\|_2^2 + \|y\|_2^2.$$

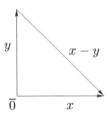

However,

$$\langle x - y, x - y \rangle = \|x\|_2^2 + \|y\|_2^2$$

if and only if

$$\langle x, x \rangle - 2\langle x, y \rangle + \langle y, y \rangle = \|x\|_2^2 + \|y\|_2^2,$$

that is, if and only if

$$\|x\|_2^2 - 2\langle x, y \rangle + \|y\|_2^2 = \|x\|_2^2 + \|y\|_2^2,$$

if and only if

$$\langle x, y \rangle = 0.$$

$\square$

A similar argument holds for orthogonal vectors in $\mathbb{R}^3$, so we may use this as a definition of orthogonality in higher dimensions.

In $\mathbb{R}^n$, the vectors x and y are **orthogonal**,* denoted by $x \perp y$, if and only if $\langle x, y \rangle = 0$. Note that this is the *standard* or *Euclidean* inner product on $\mathbb{R}^n$.

The Pythagorean theorem is easily extended to $\mathbb{R}^n$, as shown in the following theorem.

Theorem 1.4.3. *If $x, y \in \mathbb{R}^n$, then $x \perp y$ if and only if $\|x+y\|_2^2 = \|x\|_2^2 + \|y\|_2^2$.*

Proof. The points 0, x, and y form a triangle in $\mathbb{R}^n$ which lies in a plane (that is, a two-dimensional subspace of $\mathbb{R}^n$). Now the usual Pythagorean theorem in this plane gives the result.

$\square$

Note that we could have stated the previous theorem as $x \perp y$ if and only if

$$\|x - y\|_2^2 = \|x\|_2^2 + \|y\|_2^2.$$

Example 1.4.4. *(Packing the Elephant)*
Show that a three-dimensional elephant can be packed inside the unit cube of $\mathbb{R}^n$ if n is sufficiently large. In fact, our sun can be packed inside the unit cube if n is large enough.

Solution. We assume that distance in $\mathbb{R}^n$ and $\mathbb{R}^3$ is being used with the same units. For example, we assume that the length of the vector $(1, 0, 0, \ldots, 0)$ is 1 m in both $\mathbb{R}^n$ and $\mathbb{R}^3$.

Let p_i, for $i = 1, 2, 3$, be the following three points in $\mathbb{R}^{3k}$:

$$p_1 = (\underbrace{1, 1, \ldots, 1}_{k \text{ times}}, \underbrace{0, 0, \ldots, 0}_{2k \text{ times}})$$

$$p_2 = (\underbrace{0, 0, \ldots, 0}_{k \text{ times}} \underbrace{1, 1, \ldots, 1}_{k \text{ times}}, \underbrace{0, 0, \ldots, 0}_{k \text{ times}})$$

$$p_3 = (\underbrace{0, 0, \ldots, 0}_{2k \text{ times}}, \underbrace{1, 1, \ldots, 1}_{k \text{ times}}).$$

---

* Some people like to exclude the zero vector from considerations of perpendicularity. From the definition of orthogonality, the zero vector is **always** orthogonal to every other vector.

It is easily checked that

$$\langle p_1, p_2 \rangle = \langle p_1, p_3 \rangle = \langle p_2, p_3 \rangle = 0,$$

so that the segments

$$[\overline{0}, p_1], \quad [\overline{0}, p_2], \quad \text{and} \quad [\overline{0}, p_3]$$

are mutually orthogonal. Thus, they form three of the edges of a three-dimensional tetrahedron with vertices $\overline{0}$, p_1, p_2, and p_3.

This tetrahedron is entirely contained in the unit cube of $\mathbb{R}^{3k}$, and so it suffices to show that for sufficiently large k, we can pack an elephant inside this tetrahedron.

For $i = 1$, 2, 3, the length of each of the edges $[\overline{0}, p_i]$ is $\sqrt{k}$. Thus, we can make the tetrahedron as large as we like by simply increasing the magnitude of k.

$$\square$$

The notion of orthogonality can also be extended to arbitrary sets.

We say that a vector v is ***orthogonal to a set*** A if v is orthogonal to every vector determined by every pair of points p and q in A, that is, for every p and q in A, we have $\langle v, p - q \rangle = 0$.

Note that if a vector v is orthogonal to a set A, then every multiple of v is also orthogonal to A and v is orthogonal to every translate of A.

1.4.1 Nearest Points

With the Euclidean metric on $\mathbb{R}^2$, we know from the Pythagorean theorem that the hypotenuse of a right triangle is the longest side of the triangle. Thus, the line L is perpendicular (that is, orthogonal) to the segment joining $\overline{0}$ to its nearest point p in L, as in the figure below.

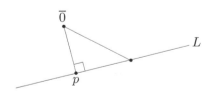

In fact, this is true no matter what the dimension of the space is.

Theorem 1.4.5. *Let L be a line in $\mathbb{R}^n$ and let p be a point on L, then p is the closest point of L to $\bar{0}$ if and only if p is orthogonal to $q - p$ for every other point q on L.*

Proof. If we let q be any point on L with $q \neq p$, then L is the line through p parallel to $v = q - p$, that is, the line through p whose equation is

$$x = p + \lambda v, \quad -\infty < \lambda < \infty.$$

Note that

$$\|p\|_2 \leq \|p + \lambda v\|_2 \quad \text{for all } \lambda$$

if and only if

$$\|p\|_2^2 \leq \|p + \lambda v\|_2^2 \quad \text{for all } \lambda,$$

that is, if and only if

$$\|p\|_2^2 \leq \|p\|_2^2 + 2\lambda \langle p, v \rangle + \lambda^2 \|v\|_2^2 \quad \text{for all } \lambda.$$

Therefore,

$$\|p\|_2 \leq \|p + \lambda v\|_2 \quad \text{for all } \lambda$$

if and only if

$$0 \leq 2\lambda \langle p, v \rangle + \lambda^2 \|v\|_2^2 \quad \text{for all } \lambda. \tag{$*$}$$

Thus, we need to show that the last inequality is true if and only if $p \perp v$.

(i) Suppose first that $\langle p, v \rangle = 0$, then $(*)$ becomes

$$\lambda^2 \|v\|_2^2 \geq 0,$$

which is true for all λ.

(ii) Conversely, suppose that $\langle p, v \rangle \neq 0$, then if we let

$$\lambda = -\frac{\langle p, v \rangle}{\|v\|_2^2},$$

then

$$2\lambda \langle p, v \rangle + \lambda^2 \|v\|_2^2 = -\frac{|\langle p, v \rangle|^2}{\|v\|_2^2} < 0$$

and $(*)$ is not true. Hence, if $(*)$ is true, then $\langle p, v \rangle = 0$.

$\square$

Corollary 1.4.6. *A line in $\mathbb{R}^n$ has a unique point of minimum Euclidean norm.*

Proof. If both p and q were points on the line with minimum norm, the previous theorem implies that

$$\langle q, q - p \rangle = 0 \qquad \text{and} \qquad \langle p, q - p \rangle = 0.$$

Subtracting, we would get $\langle q - p, q - p \rangle = 0$, which means that $\|q - p\|_2 = 0$ so that $p = q$.

$\square$

Example 1.4.7. *Given the line* $L : \; 3x_1 + 4x_2 = 12$ *in* $\mathbb{R}^2$, *find the closest point on the line* L *to the origin using*

    (a) the ℓ_2 *norm,*
    (b) the ℓ_1 *norm,*
    (c) the ℓ_∞ *norm.*

Solution. As can be seen from the figure below,

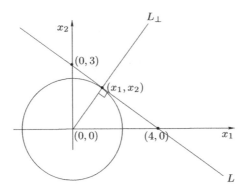

the vector equation of the line L is

$$x = (x_1, x_2) = (0, 3) + \mu \left((4, 0) - (0, 3) \right) = (0, 3) + \mu(4, -3)$$

for $-\infty < \mu < \infty$.

    (a) Using the ℓ_2 norm, the point on the line L closest to the origin is the point of intersection of L with the line $L_\perp$ through the origin in the direction $(3, 4)$. The vector equation of $L_\perp$ is

$$x = (x_1, x_2) = \bar{0} + \mu(3, 4) = \mu(3, 4), \quad -\infty < \mu < \infty,$$

and these lines intersect when

$$(0, 3) + \mu(4, -3) = \lambda(3, 4),$$

that is, when

$$4\mu = 3\lambda \quad \text{and} \quad 3 - 3\mu = 4\lambda.$$

Solving this system of equations, we get $\lambda = \frac{12}{25}$, and therefore, the point

$$(x_1, x_2) = \frac{12}{25}(3, 4) = \left(\frac{36}{25}, \frac{48}{25}\right)$$

is the point on the line L, which is closest to the origin when we use the Euclidean norm.

(b) Using the ℓ_1 norm, the closest point on L to the origin is the point at which the ℓ_1 balls centered at $(0, 0)$ first hit the line. That is, the ℓ_1 ball centered at $\bar{0}$ with radius 3 touches L at the point $(0, 3)$ and nowhere else, and any other ℓ_1 ball centered at $\bar{0}$ with a different radius either misses L or else contains $(0, 3)$.

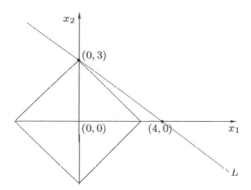

In this case, the closest point is $(0, 3)$.

(c) Using the ℓ_∞ norm, the closest point on L to the origin is the point where the ℓ_∞ balls centered at $(0, 0)$ first hit the line. That is, the smallest ℓ_∞ ball centered at $\bar{0}$ touches L at the point (x_1, x_2) where $x_1 = x_2$, as shown in the following figure. In other words, the nearest point is where the line

$$M = \{x \in \mathbb{R}^2 : x = \lambda(1, 1), \ -\infty < \lambda < \infty\}$$

meets L.

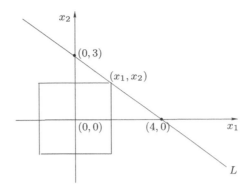

In this case, the closest point is $\left(\frac{12}{7}, \frac{12}{7}\right)$.

$\square$

From the examples given above, it should be clear now that the geometry in $\mathbb{R}^n$ depends heavily on the norm used to measure distances, and many of the familiar notions from Euclidean geometry may not look so familiar now. However, in the majority of this text, we will use the Euclidean norm, and we will see that many geometrical properties of $\mathbb{R}^n$ for $n > 2$ are the same as the familiar properties for $\mathbb{R}^2$ or $\mathbb{R}^3$.

1.4.2 Cauchy–Schwarz Inequality

An approach similar to that used in the proof of Theorem 1.4.5 yields one of the most useful relationships between the inner product and the Euclidean distance.

Theorem 1.4.8. *(The Cauchy–Schwarz Inequality)*[*]
If u and v are vectors in $\mathbb{R}^n$, then

$$|\langle u, v \rangle| \leq \|u\|_2 \|v\|_2,$$

and equality holds if and only if one of u or v is a multiple of the other.

Proof. If $|\langle u, v \rangle| = 0$, we have

$$0 = |\langle u, v \rangle| \leq \|u\|_2 \|v\|_2$$

with equality if and only if at least one of u or v is zero. Thus, the Cauchy inequality is true in this case.

---

[*] Also known as the ***Cauchy*** inequality, the ***Schwarz*** inequality, the ***Bunyakovsky*** inequality, or as the ***Cauchy–Bunyakovsky–Schwarz*** inequality.

If $|\langle u, v \rangle| \neq 0$, then for any scalar λ, we have

$$0 \leq \|u + \lambda v\|_2^2 = \|u\|_2^2 + 2\lambda\langle u, v \rangle + \lambda^2\|v\|_2^2,$$

and taking

$$\lambda = -\frac{\langle u, v \rangle}{\|v\|_2^2},$$

we get

$$0 \leq \|u\|_2^2 - 2\frac{|\langle u, v \rangle|^2}{\|v\|_2^2} + \frac{|\langle u, v \rangle|^2}{\|v\|_2^2} = \|u\|_2^2 - \frac{|\langle u, v \rangle|^2}{\|v\|_2^2},$$

so that

$$|\langle u, v \rangle|^2 \leq \|u\|_2^2 \|v\|_2^2.$$

Cauchy's inequality follows by taking nonnegative square roots.

Now note that equality holds if and only if $u + \lambda v = \overline{0}$, that is, if and only if one of u or v is a multiple of the other.

$\square$

Remark. The equality is sometimes stated as

$$\langle u, v \rangle \leq \|u\|_2 \|v\|_2,$$

and in this case, equality holds if and only if one of u or v is a nonnegative multiple of the other.

Now that we have the Cauchy–Schwarz inequality, we can give a simple proof of the **_triangle inequality_** in $\mathbb{R}^n$ with the Euclidean distance.

Theorem 1.4.9. *(The Triangle Inequality)**
For any vectors u and v in $\mathbb{R}^n$, we have

$$\|u + v\|_2 \leq \|u\|_2 + \|v\|_2$$

with equality if and only if one of u or v is a nonnegative multiple of the other.

Proof. If u and v are vectors from $\mathbb{R}^n$, then

$$\begin{aligned}
\|u + v\|_2^2 &= \|u\|_2^2 + 2\langle u, v \rangle + \|v\|_2^2 \\
&\leq \|u\|_2^2 + 2\|u\|_2\|v\|_2 + \|v\|_2^2 \\
&= (\|u\|_2 + \|v\|_2)^2,
\end{aligned}$$

---

* Sometimes known as **_Minkowski's_** inequality.

so that

$$\|u + v\|_2 \le \|u\|_2 + \|v\|_2.$$

Note that equality holds in the inequality above if and only if equality holds in the Cauchy inequality, that is, if and only if one of u or v is a nonnegative multiple of the other.

□

Note. If $u = (u_1, u_2, \ldots, u_n)$ and $v = (v_1, v_2, \ldots, v_n)$ are points in $\mathbb{R}^n$, then the coordinatized versions of the Cauchy–Schwarz inequality and Minkowski's inequality are given by

Cauchy–Schwarz Inequality:

$$\sum_{k=1}^{n} |u_k v_k| \le \left(\sum_{k=1}^{n} |u_k|^2 \right)^{\frac{1}{2}} \left(\sum_{k=1}^{n} |v_k|^2 \right)^{\frac{1}{2}}.$$

Triangle Inequality:

$$\left(\sum_{k=1}^{n} |u_k + v_k|^2 \right)^{\frac{1}{2}} \le \left(\sum_{k=1}^{n} |u_k|^2 \right)^{\frac{1}{2}} + \left(\sum_{k=1}^{n} |v_k|^2 \right)^{\frac{1}{2}}.$$

The next example shows when we can have equality in the triangle inequality in each of the usual norms.

Example 1.4.10. *Given $p = (1, 1)$ and $q = (-1, -1)$, find all points x in $\mathbb{R}^2$ such that*

(a) $\|p - x\|_2 + \|q - x\|_2 = \|p - q\|_2$,
(b) $\|p - x\|_1 + \|q - x\|_1 = \|p - q\|_1$,
(c) $\|p - x\|_\infty + \|q - x\|_\infty = \|p - q\|_\infty$.

Solution.

(a) Clearly, if x is any point on the line segment joining p and q, then

$$x = (1 - \mu)p + \mu q$$

for some $0 < \mu < 1$, and, therefore,

$$\|p - x\|_2 + \|q - x\|_2 = \mu\|p - q\|_2 + (1 - \mu)\|p - q\|_2 = \|p - q\|_2.$$

Conversely, suppose that $x \in \mathbb{R}^2$ is a point in the plane for which

$$\|p - q\|_2 = \|p - x + x - q\|_2 = \|p - x\|_2 + \|q - x\|_2,$$

then we have equality in the triangle inequality, which can happen if and only if one of $p - x$ and $x - q$ is a nonnegative multiple of the other, that is, if and only if

$$p - x = \mu(x - q)$$

for some $\mu \geq 0$, that is, if and only if

$$x = \frac{1}{1 + \mu}p + \frac{\mu}{1 + \mu}q.$$

Thus, we have equality in the triangle inequality if and only if x is on the line segment between p and q, so the set we want is the line segment $[p, q]$.

(b) If $x = (x_1, x_2) \in \mathbb{R}^2$ is any point with $p_1 \leq x_1 \leq q_1$ and $p_2 \leq x_2 \leq q_2$, then

$$\begin{aligned}
\|p - x\|_1 + \|q - x\|_1 &= |p_1 - x_1| + |p_2 - x_2| + |q_1 - x_1| + |q_2 - x_2| \\
&= (x_1 - p_1) + (x_2 - p_2) + (q_1 - x_1) + (q_2 - x_2) \\
&= q_1 - p_1 + q_2 - p_2 \\
&= |p_1 - q_1| + |p_2 - q_2| \\
&= \|p - q\|_1,
\end{aligned}$$

that is,

$$\|p - x\|_1 + \|q - x\|_1 = \|p - q\|_1.$$

On the other hand, if $x = (x_1, x_2) \subset \mathbb{R}^2$ is any point such that

$$\|p - x\|_1 + \|q - x\|_1 = \|p - q\|_1,$$

then from the triangle inequality for real numbers, we have

$$|p_1 - q_1| \leq |p_1 - x_1| + |q_1 - x_1| \tag{$*$}$$

and

$$|p_2 - q_2| \leq |p_2 - x_2| + |q_2 - x_2|. \tag{$**$}$$

Suppose that either $x_1 \notin [p_1, q_1]$ or $x_2 \notin [p_2, q_2]$, then at least one of the following is true (the reader should check this!)

$$|p_1 - x_1| > |p_1 - q_1|, \qquad |q_1 - x_1| > |p_1 - q_1|,$$
$$|p_2 - x_2| > |p_2 - q_2|, \qquad |q_2 - x_2| > |p_2 - q_2|,$$

and we would have strict inequality in at least one (or both) of the inequalities (∗) and (∗∗) above. Adding (∗) and (∗∗), we would have

$$|p_1 - q_1| + |p_2 - q_2| < |p_1 - x_1| + |p_2 - x_2| + |q_1 - x_1| + |q_2 - x_2|,$$

that is, $\|p - q\|_1 < \|p - x\|_1 + \|q - x\|_1$, which is a contradiction. Therefore, $x_1 \in [p_1, q_1]$ and $x_2 \in [p_2, q_2]$, and in this case, the set we want is the closed box

$$B = \{x = (x_1, x_2) : p_1 \le x_1 \le q_1, p_2 \le x_2 \le q_2\}.$$

(c) If $x \in [p, q]$, then $x = (1 - \mu)p + \mu q$ for some $0 < \mu < 1$, so that

$$\|x - p\|_\infty = \mu \|p - q\|_\infty \qquad \text{and} \qquad \|x - q\|_\infty = (1 - \mu)\|p - q\|_\infty,$$

and adding, we get

$$\|x - p\|_\infty + \|x - q\|_\infty = \mu \|p - q\|_\infty + (1 - \mu)\|p - q\|_\infty = \|p - q\|_\infty.$$

Conversely, if $x \notin [p, q]$, since the line joining $p = (-1, -1)$ and $q = (1, 1)$ is the line $x_2 = x_1$, then the smallest ℓ_∞ balls centered at p and q containing x overlap, as in the figure below.

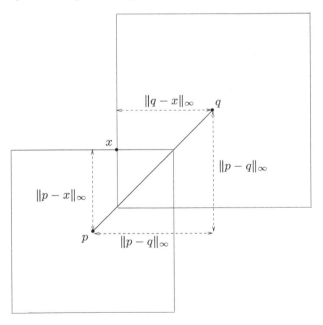

It is clear from the figure that

$$\|p - x\|_\infty + \|q - x\|_\infty > \|p - q\|_\infty.$$

In fact,

$$\|p - x\|_\infty + \|q - x\|_\infty = \|p - q\|_\infty$$

if and only if the point x lies on the line segment between p and q, that is,

$$x = (1 - \mu)p + \mu q$$

for some $0 \leq \mu \leq 1$. The set we want is the line segment $[p, q]$.

$\square$

1.4.3 Problems

In the following exercises, assume that "distance" means "Euclidean distance" unless otherwise stated.

1. (a) The **unit cube** in $\mathbb{R}^n$ is the set of points

$$\{\, x = (\alpha_1, \alpha_2, \ldots, \alpha_n) : |\alpha_i| \leq 1, \; i = 1, 2, \ldots, n \,\}.$$

   Draw the unit cube in $\mathbb{R}^1$, $\mathbb{R}^2$, and $\mathbb{R}^3$.
   (b) What is the length of the longest line segment that you can place in the unit cube of $\mathbb{R}^n$?
   (c) What is the radius of the smallest Euclidean ball that contains the unit cube of $\mathbb{R}^n$?
2. (a) Let L be the straight line through $\bar{0}$ parallel to the vector $v = (1, 3, -1, 2)$. Find the two points where the line enters and exits the unit cube.
   Hint. Solve a similar problem in $\mathbb{R}^2$ first.
   *(b) Let L be the straight line through the point $p = (\frac{1}{3}, \frac{1}{2}, -\frac{1}{3}, \frac{1}{2})$ parallel to the vector $v = (2, -1, 2, 1)$. Find the two points where the line enters and exits the unit cube.
3. Find the distance between the points $(1, -2)$ and $(-2, 3)$ using
   (a) the ℓ_1 metric,
   (b) the "sup" metric,
   (c) the Euclidean metric.
4. Let $\| \cdot \|_1$, $\| \cdot \|_2$, and $\| \cdot \|_\infty$ denote, respectively, the ℓ_1, Euclidean, and "sup" norms. Identify all those points x in $\mathbb{R}^n$ that have the property

$$\|x\|_1 = \|x\|_2 = \|x\|_\infty.$$

   Hint. Try this for $\mathbb{R}^2$ first.
5. Show that a positive homothet of a closed ball is a closed ball.

1.5 CONVEX SETS

In this section, we give a brief introduction to convex sets, plus some examples.

A subset A of $\mathbb{R}^n$ is said to be **convex** if and only if whenever x and y are two points from A, then the entire segment $[x, y]$ is a subset of A; equivalently, A is convex if and only if $x, y \in A$ and $0 < \lambda < 1$ imply that $(1 - \lambda)x + \lambda y \in A$ also.

Note that a convex set does not have any holes, dimples, or bumps.

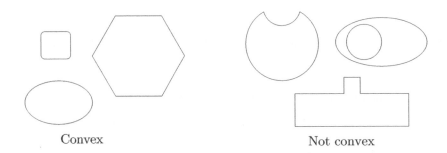

Convex   Not convex

The sets depicted on the left in the above figure are convex, while those on the right are not convex.

Note. In order to verify that a set is convex using the definition, we use the following strategy:

Choose two arbitrary points in the set and show that the segment joining the points is in the set.

Example 1.5.1. *The closed unit ball in* $\mathbb{R}^n$

$$\overline{B}(\overline{0}, 1) = \{\, x \in \mathbb{R}^n : \|x\| \le 1 \,\}$$

is a convex set.

Proof. Note that we did not specify any particular norm, so we should be able to prove this for *any* norm on $\mathbb{R}^n$.

Let $x, y \in \overline{B}(\overline{0}, 1)$, then for any point $z \in [x, y]$, we have

$$z = (1 - \lambda)x + \lambda y$$

for some scalar λ with $0 \le \lambda \le 1$. We also want to show that $z \in \overline{B}(\overline{0}, 1)$; that is, $\|z\| \le 1$.

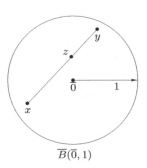

$\overline{B}(\overline{0}, 1)$

This follows from the triangle inequality:

$$\|z\| = \|(1 - \lambda)x + \lambda y\|$$
$$\leq \|(1 - \lambda)x\| + \|\lambda y\|$$
$$= |1 - \lambda| \cdot \|x\| + |\lambda| \cdot \|y\|,$$

and since $0 \leq \lambda \leq 1$, then $|1 - \lambda| = 1 - \lambda$ and $|\lambda| = \lambda$, so that

$$\|z\| \leq (1 - \lambda)\|x\| + \lambda\|y\| \leq (1 - \lambda) \cdot 1 + \lambda \cdot 1 = 1.$$

$\square$

Exercise 1.5.2. *Show that the open unit ball in $\mathbb{R}^n$*

$$B(\bar{0}, 1) = \{\, x \in \mathbb{R}^n \,:\, \|x\| < 1 \,\}$$

is a convex set.

Note. It is not difficult to see that the two extreme cases $A = \emptyset$ and $A = \mathbb{R}^n$ are both convex. In fact, if $\emptyset$ were not convex, then from the definition, we could find $x, y \in \emptyset$ such that $[x, y] \not\subset \emptyset$. Since there are no points in $\emptyset$, we cannot find such an x and y, so $\emptyset$ must be convex. We say that $\emptyset$ is convex **vacuously**.

The next theorem shows that convexity is preserved under intersection and will give us another method for showing that a set is convex. Note that the family $\mathcal{F}$ in the theorem may be finite or infinite.

Theorem 1.5.3. *Suppose that $\mathcal{F}$ is a nonempty family of convex subsets of $\mathbb{R}^n$, then the set*

$$C = \bigcap \{\, A \,:\, A \in \mathcal{F} \,\}$$

is convex.

Proof. Let $x, y \in C$. We will show that $[x, y] \subset C$.

Since $x \in C$, then $x \in A$ for every $A \in \mathcal{F}$, and since $y \in C$, then $y \in A$ for every $A \in \mathcal{F}$.

Since each $A \in \mathcal{F}$ is convex, then $[x, y] \subset A$ for each $A \in \mathcal{F}$, and, therefore,

$$[x, y] \subset \bigcap \{\, A \,:\, A \in \mathcal{F} \,\}.$$

$\square$

Remark. As mentioned earlier, this theorem provides another method of show-ing that a set is convex, namely, we show that the set is the intersection of a family of sets, all of which are known to be convex. At this point, illustrative examples would be highly contrived, and we give one such example.

Example 1.5.4. *If p and q are distinct points in $\mathbb{R}^n$, show that the segment*

$$[p, q] = \{\, x \in \mathbb{R}^n : x = (1 - \lambda)p + \lambda q, \ 0 \le \lambda \le 1 \,\}$$

is convex.

Proof. Let X be the subset of $\mathbb{R}^n$ formed by intersecting all of the closed Euclidean balls that contain p and q. We will show that $X = [p, q]$.

Suppose that p and q are contained in a closed Euclidean ball centered at x_0 with radius $r > 0$, and let

$$z = (1 - \mu)p + \mu q,$$

where $0 \le \mu \le 1$, be any point on the line segment joining p and q, then

$$
\begin{aligned}
\|z - x_0\|_2 &= \|(1 - \mu)p + \mu q - x_0\|_2 \\
&= \|(1 - \mu)p + \mu q - (1 - \mu)x_0 - \mu x_0\|_2 \\
&\le (1 - \mu)\|p - x_0\|_2 + \mu\|q - x_0\|_2 \\
&\le (1 - \mu)r + \mu r \\
&= r.
\end{aligned}
$$

That is, z is in the closed ball also. Therefore, any closed Euclidean ball that contains p and q also contains the line segment joining p and q, that is, $[p, q]$. Hence, the intersection of all such closed Euclidean balls contains the entire segment $[p, q]$, and $[p, q] \subseteq X$.

Now suppose that x is a point that is not on the line segment $[p, q]$. If p, q, and x are collinear, then by taking the center on a line bisecting $[p, q]$ and a small enough radius, as in the figure below on the left, we can find a closed Euclidean ball containing p and q, but not containing x. Similarly, if p, q, and x are not collinear, then by taking the center on a line in a plane bisecting $[p, q]$ and a large enough radius, as in the figure below on the right, we can find a closed Euclidean ball containing p and q, but not containing x.

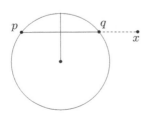 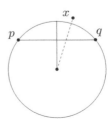

In either case, since x is not in this closed ball, it cannot possibly be in the intersection of all closed Euclidean balls containing p and q. Thus, if $x \notin [p, q]$, then $x \notin X$, and the contrapositive of this statement is true, that is, if $x \in X$, then $x \in [p, q]$. Therefore, $X \subseteq [p, q]$.

Combining these two containments, we get $X = [p, q]$, and since X is the intersection of a family of convex sets, then $[p, q]$ is convex.

$\square$

Exercise 1.5.5. *Show* $[p, q]$ *is convex directly from the definition of convexity.*

1.6 HYPERPLANES AND LINEAR FUNCTIONALS

1.6.1 Linear Functionals

Let $f : \mathbb{R}^n \longrightarrow \mathbb{R}$, then we say that f is a ***linear functional*** on $\mathbb{R}^n$ if and only if

(i) $f(x + y) = f(x) + f(y)$ for all $x, y \in \mathbb{R}^n$, and
(ii) $f(\lambda x) = \lambda f(x)$ for all $\lambda \in \mathbb{R}$, $x \in \mathbb{R}^n$.

Thus, a linear functional is a real-valued function on $\mathbb{R}^n$ that is both ***additive*** and ***homogeneous***.

Note. The notation $f = 0$ means that $f(x) = 0$ for all $x \in \mathbb{R}^n$, that is, f is identically zero. Thus, if f is a linear functional such that $f \neq 0$, then there is an $a \in \mathbb{R}^n$ such that $f(a) \neq 0$.

Example 1.6.1. *The following are examples of linear functionals:*

(1) $f(x_1, x_2) = 3x_1 + 4x_2$
(2) $f(x, y, z) = 3x + \sqrt{2}y - 52z$
(3) $f(u_1, u_2, u_3, u_4) = -1.0u_1 + 2.5u_2 - 7.3u_3 - 9.9u_4$
(4) $f(w, x, y, z) = 0$
(5) $f(x_1, x_2, x_3, x_4) = x_1 - x_3$.

Any missing terms are understood as being zero, so (5) is the same as

$$f(x_1, x_2, x_3, x_4) = x_1 + 0x_2 - x_3 + 0x_4.$$

The following are *not* linear functionals:

(1) $f(x_1, x_2) = 3x_1 + 4\sqrt{x_2}$

(2) $f(x, y, z) = \dfrac{3}{x} + \dfrac{\sqrt{2}}{y} - \dfrac{52}{z}$

(3) $f(u_1, u_2, u_3, u_4) = -1.0u_1^2 + 2.5u_2^2 - 7.3u_3^2 - 9.9u_4^2$

(4) $f(x_1, x_2, x_3, x_4) = x_1(1 - x_3)$.

$\square$

Representation of Linear Functionals

The notion of a linear functional occurs in many places in mathematics, and it is often very important to describe every possible type of linear functional that can arise in a given setting. Clearly, any function $f : \mathbb{R}^n \longrightarrow \mathbb{R}$ of the form

$$f(x_1, x_2, \ldots, x_n) = \alpha_1 x_1 + \alpha_2 x_2 + \cdots + \alpha_n x_n,$$

where $\alpha_1, \alpha_2, \ldots, \alpha_n$ are scalars, is a linear functional on $\mathbb{R}^n$. We will show in the following theorem* that these are the *only* ones!

Theorem 1.6.2. *If f is a nonzero linear functional on $\mathbb{R}^n$, then there is a unique vector $a \in \mathbb{R}^n$, $a \neq \overline{0}$, such that*

$$f(x) = \langle a, x \rangle$$

for all $x \in \mathbb{R}^n$.

Proof. Let $\{ e_1, e_2, \ldots, e_n \}$ be the standard basis vectors for $\mathbb{R}^n$, that is,

$$e_1 = (1, 0, 0, \ldots, 0), \ e_2 = (0, 1, 0, \ldots, 0), \ \ldots, \ e_n = (0, 0, 0, \ldots, 1),$$

where for $k = 1, 2, \ldots, n$, the basis vector e_k has a 1 in the kth coordinate and 0 elsewhere.

Let $x = (x_1, x_2, \ldots, x_n)$ be an arbitrary vector in $\mathbb{R}^n$, then we can write

$$x = \sum_{k=1}^{n} x_k e_k,$$

and since f is a linear functional, then

$$f(x) = \sum_{k=1}^{n} f(e_k) x_k = \langle a, x \rangle,$$

---

* This theorem is called the ***Riesz Representation Theorem***.

where a is the vector

$$a = (f(e_1), f(e_2), \ldots, f(e_n)).$$

Also, since f is nonzero, then at least one of the scalars $f(e_k)$ is nonzero, that is, $a \neq \overline{0}$.

Now we note that the vector a depends only on the linear functional f, so we have

$$f(x) = \langle a, x \rangle$$

for all $x \in \mathbb{R}^n$.

The fact that this representation of f is unique, that is, there is only one vector a that represents it, stems from the fact that there is only one way to write a vector x as a linear combination of the basis vectors.

$\square$

It is usual to identify a linear functional with the vector of constant terms that are used to define it. For example, if f is defined by the equation

$$f(x_1, x_2, x_3) = 3x_1 - 2x_2 + 7x_3,$$

then we will say that f is **represented** by $(3, -2, 7)$. If g is the linear functional

$$g(u_1, u_2, u_3, u_4) = u_1 - 2u_3 + u_4,$$

then we would say that g is represented by $(1, 0, 2, 1)$.

Thus, every n-dimensional vector $(\alpha_1, \alpha_2, \ldots, \alpha_n)$ gives rise to a unique linear functional on $\mathbb{R}^n$, and conversely, given any linear functional on $\mathbb{R}^n$, there is a unique n-dimensional vector that represents it.

Sums and Multiples of Linear Functionals

If f and g are two linear functionals on $\mathbb{R}^n$, we define the **sum** of f and g to be the function $f + g$ whose value is

$$(f + g)(x) = f(x) + g(x)$$

for all $x \in \mathbb{R}^n$, and if λ is a scalar, we define the **scalar multiple** λf to be the function whose value is

$$(\lambda f)(x) = \lambda \cdot f(x)$$

for all $x \in \mathbb{R}^n$.

It is easily verified that $f + g$ and λf are linear functionals on $\mathbb{R}^n$ and that with these pointwise definitions of addition and scalar multiplication, the set of all linear functionals on $\mathbb{R}^n$ is a real vector space (called the ***dual space***). The next theorem shows that we can identify it with $\mathbb{R}^n$, and the proof is left as an exercise.

Theorem 1.6.3. *If the linear functional f is represented by $a \in \mathbb{R}^n$ and the linear functional g is represented by $b \in \mathbb{R}^n$, and if λ is any scalar, then the linear functional $f + g$ is represented by $a + b$ and the linear functional λf is represented by λa.*

Example 1.6.4. *If f and g are defined by*

$$f(x, y, z) = 3x + y - 4x$$
$$g(x, y, z) = -x + 2y + z,$$

find the vectors that represent the functionals f, g, and $2f - g$.

Solution. From the definitions of f and g,

f is represented by $(3, 1, -4)$ and
g is represented by $(-1, 2, 1)$,

so that $2f - g$ is represented by

$$2(3, 1, -4) - (-1, 2, 1) = (6, 2, -8) - (-1, 2, 1)$$
$$= (6 - (-1), 2 - 2, -8 - 1)$$
$$= (7, 0, -9).$$

$\square$

Because the identification between vectors and linear functionals is so strong, it is usual to abuse the language and say that f *is* $(3, 1, -4)$ instead of saying that f *is represented by* the vector $(3, 1, -4)$.*

Example 1.6.5. *Let S be the set of all points in $\mathbb{R}^4$ whose fourth coordinate is zero. Find a linear functional f on $\mathbb{R}^4$ and a scalar β such that x is in S if and only if $f(x) = \beta$.*

Solution. We have to produce the linear functional

$$f(x_1, x_2, x_3, x_4) = \alpha_1 x_1 + \alpha_2 x_2 + \alpha_3 x_3 + \alpha_4 x_4$$

and the scalar β such that $x \in S$ if and only if $f(x) = \beta$. We will try to guess what f and β should be, and then we will show that our guess is correct.

---

* Woe is us! Now we can think of $(3, 1, -4)$ in *three* different ways.

First, we note that the zero vector is in S because its fourth coordinate is 0. Thus, no matter what f we try, we will always get $f(\overline{0}) = 0$, and it seems reasonable to guess that $\beta = 0$.

Next we are going to guess what each α_i should be. Notice that S contains the vector $(1, 0, 0, 0)$ so that $f(1, 0, 0, 0) = \alpha_1$. Since $f(x) = 0$ for all x in S, we have $\alpha_1 = 0$. If we repeat this argument using the points $(0, 1, 0, 0)$ and $(0, 0, 1, 0)$, we find that α_2 and α_3 must also be zero. However, we cannot use the same argument for α_4 because $(0, 0, 0, 1)$ is not in S. In fact, we cannot determine what α_4 must be, so we will guess that it can be any nonzero number.

Thus, we conclude that

$$f(x_1, x_2, x_3, x_4) = \alpha_4 x_4$$

where α_4 is any nonzero number and that $\beta = 0$.

This does not finish the solution—all we have done so far is to produce what seems like a reasonable guess. To complete the solution, we have to show that the guess is correct.

Note. A comment is worthwhile here. When you are actually writing a solution, you do not need to tell the reader how you guessed the answer. You are only obliged to show that your guess is correct. Here is a completely acceptable solution.

We will show that $f = (0, 0, 0, \alpha_4)$ and that $\beta = 0$. To check this, note that every point in S is of the form $(x_1, x_2, x_3, 0)$, so that

$$f(x_1, x_2, x_3, 0) = 0x_1 + 0x_2 + 0x_3 + \alpha_4 \cdot 0 = 0$$

for any real number α_4, which completes the proof.

$\square$

Geometry of Linear Functionals

If f is a linear functional on $\mathbb{R}^n$ and α is a scalar, then the set of all points $x \in \mathbb{R}^n$ such that $f(x) = \alpha$ is denoted by $f^{-1}(\alpha)$, that is,

$$f^{-1}(\alpha) = \{\, x \in \mathbb{R}^n \ : \ f(x) = \alpha \,\}.$$

The set $f^{-1}(\alpha)$ is called the ***counterimage*** of α under f or the ***inverse image*** of α under f.

Note that $f^{-1}(\alpha)$ consists of *all* points $x \in \mathbb{R}^n$ such that $f(x) = \alpha$, and it should be stressed that $f^{-1}(\alpha)$ is a *set*. The notation is **not** meant to imply that f is invertible! In particular, one must *avoid thinking that $f^{-1}(f(x)) = x$.**

If f is a nonzero linear functional on $\mathbb{R}^n$, then the **kernel** of f, denoted by $\ker(f)$, is the set of all $x \in \mathbb{R}^n$ such that $f(x) = 0$, that is,

$$\ker(f) = f^{-1}(0) = \{\, x \in \mathbb{R}^n \, : \, f(x) = 0 \,\}.$$

The next theorem shows that this is actually a subspace of $\mathbb{R}^n$.

Theorem 1.6.6. *If f is a linear functional on $\mathbb{R}^n$ and $f \neq 0$, then*

$$f^{-1}(0) = \ker(f) = \{\, x \in \mathbb{R}^n \, : \, f(x) = 0 \,\}$$

is a subspace of $\mathbb{R}^n$.

Proof. If $x, y \in f^{-1}(0)$, then

$$f(x + y) = f(x) + f(y) = 0 + 0 = 0,$$

so $x + y \in f^{-1}(0)$.

If $\lambda \in \mathbb{R}$ and $x \in f^{-1}(0)$, then

$$f(\lambda x) = \lambda f(x) = \lambda \cdot 0 = 0,$$

so $\lambda x \in f^{-1}(0)$.

Therefore, $f^{-1}(0)$ is closed under addition and scalar multiplication and hence is a subspace.

$\square$

Note. If f is a linear functional on $\mathbb{R}^n$ and $f \neq 0$, then there is an $a \in \mathbb{R}^n$ such that $f(a) \neq 0$, so that $a \notin f^{-1}(0)$ and $f^{-1}(0) \subsetneq \mathbb{R}^n$, that is, $f^{-1}(0)$ is a *proper subspace* of $\mathbb{R}^n$.

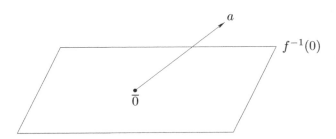

---

* The only thing that you can say is that $x \in f^{-1}(f(x))$.

The next theorem is reminiscent of the orthogonal projection theorem in linear algebra.

Theorem 1.6.7. *Let f be a nonzero linear functional on $\mathbb{R}^n$ and let $a \in \mathbb{R}^n$ be a vector such that $f(a) \neq 0$, then any point $p \in \mathbb{R}^n$ can be written uniquely as*

$$p = \lambda a + x$$

where $\lambda \in \mathbb{R}$ and $x \in f^{-1}(0)$.

Proof. Consider the vector $x = p - \lambda a$. Since f is a linear functional,

$$f(x) = f(p) - \lambda f(a),$$

and since $f(a) \neq 0$, then $f(x) = 0$ when $\lambda = \dfrac{f(p)}{f(a)}$. Therefore,

$$x_0 = p - \frac{f(p)}{f(a)} a \in f^{-1}(0),$$

so that

$$p = \frac{f(p)}{f(a)} a + x_0,$$

where $x_0 \in f^{-1}(0)$.

To show that this representation is unique, suppose that

$$p = x_1 + \mu a,$$

where $\mu \in \mathbb{R}$ and $x_1 \in f^{-1}(0)$, then

$$f(p) = f(x_1) + \mu f(a) = 0 + \mu f(a) = \mu f(a),$$

so that

$$\mu = \frac{f(p)}{f(a)} \qquad \text{and} \qquad \mu = \lambda.$$

Therefore,

$$p = x_1 + \frac{f(p)}{f(a)} a,$$

and since

$$p = x_0 + \frac{f(p)}{f(a)} a,$$

then $x_1 = x_0$.

$\square$

Note that in the theorem, a can be *any* fixed vector with $f(a) \neq 0$.

Now let

$$\mathcal{B} = \{\, e_1, e_2, \ldots, e_k \,\}$$

be a basis for the subspace $f^{-1}(0)$, then the set

$$\mathcal{B}' = \{\, a, e_1, e_2, \ldots, e_k \,\}$$

is linearly independent since $a \notin f^{-1}(0)$, and the set also spans $\mathbb{R}^n$ by the previous theorem. Therefore, it forms a basis for $\mathbb{R}^n$, and hence, $n = k + 1$, that is, $k = n - 1$. Thus, we have shown the following theorem.

Theorem 1.6.8. *If f is a nonzero linear functional on $\mathbb{R}^n$, then $f^{-1}(0)$ is a subspace of dimension $n - 1$, that is, $f^{-1}(0)$ is a subspace of* **codimension** *1.*

1.6.2 Hyperplanes

In $\mathbb{R}^2$, the set of points satisfying a linear equation such as

$$3x_1 + 4x_2 = 7$$

is a straight line.

In $\mathbb{R}^3$, the set of points satisfying a linear equation such as

$$2x_1 - 6x_2 - x_3 = 5$$

is a plane.

More generally, if $\alpha_1, \alpha_2, \ldots, \alpha_n$ and β are constant scalars and if at least one α_i is nonzero, then the set of all points $(x_1, x_2, \ldots, x_n) \in \mathbb{R}^n$ that satisfy the linear equation

$$\alpha_1 x_1 + \alpha_2 x_2 + \cdots + \alpha_n x_n = \beta \qquad (*)$$

is called a *hyperplane* in $\mathbb{R}^n$.

In other words, a hyperplane is a set H_β in $\mathbb{R}^n$ defined by

$$H_\beta = \{\, (x_1, x_2, \ldots, x_n) \in \mathbb{R}^n \, : \, \alpha_1 x_1 + \alpha_2 x_2 + \cdots + \alpha_n x_n = \beta \,\}. \quad (**)$$

In terms of linear functionals, we can rewrite equation $(*)$ as

$$f(x) = \beta$$

and we can rewrite equation $(**)$ as

$$H_\beta = f^{-1}(\beta) = \{ x \in \mathbb{R}^n : f(x) = \beta \}$$

where x is the point $(x_1, x_2, \ldots, x_n)$ and f is a nonzero linear functional on $\mathbb{R}^n$ represented by the vector $(\alpha_1, \alpha_2, \ldots, \alpha_n)$.*

Thus, in $\mathbb{R}^2$, a hyperplane is a line, and in $\mathbb{R}^3$, a hyperplane is a plane. The hyperplane in Example 1.6.5 is readily identified with $\mathbb{R}^3$.

In all three cases, the hyperplane is one dimension less than the dimension of the space in which it lives. We know from Theorem 1.6.8 that in $\mathbb{R}^n$, the hyperplane

$$f^{-1}(0) = \{(x_1, x_2, \ldots, x_n) : \alpha_1 x_1 + \alpha_2 x_2 + \cdots + \alpha_n x_n = 0\}$$

is actually a subspace of dimension $n - 1$.

Note. We can also see this as follows: in $\mathbb{R}^n$, let H be the hyperplane $f^{-1}(0)$, where f is a nonzero linear functional.

To see that the subspace H has dimension $n - 1$, note that by definition, H is the set of all solutions $(x_1, x_2, \ldots, x_n)$ to the equation

$$\alpha_1 x_1 + \alpha_2 x_2 + \cdots + \alpha_n x_n = 0,$$

where at least one α_k is nonzero.

However, we know from the theory of linear equations in linear algebra that the solution space to this equation contains $n - 1$ linearly independent vectors and no more than $n - 1$ linearly independent vectors, which is to say that H has dimension $n - 1$.

Example 1.6.9. *Show that if $S = \{v_1, v_2, \ldots, v_{n-1}\}$ is a linearly independent set of vectors in $\mathbb{R}^n$, then the linear subspace V spanned by S is a hyperplane.*

Solution. Perhaps the quickest way to see this is to use determinants. Suppose that the vectors v_k are

$$v_1 = (\alpha_{1,1}, \alpha_{1,2}, \ldots, \alpha_{1,n})$$
$$v_2 = (\alpha_{2,1}, \alpha_{2,2}, \ldots, \alpha_{2,n})$$
$$\vdots$$
$$v_{n-1} = (\alpha_{n-1,1}, \alpha_{n-1,2}, \ldots, \alpha_{n-1,n}).$$

---

* Note that the hyperplane H_β depends on both the vector $(\alpha_1, \alpha_2, \ldots, \alpha_n) \in \mathbb{R}^n$ *and the* scalar β.

Let x be the vector $(x_1, x_2, \ldots, x_n)$, and consider the determinantal equation

$$\begin{vmatrix} x_1 & x_2 & \cdots & x_n \\ \alpha_{1,1} & \alpha_{1,2} & \cdots & \alpha_{1,n} \\ \alpha_{2,1} & \alpha_{2,2} & \cdots & \alpha_{2,n} \\ \vdots & \vdots & \ddots & \vdots \\ \alpha_{n-1,1} & \alpha_{n-1,2} & \cdots & \alpha_{n-1,n} \end{vmatrix} = 0. \qquad (*)$$

When we expand this in terms of cofactors of the first row, we obtain

$$A_{11}x_1 + A_{12}x_2 + \cdots + A_{1n}x_n = 0,$$

where A_{ij} denotes the cofactor of the ith row and jth column. Since the cofactors A_{ij} are constants, we recognize $(*)$ as the equation of a hyperplane. It is clear that this hyperplane contains the vectors v_k. For example, to see that v_1 satisfies $(*)$, we substitute v_1 for x in the left side of $(*)$ and get

$$\begin{vmatrix} \alpha_{1,1} & \alpha_{1,2} & \cdots & \alpha_{1,n} \\ \alpha_{1,1} & \alpha_{1,2} & \cdots & \alpha_{1,n} \\ \alpha_{2,1} & \alpha_{2,2} & \cdots & \alpha_{2,n} \\ \vdots & \vdots & \ddots & \vdots \\ \alpha_{n-1,1} & \alpha_{n-1,2} & \cdots & \alpha_{n-1,n} \end{vmatrix}$$

which must be zero, since it contains two identical rows.

$\square$

In the next theorem, we prove that the hyperplane

$$H_\beta = f^{-1}(\beta),$$

where $\beta \neq 0$, is a translate of the subspace $f^{-1}(0)$ and so $f^{-1}(\beta)$ must have the same "dimension" as $f^{-1}(0)$.

Theorem 1.6.10. *Let f be a nonzero linear functional on $\mathbb{R}^n$, $\beta \in \mathbb{R}$, with $\beta \neq 0$, and let*

$$H_\beta = f^{-1}(\beta).$$

If $a \in \mathbb{R}^n$ is any point such that $f(a) = \beta$, then

$$H_\beta = a + f^{-1}(0),{}^*$$

that is, the hyperplane H_β is just a translate of a subspace of dimension $n - 1$.

---

* Because $\{a\} \cap f^{-1}(0) = \emptyset$, this is sometimes written as $a \oplus f^{-1}(0)$ and called a ***direct sum***.

Proof. Let $x \in H_\beta$, then

$$f(x) = \beta = f(a),$$

so that

$$f(x - a) = \beta - \beta = 0,$$

that is,

$$x - a \in f^{-1}(0).$$

Therefore, $H_\beta \subseteq a + f^{-1}(0)$.

Conversely, if $x \in a + f^{-1}(0)$, then

$$x = a + z,$$

where $z \in f^{-1}(0)$, that is, $f(z) = 0$. Therefore,

$$f(x) = f(a) = \beta.$$

Hence, $x \in H_\beta$ and $a + f^{-1}(0) \subseteq H_\beta$.

□

Remark. Hyperplanes in infinite dimensional spaces are defined as being *maximal proper subspaces*, or *translates of maximal proper subspaces*. Theorems 1.6.8 and 1.6.10 show that in $\mathbb{R}^n$ at least, this definition and the one we gave are equivalent.

Note. Summarizing all this, any hyperplane

$$H_\beta = \{\, x \in \mathbb{R}^n \; : \; f(x) = \beta \,\}$$

can be written as

$$H_\beta = \{\, x \in \mathbb{R}^n \; : \; \langle p, x \rangle = \beta \,\}$$

for some fixed $p \neq \overline{0}$ in $\mathbb{R}^n$,

and

$$f^{-1}(0) = \{\, x \in \mathbb{R}^n \; : \; \langle p, x \rangle = 0 \,\}.$$

Thus, $f^{-1}(0)$ is the subspace of all vectors orthogonal to p, and H_β is just a translate of this subspace, as shown in the following figure.

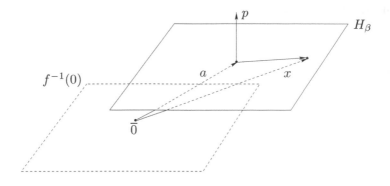

Here

$$H_\beta = a + f^{-1}(0),$$

where $\langle p, a \rangle = \beta$, so that $x \in H_\beta$ if and only if

$$f(x) = f(a + x - a) = f(a) + f(x - a) = \langle p, a \rangle + \langle p, x - a \rangle = \beta + 0 = \beta.$$

Example 1.6.11. *If $p = (4, -2)$ and $\beta = 1$, sketch the hyperplane in $\mathbb{R}^2$ determined by p and β.*

Solution. The hyperplane is the set

$$H_\beta = \{ (x, y) \in \mathbb{R}^2 : 4x - 2y = 1 \},$$

since in $\mathbb{R}^2$, a hyperplane is just a line, a translate of a subspace of dimension 1. The line passes through the point $a = (\frac{1}{4}, 0)$ and is perpendicular to the vector $p = (4, -2)$. The hyperplane is sketched below.

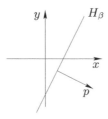

□

Example 1.6.12. *If $p = (4, -2, 3)$ and $\beta = 1$, sketch the hyperplane in $\mathbb{R}^3$ determined by p and β.*

Solution. The hyperplane is the set

$$H_\beta = \{\, (x, y, z) \in \mathbb{R}^3 \ : \ 4x - 2y + 3z = 1 \,\},$$

since in $\mathbb{R}^3$, a hyperplane is just a plane, a translate of a subspace of dimension 2. The plane passes through the point $a = (\frac{1}{4}, 0, 0)$ and is perpendicular to the vector $p = (4, -2, 3)$. The portion of the hyperplane in one octant is sketched below.

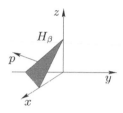

Halfspaces

A hyperplane determines two **halfspaces**, one on each side of the hyperplane, and the most fundamental property of the hyperplane is that it divides the space $\mathbb{R}^n$ into three disjoint parts:

- the hyperplane itself

$$H = \{\, x \in \mathbb{R}^n \ : \ f(x) = \beta \,\}$$

- an **open halfspace** to one side of the hyperplane

$$H^+ = \{\, x \in \mathbb{R}^n \ : \ f(x) > \beta \,\}$$

- the open halfspace on the other side of the hyperplane

$$H^- = \{\, x \in \mathbb{R}^n \ : \ f(x) < \beta \,\}.$$

If the hyperplane is adjoined to either of the open halfspaces, the result is a set of the type

$$\{\, x \in \mathbb{R}^n \ : \ f(x) \le \beta \,\} \quad \text{or} \quad \{\, x \in \mathbb{R}^n \ : \ f(x) \ge \beta \,\}.$$

Such sets are called **closed halfspaces**.

A hyperplane H separates $\mathbb{R}^n$ in the following sense.

Theorem 1.6.13. *If the point x is in one of the open halfspaces determined by a hyperplane H and y is in the other open halfspace, then the line segment (x, y) intersects the hyperplane H at precisely one point z.*

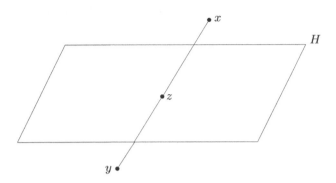

Proof. Suppose that the hyperplane H is given by

$$H = \{\, z \in \mathbb{R}^n \; : \; \langle p, z \rangle = \lambda \,\}$$

for some scalar λ and some nonzero vector $p \in \mathbb{R}^n$. Since x and y are in different halfspaces, we may assume that

$$\alpha = \langle p, x \rangle < \lambda < \langle p, y \rangle = \beta.$$

We have to show that there is a point $w \in (x, y)$ such that $\langle p, w \rangle = \lambda$.

Let $w = (1 - \mu)x + \mu y$ be a typical point on the line through x and y. We want to find a scalar μ with $0 < \mu < 1$ such that

$$\langle p, (1 - \mu)x + \mu y \rangle = \lambda,$$

that is,

$$(1 - \mu)\langle p, x \rangle + \mu \langle p, y \rangle = \lambda,$$

or

$$(1 - \mu)\alpha + \mu\beta = \lambda.$$

Solving this equation for μ, we have

$$\mu = \frac{\lambda - \alpha}{\beta - \alpha},$$

and since $\alpha < \lambda < \beta$, then $0 < \mu < 1$, that is, w is in the line segment (x, y).

$\square$

Example 1.6.14. *Given $f = (1, -2, -1, 4)$, show that the hyperplane $f^{-1}(10)$ misses the closed unit ball.*

Solution. Since the norm has not been specified, we assume that we are dealing with the Euclidean ball. We will solve this problem by showing that the ball lies to one side of the hyperplane, that is, we will show that the ball is contained in the open halfspace $\{\, x \in \mathbb{R}^4 : f(x) < 10 \,\}.$*

Now, a typical point of the unit ball $x = (x_1, x_2, x_3, x_4)$ has

$$x_1^2 + x_2^2 + x_3^2 + x_4^2 \leq 1.$$

Consequently, we know for certain that for each coordinate x_k, the absolute value $|x_k|$ must be no greater than 1. Thus,

$$\begin{aligned} f(x) &= x_1 - 2x_2 - x_3 + 4x_4 \\ &\leq |x_1| + 2|x_2| + |x_3| + 4|x_4| \\ &\leq 1 + 2 + 1 + 4, \end{aligned}$$

which shows that $f(x) < 10$ and

$$\overline{B}(\overline{0}, 1) \subsetneq \{\, x \in \mathbb{R}^4 : f(x) < 10 \,\}.$$

$\square$

Example 1.6.15. *Given that $f = (3, -1, 0, 2)$, find the point where the line through $(1, 0, 0, 2)$ parallel to $(1, 1, -1, 3)$ intersects the hyperplane $f^{-1}(1)$.*

Solution. The problem asks us to find the point x on the line for which $f(x) = 1$.

A typical point $x = (x_1, x_2, x_3, x_4)$ on the line can be written as

$$(x_1, x_2, x_3, x_4) = (1, 0, 0, 2) + \mu(1, 1, -1, 3) = (1 + \mu, \mu, -\mu, 2 + 3\mu)$$

for some scalar μ.

Therefore,

$$f(x) = f(1 + \mu, \mu, -\mu, 2 + 3\mu) = 3(1 + \mu) - 1\mu + 0 + 2(2 + 3\mu) = 7 + 6\mu,$$

and setting this equal to 1 yields $\mu = -1$. Hence, the point of intersection of the line and hyperplane is

$$x = (1 + \mu, \mu, -\mu, 2 + 3\mu) = (0, -1, 1, -1).$$

$\square$

---

* Can you see why we know immediately that the unit ball cannot be contained in the other open halfspace?

Example 1.6.16. *Given that x and y are points in $\mathbb{R}^n$ and that f is a linear functional with $f(x) = 3$ and $f(y) = 9$, find the hyperplane determined by f that contains the midpoint of the straight line segment joining x and y.*

Solution. The midpoint of the segment is

$$z = \tfrac{1}{2}x + \tfrac{1}{2}y,$$

and hence, the value of the linear functional f at the midpoint is

$$f(z) = f\left(\tfrac{1}{2}x + \tfrac{1}{2}y\right) = \tfrac{1}{2}f(x) + \tfrac{1}{2}f(y) = \tfrac{1}{2} \cdot 3 + \tfrac{1}{2} \cdot 9 = 6.$$

Therefore, the hyperplane is $f^{-1}(6)$.

$\square$

We will need the following theorem. Although it is possible to prove it at this point, a simpler proof will be given later in the text.

Theorem 1.6.17. *Given the hyperplane $H = f^{-1}(\beta)$ in $\mathbb{R}^n$, where f is represented by p, the point on H that has minimum Euclidean norm is the point where the straight line through $\overline{0}$ in the direction of p intersects H.*

Proof. Suppose that the line through $\overline{0}$ in the direction of p intersects the hyperplane H at x_0, as depicted in the figure below.

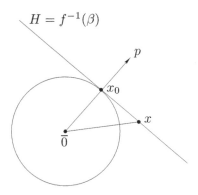

Let x be an arbitrary point in H, then

$$\langle p, x - x_0 \rangle = \langle p, x \rangle - \langle p, x_0 \rangle = \beta - \beta = 0,$$

so that the vector $x - x_0$ is orthogonal to p, and from the Pythagorean theorem,

$$\|x\|_2^2 = \|x_0\|_2^2 + \|x - x_0\|_2^2 \geq \|x_0\|_2^2,$$

so that

$$\|x\|_2 \geq \|x_0\|_2$$

for all $x \in H$.

Therefore, x_0 is the point in the hyperplane H that has minimum Euclidean norm.

$\square$

Corollary 1.6.18. *In the hyperplane $H = f^{-1}(\beta)$, where f is represented by p, the point closest to $\bar{0}$ is*

$$x_0 = \frac{\beta}{\|p\|_2^2} \, p$$

and $\|x_0\|_2 = \dfrac{|\beta|}{\|p\|_2}$.

Proof. Since

$$f(x_0) - \frac{\beta}{\|p\|_2^2} f(p) = \frac{\beta}{\|p\|_2^2} \langle p, p \rangle = \beta,$$

by the previous theorem, the point x_0 is precisely the point in $H = f^{-1}(\beta)$, which is closest to $\bar{0}$.

Also,

$$\|x_0\|_2 = \frac{|\beta|}{\|p\|_2^2} \|p\|_2 = \frac{|\beta|}{\|p\|_2}.$$

$\square$

The word "normal" in geometry is also synonymous with perpendicularity. If H is the hyperplane $f^{-1}(\beta)$, the vector p representing the nonzero linear functional f is often called a ***normal vector*** to the hyperplane. Of course, any nonzero multiple of p is also a normal vector.

For example, recall that for the plane

$$ax + by + cz = d$$

in $\mathbb{R}^3$, the distance from the plane to the origin is given by the formula

$$\rho = \frac{|d|}{\sqrt{a^2 + b^2 + c^2}},$$

as in the corollary above. You may also recall that the equation

$$\frac{ax + by + cz}{\sqrt{a^2 + b^2 + c^2}} = \frac{d}{\sqrt{a^2 + b^2 + c^2}}$$

is sometimes called the *normal form* of the equation of the plane $ax + by + cz = d$.

Example 1.6.19. *Find the point $z_0 = (x_0, y_0)$ on the line*

$$L = \{\, z = (x, y) \in \mathbb{R}^2 \; : \; 3x - 2y = 1 \,\}$$

in $\mathbb{R}^2$ that is closest to the origin and find the norm of z_0.

Solution. Since no distance function is specified, we assume that the Euclidean distance is to be used, and, hence, the previous theorem will apply.

The line L is really the hyperplane $H = f^{-1}(1)$, where f is represented by the vector $p = (3, -2)$. According to the theorem, the point z in question is where the line through $\bar{0} = (0, 0)$ and $p = (3, -2)$ intersects H.

Now, a typical point on the line through $(0, 0)$ and $(3, -2)$ is

$$z = (x, y) = (0, 0) + \mu(3, -2) = (3\mu, -2\mu)$$

and we are looking for a value of $\mu \in \mathbb{R}$ such that

$$f(x, y) = \langle p, z \rangle = 3 \cdot (3\mu) - 2 \cdot (-2\mu) = 13\mu = 1.$$

This yields $\mu = \frac{1}{13}$, and the point z_0 we want is

$$z_0 = (3\mu, -2\mu) = \left(\tfrac{3}{13}, -\tfrac{2}{13}\right),$$

with norm

$$\|z_0\|_2 = \left\|\left(\tfrac{3}{13}, -\tfrac{2}{13}\right)\right\| = \sqrt{\left(\tfrac{3}{13}\right)^2 + \left(-\tfrac{2}{13}\right)^2} = \tfrac{1}{\sqrt{13}}.$$

Remark. In the preceding example, the distance from z_0 to $\bar{0}$ is $\frac{1}{\sqrt{13}}$. This means that the hyperplane H is tangent to the closed ball $\overline{B}(\bar{0}, \frac{1}{\sqrt{13}})$ at the point $\left(\frac{3}{13}, -\frac{2}{13}\right)$. The next example uses the correspondence between tangency and points of minimum ℓ_2 norm.

Example 1.6.20. *Let f be the linear functional on $\mathbb{R}^4$ with the Euclidean norm represented by the vector $p = (-1, 2, -1, 1)$. Find the point on the closed unit ball in $\mathbb{R}^4$, where f attains its maximum value and find that maximum value.*

Solution. First, we make sure the question is clear. For every point x in $\overline{B}(\bar{0}, 1)$, $f(x)$ has a specific value. What we want to find is the point where this value is a maximum.

Next, we interpret this geometrically. Suppose that the maximum happens to be β and that it occurs at the point x_0. Then $f(x_0) = \beta$, and for every other point y in $\overline{B}(\overline{0}, 1)$, we must have $f(y) < \beta$.

However, this would mean that the hyperplane H, whose equation is $f(x) = \beta$, touches the closed unit ball at precisely x_0. Furthermore, we know that in this situation, the point x_0 is where the line through $\overline{0}$ would intersect the hyperplane H. Thus, if we can find x_0, then we will have found the solution.

There is something else about x_0 that we can exploit. Since x_0 is the point of tangency to the unit ball, it cannot be inside the ball, and therefore, $\|x_0\|_2 = 1$. Thus, the problem reduces to the following: find the point where the line through $\overline{0}$ and p intersects the unit sphere. There will be two such points, and one of them will be the one we want.

Now we proceed with the solution. We want to find the points x in $\mathbb{R}^4$ where the line through $\overline{0} = (0, 0, 0, 0)$ and $p = (-1, 2, -1, 1)$ pierces the unit sphere

$$S = \{\, x \in \mathbb{R}^4 : \|x\|_2 = 1 \,\}.$$

The line has the equation

$$x = (1 - \mu)\overline{0} + \mu(-1, 2, -1, 1) = \mu(-1, 2, -1, 1), \quad -\infty < \mu < \infty$$

and we want to find values of μ such that $\|x\|_2 = 1$.

Now,

$$\|x\|_2 = \|\mu(-1, 2, -1, 1)\| = |\mu|\sqrt{(-1)^2 + 2^2 + (-1)^2 + 1^2}$$

so that

$$\|x\|_2 = \sqrt{7}\,|\mu|,$$

and, therefore, $\|x\|_2 = 1$ when $|\mu| = \frac{1}{\sqrt{7}}$. The two places where the line intersects the sphere are

$$\frac{1}{\sqrt{7}}(-1, 2, -1, 1) \quad \text{and} \quad -\frac{1}{\sqrt{7}}(-1, 2, -1, 1).$$

Checking the value of f at each of these points, we have

$$f\left(\frac{1}{\sqrt{7}}(-1, 2, -1, 1)\right) = \sqrt{7} \quad \text{and} \quad f\left(-\frac{1}{\sqrt{7}}(-1, 2, -1, 1)\right) = -\sqrt{7}.$$

Therefore, f attains a maximum value of $\sqrt{7}$ on the closed unit ball, and it attains that value at the point

$$x_0 = \frac{1}{\sqrt{7}}(-1, 2, -1, 1).$$

$\square$

Example 1.6.21. *If p is a point in the unit sphere in $\mathbb{R}^n$, then the hyperplane H whose equation is $\langle p, x \rangle = 1$ is tangent to the unit sphere at p, and only at the point p (here the Euclidean norm is being used).*

Solution. Since p is on the sphere, then $\|p\|_2 = 1$. Taking $x = p$ in the equation of the hyperplane, we have

$$\langle p, x \rangle = \langle p, p \rangle = \|p\|^2 = 1,$$

and thus $p \in H$.

Now let $q \in S\left(\overline{0}, 1\right)$, where $q \neq p$. We want to show that $x = q$ does not satisfy the equation for H. We recall that in the triangle inequality

$$\langle x, y \rangle \leq \|x\|_2 \cdot \|y\|_2,$$

and equality holds if and only if one of x or y is a nonnegative multiple of the other.

The only nonnegative multiple of p that belongs to $S\left(\overline{0}, 1\right)$ is p itself, and hence, neither p nor q is a nonnegative multiple of the other, so that

$$\langle p, q \rangle < \|p\|_2 \cdot \|q\|_2 = 1 \cdot 1 = 1.$$

Therefore, $q \notin H$.

$\square$

Example 1.6.22. *Let S be the unit sphere (in the Euclidean norm) in $\mathbb{R}^n$. Suppose that H is the hyperplane whose equation is $\langle p, x \rangle = \alpha$, where $0 < \alpha < 1$. Let $T = S \cap H$. Show that T is a sphere in H, that is, show that there is some point q in H and some constant δ such that $\|x - q\|_2 = \delta$ for all $x \in T$.*

Solution. The vector p is orthogonal to the hyperplane H, and if $x \in T = S \cap H$, then

$$\alpha = \langle p, x \rangle \leq \|p\|_2 \cdot \|x\|_2 = \|p\|_2,$$

so that $\|p\|_2 \geq \alpha$.

The equation of the line L through $\overline{0}$ in the direction of p is

$$L: \quad x = \mu p, \quad 0 < \mu < \infty$$

and this line intersects H at the point $x = \mu p$ where

$$\langle p, x \rangle = \mu \langle p, p \rangle = \alpha,$$

so that

$$\mu = \frac{\alpha}{\langle p, p \rangle} = \frac{\alpha}{\|p\|_2^2}.$$

Thus, the point in H closest to $\overline{0}$ is the point

$$q = \frac{\alpha}{\|p\|_2^2} \cdot p$$

so that $\|q\|_2 = \dfrac{\alpha}{\|p\|_2}$, and hence, $\|q\|_2 \leq 1$.

Now let $x \in T = S \cap H$ and let $z = x - q$, then z is orthogonal to p, since

$$\langle z, p \rangle = \langle x - q, p \rangle = \langle x, p \rangle - \langle q, p \rangle = \alpha - \alpha = 0,$$

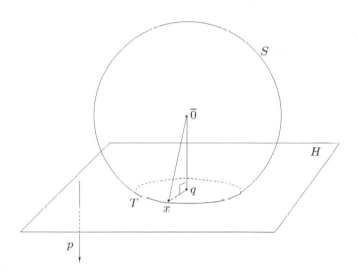

and from the Pythagorean theorem, we have

$$\|x - q\|_2^2 + \|q\|_2^2 = \|x\|_2^2 = 1,$$

that is, $\|x - q\|_2^2 = 1 - \dfrac{\alpha^2}{\|p\|_2^2}$ for all $x \in T = S \cap H$. Taking

$$\delta = \sqrt{1 - \frac{\alpha^2}{\|p\|_2^2}},$$

then $0 \leq \delta < 1$ and

$$\|x - q\|_2 = \delta$$

for all $x \in T = S \cap H$, that is, T is a sphere in H.

$\square$

1.6.3 Problems

In the following exercises, unless otherwise stated, assume that the closed unit ball is the closed unit ball in the Euclidean norm.

*1. Find a hyperplane $H = f^{-1}(1)$ in $\mathbb{R}^4$ that is tangent to the unit cube at the point $\left(1, \frac{1}{2}, \frac{1}{3}, \frac{1}{4}\right)$. Verify your answer.

2. Draw each of the following hyperplanes in $\mathbb{R}^2$:

   (a) The hyperplane through the point $(1, 3)$ that is perpendicular to the line through $(0, 0)$ and $(1, 3)$,

   (b) The hyperplane that is tangent to the unit sphere at the point $\left(\frac{\sqrt{3}}{2}, \frac{1}{2}\right)$,

   (c) The hyperplane whose equation is $-\alpha_1 + 2\alpha_2 = 3$,

   (d) $f^{-1}(0)$ where f is represented by $p = (3, 1)$,

   (e) $f^{-1}(1)$ where f is represented by $p = (3, 1)$,

   (f) $f^{-1}(2)$ where f is represented by $p = (3, 1)$.

3. Find an equation for the hyperplane of

   (a) Problem 2 (a),

   (b) Problem 2 (b).

4. Find the point of intersection of the plane

$$H = \left\{ (x_1, x_2, x_3) \in \mathbb{R}^3 \ : \ 2x_1 - 3x_2 + x_3 = 2 \right\}$$

   with the line through $(1, 0, 1)$ and $(-2, 1, 2)$.

5. Given the linear functional $f(x_1, x_2) = 4x_1 - 3x_2$, find

   (a) the point x on the closed unit ball where $f(x)$ is a maximum,

   (b) the point x in the hyperplane $f^{-1}(2)$ that is closest to the origin,

   (c) the point x in the hyperplane $f^{-1}(3)$ that is closest to the origin.

6. Given the linear functional $f(x_1, x_2, x_3) = 2x_1 - 3x_2 + x_3$, find

   (a) the point x on the unit ball where $f(x)$ is a maximum,

   (b) the point x in the hyperplane $f^{-1}(2)$ that is closest to the origin,

   (c) the point x in the hyperplane $f^{-1}(3)$ that is closest to the origin.

7. Let f be the linear functional on $\mathbb{R}^3$ represented by the vector $p = (3, -2, -3)$ and let S be the set

$$S = \left\{ (1, 1, -2), \ (-3, 4, 1), \ (60, 10, 15), \ (-8, -2, 4), \ (0, 1, 1) \right\}.$$

   (a) Determine which points of S are on the same side of $f^{-1}(0)$,

   (b) Which point or points of S are closest to $f^{-1}(0)$?

(c) Which points of S are on the same side of $f^{-1}(8)$ as the origin?

(d) Find the point or points of S that are closest to $f^{-1}(8)$.

8. Given the line $L = \{(x, y) : 3x + 4y = 5\}$ in $\mathbb{R}^2$, find the point on L of minimum norm in each of the following cases, and draw a figure with the appropriate unit ball:

(a) for the Euclidean norm,

(b) for the ℓ_1 norm,

(c) for the ℓ_∞ norm.

9. Given that H is the hyperplane $f^{-1}(2)$, and given that $g = 4f$, find β such that $g^{-1}(\beta)$ is exactly the same as H.

10. Let L be the line

$$L = \{x \in \mathbb{R}^n : x = p + \mu q, \ -\infty < \mu < \infty\},$$

where p and q are distinct points in $\mathbb{R}^n$, and let f be a linear functional on $\mathbb{R}^n$ such that $f(p) = 3$ and $f(q) = -1$. Find

(a) the point where L intersects the hyperplane $f^{-1}(1)$; in other words, find the scalar μ such that $f(p + \mu q) = 1$,

(b) the scalar β such that the hyperplane $f^{-1}(\beta)$ intersects the line at the point $x = p + \mu q$ where $\mu = 3.4$.

11. Let L be the line

$$L = \{x \in \mathbb{R}^n : x = \mu p + (1 - \mu)q, \ -\infty < \mu < \infty\},$$

where p and q are distinct points in $\mathbb{R}^n$, and let f be a linear functional on $\mathbb{R}^n$ such that $f(p) = 6$ and $f(q) = 1$. Find

(a) the point where L intersects the hyperplane $f^{-1}(-2)$,

(b) the scalar β such that the hyperplane $f^{-1}(\beta)$ passes through the midpoint of the line segment joining p and q.

12. (a) If a hyperplane in $\mathbb{R}^n$, where $n > 1$, meets a straight line at two distinct points, show that the hyperplane contains the straight line.
    Hint. Let H be the hyperplane $f^{-1}(\alpha)$ and let L be a straight line that intersects H at two distinct points p and q.

(b) Consequently, show that a hyperplane H and a straight line L must be related in exactly one of the following ways:

   (i) H and L intersect in exactly one point,

   (ii) $L \subset H$,

   (iii) L misses H.

13. Given that $H = f^{-1}(1)$, where the linear functional f on $\mathbb{R}^4$ is represented by the vector $p = (1, 0, 1, -1)$, find

    (a) a line L_1 through $\overline{0}$ that intersects H in exactly one point,

    (b) a line L_2 through $\overline{0}$ that misses H.

14. If a hyperplane $f^{-1}(\alpha)$ misses the straight line L, then for some scalar β, the hyperplane $f^{-1}(\beta)$ contains L.
    Hint. Let p and q be points on the line. If $f(p) \neq f(q)$, then for every real number α, there is a unique solution δ to the equation $\delta f(p) + (1 - \delta)f(q) = \alpha$ (why?). Thus, $f^{-1}(\alpha)$ would intersect L (why?). Therefore, $f(p)$ and $f(q)$ must be the same. Let $\beta = f(p)$.

15. If the hyperplane $H = f^{-1}(\alpha)$ intersects the straight line L in exactly one point, then for every scalar β, the hyperplane $H_\beta = f^{-1}(\beta)$ intersects L in exactly one point.
    Hint. Conclude that this must happen because of what we know from Problems 12 and 14.

*16. (a) In $\mathbb{R}^3$, the intersection of the closed unit ball with a plane is either the empty set, a single point, or a disk (which is like a closed ball from $\mathbb{R}^2$ embedded in $\mathbb{R}^3$).

    (b) List (without proof) the possible intersections in $\mathbb{R}^2$ of the closed unit ball with a straight line.

    (c) List (without proof) the possible intersections in $\mathbb{R}^4$ of the closed unit ball with a hyperplane.

17. Prove Theorem 1.6.3.

18. If the ℓ_1 or ℓ_∞ norm is used, show that a line may have infinitely many points of minimum norm.

19. Show that a hyperplane in $\mathbb{R}^n$ has a unique point of minimum norm.

20. Use the Cauchy–Schwarz inequality to show that the triangle inequality holds.
    Hint. $\|u + v\|^2 = ?$

21. Show that if f is a linear functional on $\mathbb{R}^n$ and f is represented by the vector $p \in \mathbb{R}^n$, then

$$\|p\| = \max\{\, f(x) \,:\, x \in B \,\} = \max\{\, \langle p, x \rangle \,:\, x \in B \,\},$$

where B is the closed unit ball in $\mathbb{R}^n$.

22. Let f be the linear functional on $\mathbb{R}^4$ represented by $p = (-1, 1, 1, -3)$. Find the point of the hyperplane $f^{-1}(0)$ that is closest to the point $x = (3, -2, 2, 1)$.

23. Develop a general formula for the point q on the hyperplane

$$H_\beta = \{\, x \in \mathbb{R}^n \,:\, \langle p, x \rangle = \beta \,\}$$

that is closest to the point x_0. Assume that $p \neq \overline{0}$.

2 TOPOLOGY

2.1 INTRODUCTION

In this chapter, we give a quick overview of the basic parts of general topology
or point set topology that are needed in the text. The fundamental ideas
concentrate on the notions of *open sets*, *closed sets*, and *compact sets*. There
are several ways to characterize these sets, and these characterizations should
be thoroughly understood. We should emphasize that many of the results
concerning compactness stated here are specific to $\mathbb{R}^n$.

The following is some of the terminology that we introduce:

*Open set, closed set, compact set, accumulation point, limit point, interior
point, bounded set, boundary, boundary points, interior, closure, sequences,
subsequences, convergent sequences, Cauchy sequences, topology, norm topology, supremum and infimum of a set of real numbers.*

In basic line and circle geometry of the plane, you may have used a sentence
such as

"*Let L be a line through the point p that is tangent to the circle C.*"

This sentence is meaningful to us because we all know how to use a straightedge and compass to actually construct a tangent line.

Geometry of Convex Sets, First Edition. I. E. Leonard and J. E. Lewis.
© 2016 John Wiley & Sons, Inc. Published 2016 by John Wiley & Sons, Inc.

We want to be able to make statements such as this about more general types of sets. For example, there seems to be no doubt that we can find a line through p that is tangent to the set shown in the figure on the right.

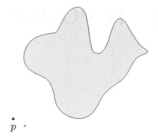

$\overset{\bullet}{p}$ ·

Unfortunately, as the figures below show, there are sets for which a statement of this type is not always meaningful. We would like to characterize those sets for which it is.

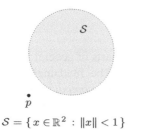

$S = \{\, x \in \mathbb{R}^2 \,:\, \|x\| < 1 \,\}$

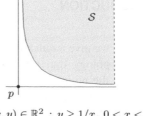

$S = \{\, (x, y) \in \mathbb{R}^2 \,:\, y \geq 1/x,\ 0 < x < \infty \,\}$

The characterization, which is described in detail later, is called *compactness*. Circles and sets, such as the first one above, are compact, while sets such as those in the figure directly above are not. It should be pointed out that the notion of compactness is applicable to questions that are far more complex than the question as to whether or not a tangent line can be found. It is not an exaggeration to say that compactness is among the most useful concepts in all of mathematics. Its application to tangent lines occupies only a very few pages in the story.

Let us try to convince ourselves, in a rather informal way, that a tangent line really does exist for the figure on the right. Of course, we would not expect to be able to use a straightedge and compass to construct the line. Instead, we will use a sort of "thought experiment." It goes like this: First, imagine that we have passed a line through p that cuts the set, like the line L in the figure.

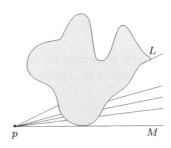

Now let us rotate the line about p in a clockwise direction. Continue rotating the line as long as the line still cuts the set. It is clear that we cannot continue this forever, because the line will eventually miss the set entirely. Thus, there must be some limiting position, like the line M in the above figure. This limiting line will be the tangent line.

If we repeat the process in the figure on the right, we still end up producing a limiting line, but in this case, the limiting line does not touch the set. To check this assertion, note that the limiting line M in the figure is parallel to the x-axis. Now, the point on M that is closest to the origin is at a distance 1 from the origin.

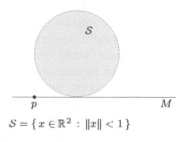

$$S = \{\, x \in \mathbb{R}^2 \,:\, \|x\| < 1 \,\}$$

The set S is the open unit ball in $\mathbb{R}^2$, and so everything in the set is strictly closer to the origin than 1 unit. Thus, there can be no points common to both the set and the limiting line. The problem with the set $S = \{\, x \in \mathbb{R}^2 \,:\, \|x\| < 1 \,\}$ is perhaps obvious: almost everyone will say that the process fails because the set is missing its "boundary."

However, the set $S = \{\, (x, y) \in \mathbb{R}^2 \,:\, y \geq 1/x,\ 0 < x < \infty \,\}$ shows that simply having a boundary is not sufficient. Trying the same process here produces the x-axis as the limiting line, and the set in question does not touch this axis even though it does have a boundary. The failure of the process is even more spectacular when we try to find a tangent line through the point $q = (-1, -1)$, as in the figure below.

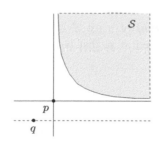

$$S = \{\, (x, y) \in \mathbb{R}^2 \,:\, y \geq 1/x,\ 0 < x < \infty \,\}$$

In this case, as we rotate the line, before reaching the limiting position, the line always intersects the interior of the set. The limiting line, however, not only misses the set, but no point on the limiting line is even close to the set.

It may be obvious why the set S in the above figure is causing a problem—the set contains points that are arbitrarily far from the origin.

2.2 INTERIOR POINTS AND OPEN SETS

The discussion so far has used some terms whose meanings are perhaps only intuitively understood. In addition to *compactness*, we have referred to the *boundary* and the *interior* of a set. We have also talked about a limiting line without giving a concrete explanation of what we really mean by the word *limit*. These sound like ordinary English words, but in fact they all have very precise mathematical meanings. It is common practice to name a mathematical concept with an English word, and often a word is chosen whose English meaning is fairly close to the mathematical meaning. However, the English meaning of the word hardly ever conveys the exact mathematical meaning. Whenever the word is used in a mathematical setting, it is always the mathematical meaning that is understood.

In this and the following sections, we define some of the topological notions that have spread through the rest of mathematics. If you are not already familiar with these, you might find it interesting to reread the introduction after you have learned them.

There are many different norms on $\mathbb{R}^n$, and the following definitions would at first seem to depend upon which particular norm we are using. We will see shortly that this is not the case. If this concerns you, consider for the moment that the norm $\| \cdot \|$ is the ℓ_2 norm, and the balls $B(x, \delta)$ are the balls derived from that norm. If you examine the proofs, you will see that the discussions would be equally valid if we were to use the ℓ_1 or ℓ_∞ norm. In fact, we use only the general properties that are shared by all norms, namely, *nonnegativity*, *homogeneity*, and the *triangle inequality*.

A point x is an **interior point** of a set $S \subset \mathbb{R}^n$ if S contains some open ball of positive radius centered at x, that is, there exists a $\delta > 0$ such that $B(x, \delta) \subset S$.

For example,

(a) In $\mathbb{R}$, the point 0 is an interior point of the interval $[-1, 1]$ since, for example, the open ball

$$B(0, 1) = \{ x \in \mathbb{R} : |x| < 1 \} = \{ x \in \mathbb{R} : -1 < x < 1 \}$$

centered at 0 with radius 1 is contained in $[-1, 1]$.

(b) In $\mathbb{R}$, the point 1 is **not** an interior point of $[-1, 1]$ since, for example, given any $\delta > 0$, the open ball centered at 1 with radius δ,

$$B(1, \delta) = \{ x \in \mathbb{R} : |x - 1| < \delta \} = \{ x \in \mathbb{R} : 1 - \delta < x < 1 + \delta \},$$

contains the point $1 + \frac{\delta}{2}$, which is not in $[-1, 1]$.

(c) In $\mathbb{R}^n$ with any norm, a finite set has no interior points. To see this, suppose that

$$F = \{ x_1, x_2, \dots, x_k \}$$

is a finite subset of $\mathbb{R}^n$, and let $\delta > 0$ be any positive number. If $1 \leq i \leq k$, then

$$B(x_i, \delta) = \{ x \in \mathbb{R}^n : \|x - x_i\| < \delta \}$$

contains infinitely many points of $\mathbb{R}^n$ and therefore contains points other than points of F. Hence, $B(x_i, \delta) \not\subset F$ for any $\delta > 0$, and x_i is not an interior point of F.

(d) In $\mathbb{R}$, an argument similar to that in (c) can be used to show that the set of positive integers $\mathbb{N} = \{ 1, 2, 3, \dots \}$ has no interior points.

It should be mentioned that interior points can also be described in terms of *convergent sequences*.

A sequence $\{x_k\}_{k \geq 1}$ of points in $\mathbb{R}^n$ *converges* to a point x_0 in $\mathbb{R}^n$ if given any $\epsilon > 0$, there exists a positive integer k_0 such that

$$\|x_k - x_0\| < \epsilon$$

for all $k \geq k_0$, and we say the sequence is *convergent*. A sequence that does not converge is said to *diverge* or to be *divergent*.

Theorem 2.2.1. *Given a set $A \subset \mathbb{R}^n$, a point $x_0 \in \mathbb{R}^n$ is an interior point of A if and only if given any sequence $\{x_k\}_{k \geq 1}$ of points in $\mathbb{R}^n$ that converges to x_0, the sequence $\{x_k\}_{k \geq 1}$ is in A* **eventually,** *that is, there exists an integer k_0 such that x_k is in A for all $k \geq k_0$.*

Proof. If x_0 is an interior point of A, then there exists a positive number $\delta > 0$ such that $B(x_0, \delta) \subset A$. If $\{x_k\}_{k \geq 1}$ is a sequence in $\mathbb{R}^n$ that converges to x_0, then given $\delta > 0$, there exists an integer k_0 such that

$$\|x_k - x_0\| < \delta$$

for all $k \geq k_0$. Thus, $x_k \in B(x_0, \delta)$ for all $k \geq k_0$, so that $x_k \in A$ for all $k \geq k_0$, and the sequence $\{x_k\}_{k \geq 1}$ is in A eventually.

On the other hand, if x_0 is not an interior point of A, then for each positive integer $k \geq 1$, the open ball $B(x_0, \frac{1}{k})$ contains a point x_k, which is not in A. The sequence $\{x_k\}_{k \geq 1}$ converges to x_0 since

$$\|x_k - x_0\| < \frac{1}{k}$$

for each $k \geq 1$, but no point of the sequence is in A. Therefore, if every sequence converging to x_0 is eventually in A, then x_0 is an interior point of A.

$\square$

A set is said to be **open** if every point of the set is an interior point.

Example 2.2.2. *An open set $G \subset \mathbb{R}^n$ can be written as a union of open balls.*

Solution. If $x \in G$, then there exists a $\delta_x > 0$ such that $B(x, \delta_x) \subset G$, and therefore,

$$G = \bigcup \{B(x, \delta_x) : x \in G\}$$

is a union of open balls.

$\square$

Example 2.2.3. *The open unit ball is an open set; in fact, any open ball $B(x, \delta)$ is an open set.*

Solution. To show that this is the case, we must show that given any x in $B\left(\overline{0}, 1\right)$, we can find a $\delta > 0$ such that $B\left(x, \delta\right) \subset B\left(\overline{0}, 1\right)$.

If $x = \overline{0}$, then clearly $\delta = 1$ will do. If, on the other hand, $x \neq \overline{0}$, then we will show that $\delta = 1 - \|x\|$ will do. Since $x \in B\left(\overline{0}, 1\right)$, then $\|x\| < 1$, so we know that $\delta > 0$.

Referring to the figure below and using the triangle inequality, we see that if z is in $B\left(x, \delta\right)$, then

$$\|z\| = \|z - x + x\| \leq \|z - x\| + \|x\| < \delta + \|x\| = 1.$$

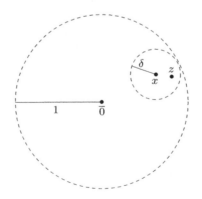

Thus, $z \in B\left(\overline{0}, 1\right)$, so $B\left(x, \delta\right) \subset B\left(\overline{0}, 1\right)$. Therefore, x is an interior point of $B\left(\overline{0}, 1\right)$, and since x was arbitrary, then $B\left(\overline{0}, 1\right)$ is an open set.

The proof for

$$B(x, \delta) = \{\, x \in \mathbb{R}^n \; : \; \|x\| < \delta \,\}$$

is similar and left as an exercise.

$\square$

Since the open sets are ultimately determined by the open balls, and since the shape of the unit ball is different for each norm, it is natural to expect that each norm may give rise to a different family of open sets.

Does it matter which norm we are using? For example, is the open unit ball in the ℓ_2 norm also an open set if we are using the ℓ_1 norm or the ℓ_∞ norm? The answer to this latter question is yes; in fact, it does not matter which norm we use. We will show this for the standard norms we are using, the ℓ_1, ℓ_2, and ℓ_∞ norms on $\mathbb{R}^n$:

$$\|x\|_1 = \sum_{k=1}^{n} |x_k|, \quad \|x\|_2 = \left(\sum_{k=1}^{n} |x_k|^2 \right)^{\frac{1}{2}}, \quad \|x\|_\infty = \max_{1 \leq k \leq n} |x_k|,$$

for $x = (x_1, x_2, \ldots, x_n) \in \mathbb{R}^n$.

Lemma 2.2.4. *For each $x \in \mathbb{R}^n$,*

(a) $\dfrac{1}{n}\|x\|_1 \leq \|x\|_2 \leq \|x\|_1$, *and*

(b) $\|x\|_\infty \leq \|x\|_2 \leq n\|x\|_\infty$.

Proof. Let $x = (x_1, x_2, \ldots, x_n) \in \mathbb{R}^n$; since all the inequalities are valid for $x = \overline{0}$, we may assume that $x \neq \overline{0}$.

Note that $|x_k|^2 \leq \sum_{i=1}^{n} |x_i|^2 = \|x\|_2^2$, so that

$$|x_k| \leq \|x\|_2 \qquad\qquad (*)$$

for $1 \leq k \leq n$.

(a) From $(*)$, summing over k, we have

$$\|x\|_1 = \sum_{k=1}^{n} |x_k| \leq n\|x\|_2.$$

Also, since $2 - 1 = 1$, from $(*)$, we have $|x_k|^{2-1} \leq \|x\|_2$, so that

$$\frac{|x_k|^2}{\|x\|_2} \leq |x_k|.$$

Summing we get

$$\frac{1}{\|x\|_2} \sum_{k=1}^n |x_k|^2 \leq \sum_{k=1}^n |x_k| = \|x\|_1,$$

that is, $\|x\|_2 \leq \|x\|_1$.

(b) Again, from $(*)$, we have $|x_k| \leq \|x\|_2$ for $1 \leq k \leq n$, so that

$$\|x\|_\infty = \max_{1 \leq k \leq n} |x_k| \leq \|x\|_2.$$

Also, we have

$$|x_k| \leq \max_{1 \leq i \leq n} |x_i| = \|x\|_\infty,$$

for $1 \leq k \leq n$, so that

$$\left(\sum_{k=1}^n |x_k|^2 \right)^{\frac{1}{2}} \leq n^{\frac{1}{2}} \|x\|_\infty \leq n \|x\|_\infty,$$

that is, $\|x\|_2 \leq n \|x\|_\infty$.

$\square$

Two norms $\|\cdot\|$ and $\|\cdot\|_*$ on $\mathbb{R}^n$ are said to be ***equivalent*** if there exist positive constants m and M such that

$$m \|x\| \leq \|x\|_* \leq M \|x\|$$

for all $x \in \mathbb{R}^n$.

Lemma 2.2.4 shows that in $\mathbb{R}^n$, the ℓ_2 norm is equivalent to both the ℓ_1 norm and the ℓ_∞ norm. And now we can show the equivalence of open sets defined by these norms.

Theorem 2.2.5. *A subset G of $\mathbb{R}^n$ is open with respect to the ℓ_2 norm if and only if it is open with respect to the ℓ_1 norm, if and only if it is open with respect to the ℓ_∞ norm.*

Proof. Suppose that $G \subset \mathbb{R}^n$ is open with respect to the ℓ_2 norm. If we let $a \in G$ be any point of G, then there exists a $\delta > 0$ such that

$$B_2(a, \delta) = \{x \in \mathbb{R}^n : \|x - a\|_2 < \delta\| \subset G.$$

Now, if $\|x - a\|_1 < \delta$, then the previous inequalities imply that $\|x - a\|_2 < \delta$, so that

$$B_1(a, \delta) = \{x \in \mathbb{R}^n \ : \ \|x - a\|_1 < \delta\} \subset B_2(a, \delta) \subset G,$$

and a is an interior point of G with respect to the ℓ_1 norm. Since $a \in G$ was arbitrary, then G is open with respect to the ℓ_1 norm.

Conversely, suppose that G is open with respect to the ℓ_1 norm and let $a \in G$ be any point of G, then there exists an $\epsilon > 0$ such that

$$B_1(a, \epsilon) = \{x \in \mathbb{R}^n \ : \ \|x - a\|_1 < \epsilon\} \subset G.$$

Now, if $\|x - a\|_2 < \epsilon/n$, then the previous inequalities imply that $\|x - a\|_1 < \epsilon$, so that

$$B_2(a, \epsilon/n) = \{x \in \mathbb{R}^n \ : \ \|x - a\|_2 < \epsilon/n\} \subset B_1(a, \epsilon) \subset G,$$

and a is an interior point of G with respect to the ℓ_2 norm. Since $a \in G$ was arbitrary, then G is open with respect to the ℓ_2 norm.

A similar argument shows that G is open with respect to the ℓ_2 norm if and only if G is open with respect to the ℓ_∞ norm and is omitted.

$\square$

In general,

Theorem 2.2.6. *In $\mathbb{R}^n$, a set is open with respect to one norm if and only if it is open with respect to all other norms.*

As you might have guessed, this theorem extends to the other topological notions.

Corollary 2.1. *All norms for $\mathbb{R}^n$ result in the same interior points, accumulation points, boundary points, closed sets, and convergent sequences.*

We will omit the proof of these results until after we have discussed compact sets and shown that the unit sphere $S(\overline{0}, 1)$ in $\mathbb{R}^n$ with the Euclidean norm is compact (see Theorem 2.6.5).

The collection of all interior points of a set A is called the **interior** of A and is usually denoted by $\text{int}(A)$. Thus,

$$\text{int}(A) = \{\, x \in A \ : \ x \text{ is an interior point of } A \,\}.$$

Example 2.2.7. *The interior of the closed unit ball is the open unit ball.*

Solution. If $x \in B(\overline{0}, 1)$, then $\|x\| < 1$ and there exists a $\delta > 0$ such that $B(x, \delta) \subset B(\overline{0}, 1)$. Since $B(x, \delta) \subset \overline{B}(\overline{0}, 1)$, then $x \in \text{int}\left(\overline{B}(\overline{0}, 1)\right)$, and therefore, $B(\overline{0}, 1) \subset \text{int}\left(\overline{B}(\overline{0}, 1)\right)$.

On the other hand, if $x \notin B(\overline{0}, 1)$, then $\|x\| \geq 1$, and for every positive number $\delta > 0$, the point $z = (1 + \delta/2)x$ is in $B(x, \delta)$ but is not in $\overline{B}(\overline{0}, 1)$. Thus, every open ball centered at x contains points not in $\overline{B}(\overline{0}, 1)$, so that x is not an interior point of $\overline{B}(\overline{0}, 1)$. Thus, we have shown that $\text{int}\left(\overline{B}(\overline{0}, 1)\right) \subset B(\overline{0}, 1)$.

$\square$

Example 2.2.8. *Show that the open halfspace*

$$H = \{ x \in \mathbb{R}^n : \langle p, x \rangle < \alpha \}$$

is an open set.

Solution. If $y \in H$ and q is the point in the hyperplane

$$H_\alpha = \{ x \in \mathbb{R}^n : \langle p, x \rangle = \alpha \}$$

nearest to y, let δ be the shortest distance from y to H_α, that is, $\delta = \|y - q\|$, then $B(y, \delta/2) \subset H$, so that y is an interior point of H. Since $y \in H$ is arbitrary, then H is open.

$\square$

Exercise 2.2.9. *Show that in $\mathbb{R}^n$, $\text{int}(S(\overline{0}, 1)) = \emptyset$.*

We have the following theorem:

Theorem 2.2.10. *A subset A of $\mathbb{R}^n$ is open if and only if $\text{int}(A) = A$.*

Proof. A is open if and only if every point of A is an interior point of A.

$\square$

And it seems reasonable that:

Theorem 2.2.11. *If $A \subset \mathbb{R}^n$, then $\text{int}(A)$ is an open set.*

Proof. If $a \in \text{int}(A)$, then we have to show that a is an interior point of $\text{int}(A)$. Since $a \in \text{int}(A)$, there exists a $\delta > 0$ such that $B(a, \delta) \subset A$. We will show that every point of $B(a, \delta)$ is in $\text{int}(A)$.

Let $y \in B(a, \delta)$ and let $\epsilon = \delta - \|y - a\|$, then $\epsilon > 0$ and $\|y - z\| < \epsilon$ for all $z \in B(y, \epsilon)$.

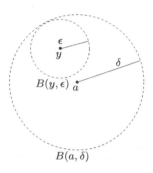

$$B(a, \delta)$$

Thus, $\|y - z\| < \delta - \|y - a\|$ for all $z \in B(y, \epsilon)$, so that

$$\|a - z\| = \|a - y + y - z\| \leq \|a - y\| + \|y - z\| < \delta$$

for all $z \in B(y, \epsilon)$, and

$$B(y, \epsilon) \subset B(a, \delta) \subset A,$$

that is, $y \in \operatorname{int}(A)$ for all $y \in B(a, \delta)$.

Hence, $B(a, \delta) \subset \operatorname{int}(A)$ so that $a \in \operatorname{int}(\operatorname{int}(A))$ and a is an interior point of $\operatorname{int}(A)$. Since a was an arbitrary point of $\operatorname{int}(A)$, then $\operatorname{int}(A)$ is open.

$\square$

Theorem 2.2.12. *For every set $A \subset \mathbb{R}^n$, the interior of A is an open set contained in A and is the largest open set contained in A, in fact,*

$$\operatorname{int}(A) = \bigcup \{U \subset \mathbb{R}^n : U \text{ is open, and } U \subset A\}.$$

Proof. Let

$$B = \bigcup \{U \subset \mathbb{R}^n : U \text{ is open, and } U \subset A\}.$$

If $x_0 \in B$, then $x_0 \in U$ for some open set $U \subset A$, and since U is open, there exists a $\delta > 0$ such that $B(x_0, \delta) \subset U \subset B$. Therefore, x_0 is an interior point of B, and since x_0 is arbitrary, then B is open.

Since $\operatorname{int}(A)$ is an open set and $\operatorname{int}(A) \subset A$, then $\operatorname{int}(A)$ is one of the sets U in the definition of B, so that $\operatorname{int}(A) \subseteq B$.

Now, if U is open and $U \subset A$, then every point of U is an interior point of A, so that $U \subset \text{int}(A)$, so that

$$B = \bigcup \{U \subset \mathbb{R}^n : U \text{ is open, and } U \subset A\} \subseteq \text{int}(A).$$

We have shown that $B \subseteq \text{int}(A)$ and that $\text{int}(A) \subseteq B$, that is, $B = \text{int}(A)$.

$\square$

Note. Since the interior of A is the union of all open sets contained in A, the interior of A is sometimes defined as the *largest open subset* of A.

2.2.1 Properties of Open Sets

We have the following result concerning open sets, and again, the norm we are using is not necessarily the Euclidean norm.

Theorem 2.2.13. *In $\mathbb{R}^n$, the following are true:*

(i) *The empty set $\emptyset$ and the whole space $\mathbb{R}^n$ are open sets.*
(ii) *The intersection of any finite number of open sets is an open set.*
(iii) *The union of any family of open sets is an open set.*

Proof. We will prove only the second part, the other parts are left as exercises.

Let U_1 and U_2 be open subsets of $\mathbb{R}^n$, and let $U_3 = U_1 \cap U_2$. If $x \in U_3$, then $x \in U_1$ and $x \in U_2$ also.

Since U_1 and U_2 are open, there exist $\delta_1 > 0$ and $\delta_2 > 0$ such that $B(x, \delta_1) \subset U_1$ and $B(x, \delta_2) \subset U_2$. Therefore, $B(x, \delta_1) \cap B(x, \delta_2) \subset U_1 \cap U_2$, and if we let $\delta = \min\{\delta_1, \delta_2\}$, then

$$B(x, \delta) \subset B(x, \delta_1) \cap B(x, \delta_2) \subset U_1 \cap U_2,$$

so that x is an interior point of $U_3 = U_1 \cap U_2$. Since x is arbitrary, then $U_1 \cap U_2$ is open.

We have shown that the intersection of two open sets is also open, and an easy induction argument will show that the intersection of any finite number of open sets is open.

$\square$

Note. The previous theorem says that the family of open sets in $\mathbb{R}^n$ form what is known as a ***topology*** for $\mathbb{R}^n$. When $\mathbb{R}^n$ is endowed with a family of sets satisfying these three statements, $\mathbb{R}^n$ is called a ***topological space***.

More specifically, a family $\mathcal{U}$ of subsets of a given set X is called a ***topology*** for X and the members of $\mathcal{U}$ are called ***open sets*** if and only if

(i) $\emptyset \in \mathcal{U}$ and $X \in \mathcal{U}$,

(ii) $U_1, U_2, \ldots, U_n \in \mathcal{U}$ implies that $U_1 \cap U_2 \cap \cdots \cap U_n \in \mathcal{U}$,

(iii) $U_\alpha \in \mathcal{U}$ for all $\alpha \in I$ implies that $\displaystyle\bigcup_{\alpha \in I} U_\alpha \in \mathcal{U}$.

There are many different topologies for $\mathbb{R}^n$, but we will only talk about those that are generated by the open balls with respect to some norm on $\mathbb{R}^n$, called ***norm topologies***.

Example 2.2.14. *The subset of $\mathbb{R}^2$ given by*

$$A = \{(x,y) : (x+1)^2 + y^2 \leq 1\} \cup \{(x,y) : (x-1)^2 + y^2 \leq 1\} \cup L,$$

*where $L = \{(x,y) : x = 0, \ -1 \leq y \leq 1\}$, is called **Schatz's Apple** and is shown below.*

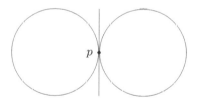

(a) *Let $p = (0,0)$, show that p has the property that every ray with endpoint p has an initial segment of the ray contained in A.*

(b) *Show that p is **not** an interior point of A.*

Solution.

(a) If $0 < x < 2$, and we let

$$y^2 = 1 - (x-1)^2 = x(2-x),$$

then the point (x,y) lies on the circle $(x-1)^2 + y^2 = 1$ (on the upper half if $y > 0$ and on the lower half if $y < 0$).

Since the vertical segment $x = 0, -1 < y < 1$ is tangent to the circle, then by convexity, the entire segment $[p, q]$, where $p = (0, 0)$ and $q = (x, \sqrt{x(2 - x)})$, lies in the set A, as in the figure below.

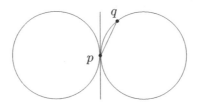

Similarly, if $-2 < x < 0$ and $y = \sqrt{-x(2 + x)}$, by convexity, the entire segment $[p, r]$, where $p = (0, 0)$ and $r = (x, \sqrt{-x(2 + x)})$ lies in the set A.

Therefore, $p = (0, 0)$ has the property that every ray with endpoint p has an initial segment of the ray contained in A.

(b) Given any $\epsilon > 0$, the ball centered at $p = (0, 0)$ with radius ϵ contains a point that is not in the set A, as in the figure below.

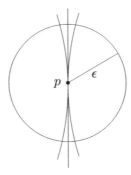

For example, the circles

$$(x - 1)^2 + y^2 = 1 \qquad \text{and} \qquad x^2 + y^2 = \epsilon^2$$

intersect when $x = \epsilon^2/2$, so the point (x_0, y_0) with

$$0 < x_0 < \epsilon^2/2 \qquad \text{and} \qquad y_0 = \epsilon\sqrt{1 - \epsilon^2/4}$$

is in the open ball with radius ϵ centered at p, but is not in the set A. Thus, p is not an interior point of the set A.

$\square$

2.3 ACCUMULATION POINTS AND CLOSED SETS

Given a subset A of $\mathbb{R}^n$, a point p in $\mathbb{R}^n$ is called an **accumulation point** of A if every open ball of positive radius centered at p contains a point of A other than p, that is, $(A \setminus \{p\}) \cap B(p, \delta) \neq \emptyset$ for every $\delta > 0$.

For example,

(a) In $\mathbb{R}$, if we let $A = (0, 1]$, then $p = 0$, $p = \frac{1}{2}$, and $p = 1$ are accumulation points of A, but $p = -\frac{1}{10}$ is **not** an accumulation point of A. In fact, the set of **all** accumulation points of A is $[0, 1]$.

(b) In $\mathbb{R}^2$, the set $A = \{ (0,0), (1,0), (0,1), (1,1) \}$ has **no** accumulation points.

(c) In $\mathbb{R}$, the set $A = \{ 2 + \frac{1}{k} : k = 1, 2, 3, \ldots \}$ has exactly **one** accumulation point, namely, the point $p = 2$.

(d) If a set has an accumulation point, then it must contain infinitely many points. Hence, the empty set $\emptyset$ and any finite subset $F = \{ x_1, x_2, \ldots, x_k \}$ have **no** accumulation points.

(e) In $\mathbb{R}$, the set of positive integers $\mathbb{N} = \{ 1, 2, 3, \ldots \}$ is an example of an infinite set that has **no** accumulation points.

Just as with interior points, accumulation points can also be described in terms of convergent sequences.

Theorem 2.3.1. *Given a set $A \subset \mathbb{R}^n$, a point $p \in \mathbb{R}^n$ is an accumulation point of A if and only if there is a sequence of points $\{p_k\}_{k \geq 1}$, with each $p_k \in A \setminus \{p\}$, such that $\lim_{k \to \infty} p_k = p$.*

Proof. If p is an accumulation point of A, then for any positive integer $k \geq 1$, the open ball $B\left(p, \frac{1}{k}\right)$ contains at least one point p_k of A different from p. Therefore, $p_k \in A$ and $p_k \neq p$ for all $k \geq 1$, and

$$\|p_k - p\| < \frac{1}{k}$$

for all $k \geq 1$, that is, $\lim_{k \to \infty} p_k = p$.

Conversely, suppose that there exists a sequence $\{p_k\}_{k \geq 1}$ with $p_k \in A$ and $p_k \neq p$ for all $k \geq 1$, such that $\lim_{k \to \infty} p_k = p$. Given any $\delta > 0$, there exists a positive integer $k_0 = k_0(\delta)$ such that

$$\|p_k - p\| < \delta$$

for all $k \geq k_0$, that is, given any $\delta > 0$, the open ball $B(p, \delta)$ contains the points p_k for $k \geq k_0$, which belong to A and which are different from p. Therefore, p is an accumulation point of A.

$\square$

Note. Given a subset $A \subset \mathbb{R}^n$, the set of all accumulation points of A is called the ***derived set*** of A and is denoted by A', and the fact that a point p either belongs to A or does not belong to A has no bearing whatsoever upon whether or not p is an accumulation point of A.

A set A is said to be ***closed*** if it contains all of its accumulation points, that is, if $A' \subseteq A$.

For example,

(a) In $\mathbb{R}^2$, the open unit ball

$$B\left(\overline{0}, 1\right) = \left\{ (x, y) \in \mathbb{R}^2 : x^2 + y^2 < 1 \right\}$$

is ***not*** a closed set. The point $(1, 0)$ is an accumulation point of $B\left(\overline{0}, 1\right)$ that is not in $B\left(\overline{0}, 1\right)$.

(b) In $\mathbb{R}^2$, the unit circle

$$S\left(\overline{0}, 1\right) = \left\{ (x, y) \in \mathbb{R}^2 : x^2 + y^2 = 1 \right\}$$

is a closed set.

(c) In $\mathbb{R}^2$, the set

$$H = \left\{ (x, y) \in \mathbb{R}^2 : x \geq 0, \ -\infty < y < \infty \right\}$$

is closed.

(d) In $\mathbb{R}^2$, the set

$$H = \left\{ (x, y) \in \mathbb{R}^2 : x > 0, \ -\infty < y < \infty \right\}$$

is ***not*** closed, since the origin is an accumulation point of the set that is not contained in the set.

(e) In $\mathbb{R}^2$, the set

$$H = \left\{ (x, 0) \in \mathbb{R}^2 : x > 0 \right\}$$

is ***not*** closed.

(f) A line in $\mathbb{R}^2$ is a closed set.

Example 2.3.2. *The closed unit ball is a closed set, in fact, any closed ball $\overline{B}(x, \delta)$ is a closed set.*

Solution. To show that this is the case, let p be any point not in $\overline{B}(\overline{0}, 1)$, we will show that p is not an accumulation point of $\overline{B}(\overline{0}, 1)$; thus, $\overline{B}(\overline{0}, 1)$ contains all of its accumulation points.

Since $p \notin \overline{B}(\overline{0}, 1)$, then we know that $\|p\| > 1$, and if we let $\delta = \|p\| - 1$, then the open ball $B(p, \delta)$ does not contain any points of $\overline{B}(\overline{0}, 1)$.

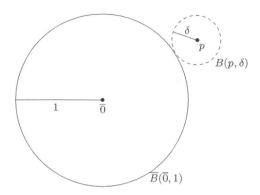

To see this, suppose that z is a point common to both $B(p, \delta)$ and $\overline{B}(\overline{0}, 1)$, from the triangle inequality,

$$\|p\| = \|(p - z) + z\| \leq \|p - z\| + \|z\| < \delta + 1 = \|p\|,$$

which is a contradiction.

Therefore, $B(p, \delta) \cap \overline{B}(\overline{0}, 1) = \emptyset$, so that p is not an accumulation point of $\overline{B}(\overline{0}, 1)$, and $\overline{B}(\overline{0}, 1)$ contains all of its accumulation points. Thus, $\overline{B}(\overline{0}, 1)$ is a closed set.

$\square$

Example 2.3.3. *If f is a nonzero linear functional on $\mathbb{R}^n$, which is represented by the nonzero vector $p \in \mathbb{R}^n$, then the hyperplane*

$$H_\alpha = f^{-1}(\alpha) = \{\, x \in \mathbb{R}^n \,:\, \langle p, x \rangle = \alpha \,\}$$

is a closed set.

Solution. We may assume without loss of generality that $\alpha \geq 0$, since if $\alpha < 0$, we simply replace p by $-p$ and obtain the same hyperplane.

Now let x_0 be an accumulation point of H_α, then given any $\epsilon > 0$, there is a point x_ϵ in H_α inside $B(x_0, \epsilon)$, and

$$\alpha = \langle p, x_\epsilon \rangle = \langle p, x_\epsilon - x_0 \rangle + \langle p, x_0 \rangle,$$

so that

$$\alpha - \langle p, x_0 \rangle = \langle p, x_\epsilon - x_0 \rangle.$$

From the Cauchy–Schwarz inequality, we have

$$|\alpha - \langle p, x_0 \rangle| = |\langle p, x_\epsilon - x_0 \rangle| \leq \|p\| \cdot \|x_\epsilon - x_0\| < \|p\| \cdot \epsilon$$

since $\|x_\epsilon - x_0\| < \epsilon$. Therefore,

$$-\|p\| \cdot \epsilon < \alpha - \langle p, x_0 \rangle < \|p\| \cdot \epsilon,$$

for all $\epsilon > 0$. Letting $\epsilon \to 0$, we have $\alpha = \langle p, x_0 \rangle$, so that $x_0 \in H_\alpha$. Thus, H_α contains all of its accumulation points and is a closed set.

We can give an alternative proof using the characterization of accumulation points in terms of sequences given in Theorem 2.3.1. If we let x_0 be an accumulation point of H_α and let $\{x_k\}_{k \geq 1}$ be a sequence of points in H_α such that $\lim_{k \to \infty} x_k = x_0$, then $\langle p, x_k \rangle = \alpha$ for all $k \geq 1$, so that

$$\lim_{k \to \infty} \langle p, x_k \rangle = \alpha.$$

Also, from the Cauchy–Schwarz inequality,

$$|\langle p, x_k \rangle - \langle p, x_0 \rangle| = |\langle p, x_k - x_0 \rangle| \leq \|p\| \cdot \|x_k - x_0\|$$

and $\lim_{k \to \infty} \|x_k - x_0\| = 0$, so that $\lim_{k \to \infty} \langle p, x_k \rangle = \langle p, x_0 \rangle$, and $\langle p, x_0 \rangle = \alpha$. Thus, $x_0 \in H_\alpha$ and H_α contains all of its accumulation points, that is, H_α is a closed set.

$\square$

Exercise 2.3.4. *Show that the closed halfspaces*

$$H_\leq = \{\, x \in \mathbb{R}^n : \langle p, x \rangle \leq \alpha \,\} \quad and \quad H_\geq = \{\, x \in \mathbb{R}^n : \langle p, x \rangle \geq \alpha \,\}$$

are both closed sets.

In $\mathbb{R}^n$ with a norm topology, there is a close relationship between the family of closed sets and the family of open sets. However, it is important to understand that being open is *not* the logical negation of being closed. The following theorem establishes the correct relationship.

Theorem 2.3.5. *A subset F of $\mathbb{R}^n$ is closed if and only if its complement $\mathbb{R}^n \setminus F$ is open.*

Proof. Suppose that F is closed and, hence, contains all of its accumulation points. If $x \notin F$, then x is not an accumulation point of F, so there exists a $\delta > 0$ such that $B(x, \delta) \cap (F \setminus \{x\}) = \emptyset$. Since $x \notin F$, this means that $B(x, \delta) \cap F = \emptyset$, that is, $B(x, \delta) \subset \mathbb{R}^n \setminus F$. Therefore, $x \in \text{int}\,(\mathbb{R}^n \setminus F)$, and since x was arbitrary, then $\mathbb{R}^n \setminus F$ is open.

Conversely, suppose that $\mathbb{R}^n \setminus F$ is open and let $x \in \mathbb{R}^n \setminus F$, then there is a $\delta > 0$ such that $B(x, \delta) \subset \mathbb{R}^n \setminus F$; thus, x is not an accumulation point of F. So we have shown that if $\mathbb{R}^n \setminus F$ is open, then F contains all of its accumulation points, so F is closed.

$\square$

Exercise 2.3.6. *Show that the open halfspaces*

$$H_< = \{\, x \in \mathbb{R}^n : \langle p, x \rangle < \alpha \,\} \quad and \quad H_> = \{\, x \in \mathbb{R}^n : \langle p, x \rangle > \alpha \,\}$$

are both open sets.

As with open sets, closed sets can be characterized in terms of sequences.

Example 2.3.7. *(Closed Sets and Sequences)*

If $F \subseteq \mathbb{R}^n$, then the following are equivalent.

(i) *F is a closed subset of $\mathbb{R}^n$.*
(ii) *If $\{x_k\}_{k \geq 1}$ is any convergent sequence of points of F, and $x_0 = \lim\limits_{k \to \infty} x_k$, then x_0 belongs to F.*

Solution. (i) implies (ii). Let $\{x_k\}_{k \geq 1}$ be a sequence of points of F and suppose that $x_0 = \lim\limits_{k \to \infty} x_k$. We have to show that $x_0 \in F$.

Suppose that $x_0 \notin F$, since F is closed, then its complement $\mathbb{R}^n \setminus F$ is open; therefore, there exists a $\delta > 0$ such that the open ball $B(x_0, \delta)$ centered at x_0 with radius δ is completely contained in $\mathbb{R}^n \setminus F$. Now since $x_0 = \lim\limits_{k \to \infty} x_k$, there is an integer $k_0 = k_0(\delta)$ such that $x_k \in B(x_0, \delta)$ for all $k \geq k_0$. However, this implies that $x_{k_0} \in \mathbb{R}^n \setminus F$, which is a contradiction, since $x_k \subset F$ for all $k \geq 1$. Thus, $x_0 \in F$.

(ii) implies (i). Suppose that whenever $\{x_k\}_{k \geq 1}$ is any convergent sequence of points of F and $x_0 = \lim\limits_{k \to \infty} x_k$, then x_0 belongs to F.

Suppose also that F is not closed. Then its complement, $U = \mathbb{R}^n \setminus F$, is not open, so there exists a point $x_0 \in U$, which is not an interior point of U; therefore for each $k \geq 1$, there is a point $x_k \in \mathbb{R}^n \setminus U = F$, such that

$$\|x_k - x_0\| < \tfrac{1}{k}$$

so that $x_0 = \lim\limits_{k \to \infty} x_k$. However, this contradicts the hypothesis (ii), since $x_k \in F$ for all $k \geq 1$ and $x_0 = \lim\limits_{k \to \infty} x_k$, but $x_0 \notin F$. Therefore, F is closed.

$\square$

From the above, it is easily seen that the family of closed sets has properties that "mirror" those of open sets.

2.3.1 Properties of Closed Sets

The following sections provide further examples of closed sets. For the most part, one checks that a set is closed directly from the definition. Sometimes, however, the following theorem is useful; again, the norm we are using is not necessarily the Euclidean norm.

Theorem 2.3.8. *In $\mathbb{R}^n$, the following are true:*

(i) *The empty set $\emptyset$ and the whole space $\mathbb{R}^n$ are closed sets.*
(ii) *The union of any finite number of closed sets is a closed set.*
(iii) *The intersection of any family of closed sets is a closed set.*

Proof.

(i) The empty set $\emptyset$ is closed since its complement $\mathbb{R}^n \setminus \emptyset$ is open, and $\mathbb{R}^n$ is closed since its complement $\emptyset = \mathbb{R}^n \setminus \mathbb{R}^n$ is open.
(ii) If F_1 and F_2 are closed sets, then

$$U = \mathbb{R}^n \setminus (F_1 \cup F_2) = (\mathbb{R}^n \setminus F_1) \cap (\mathbb{R}^n \setminus F_2)$$

is open since $\mathbb{R}^n \setminus F_1$ and $\mathbb{R}^n \setminus F_2$ are both open sets. Therefore, $F_1 \cup F_2$ is closed. Now an easy induction argument shows that the union of any finite number of closed sets is closed.
(iii) Let $\{ F_\alpha : \alpha \in I \}$ be a family of closed sets (here I is an index set and may be finite or infinite).
If $p \in \mathbb{R}^n$ is an accumulation point of $\bigcap_{\alpha \in I} F_\alpha$, then for every positive $\delta > 0$, the open ball $B(p, \delta)$ must contain some point q of $\bigcap_{\alpha \in I} F_\alpha$ with $q \neq p$. However, $q \in \bigcap_{\alpha \in I} F_\alpha$ implies that $q \in F_\alpha$ for every $\alpha \in I$.
In other words, for each $\alpha \in I$, given any positive δ, the ball $B(p, \delta)$ must contain some point $q \in F_\alpha$ with $q \neq p$. Thus, p is an accumulation point of F_α for each $\alpha \in I$.
Since each F_α is closed, this shows that $p \in \bigcap_{\alpha \in I} F_\alpha$, so that $\bigcap_{\alpha \in I} F_\alpha$ contains all of its accumulation points and is, therefore, a closed set.

$\square$

2.3.2 Boundary Points and Closed Sets

A point x_0 in $\mathbb{R}^n$ is a **boundary point** of a set $A \subset \mathbb{R}^n$ if every open ball centered at x_0 contains both points of A and points of the complement of A; that is, x_0 is a boundary point of A if and only if given any positive number $\delta > 0$,

$$A \cap B(x_0, \delta) \neq \emptyset \quad \text{and} \quad (\mathbb{R}^n \setminus A) \cap B(x_0, \delta) \neq \emptyset.$$

The collection of all boundary points of A is called the **boundary** of A and is denoted by $\mathrm{bdy}(A)$.

For example,

(a) In $\mathbb{R}^2$, the boundary of the open unit ball is the unit circle.
(b) In $\mathbb{R}^2$, the boundary of the closed unit ball is the unit circle.
(c) In $\mathbb{R}^2$, the boundary of the set

$$A = \{ (x, y) \in \mathbb{R}^2 : y = 0, \ -\infty < x < \infty \},$$

that is, the x-axis, is the x-axis.
(d) In $\mathbb{R}^n$, the boundary of a singleton set $A = \{x_0\}$, that is, a set containing a single point, is the set itself.
(e) In $\mathbb{R}^n$, the boundary of a finite set $F = \{ x_1, x_2, \ldots, x_k \}$ is the set itself.
(f) In $\mathbb{R}^n$, the boundary of the empty set $A = \emptyset$ is the empty set $\emptyset$, since no open ball contains points that belong to the empty set.

The following theorem shows that a closed set can be described as one that contains its entire boundary.

Theorem 2.3.9. *In $\mathbb{R}^n$, a set A is closed if and only if A contains all of its boundary points; that is, A is closed if and only if $\mathrm{bdy}(A) \subset A$.*

Proof. If $A \subset \mathbb{R}^n$ is not closed, then there is an accumulation point x_0 of A that does not belong to A. Since x_0 is an accumulation point of A, every open ball centered at x_0 must contain points of A.

On the other hand, every open ball centered at x_0 contains a point that does not belong to A, namely, the point x_0 itself. Thus, x_0 is a boundary point of A. We have shown that if A is not closed, it fails to contain one of its boundary points. Therefore, if $\mathrm{bdy}(A) \subset A$, then A is closed.

Conversely, If A does not contain all of its boundary points, then there exists a point $x_0 \in \mathrm{bdy}(A)$ such that $x_0 \notin A$. Since $x_0 \in \mathrm{bdy}(A)$, given any $\delta > 0$, the open ball $B(x_0, \delta)$ contains both points in A and points not in A, so x_0 is an accumulation point of A, which is not in A.

We have shown that if A does not contain all of its boundary points, then A is not closed.

Therefore, if A is closed, then $\mathrm{bdy}(A) \subset A$.

$\square$

2.3.3 Closure of a Set

Intuitively, one thinks of a nonclosed set as being one in which "some points are missing." The missing points are, of course, accumulation points. It would seem that we should be able to "repair" the set by including the accumulation points, and we have the following definition.

The union of a set A and its accumulation points A' is called the *closure* of A and is denoted by $\overline{A}$, so that

$$\overline{A} = A \cup A',$$

where A' is the derived set of A.

Calling $\overline{A}$ the *closure* of A suggests that $\overline{A}$ is a closed set, and we will prove this below.

Perhaps we should ask if there is really anything to prove. Is it not the case that if we throw in the accumulation points of A, then the resulting set must contain all of its accumulation points? To convince ourselves that there is really some work to do, let us consider the following.

A set is called *midpoint convex* if it contains the midpoint of each pair of points that belong to the set. For example, a straight line is midpoint convex, and the closed unit ball in $\mathbb{R}^n$ is midpoint convex. Of course, not all sets are midpoint convex. Suppose that we have a set A that is missing some of its midpoints. If we throw in all the midpoints, the set that we end up with will likely not be midpoint convex. For example, what happens if we start with a set A that consists of two distinct points?

The problem should now be clear: when we include the accumulation points of A, we create the larger set $\overline{A}$. However, the larger set may have more accumulation points than the original one. In other words, we have to check that $\overline{A}$ actually contains all of the accumulation points of $\overline{A}$, not just the accumulation points of A.

Theorem 2.3.10. *The closure of a set A is a closed set.*

Proof. Suppose that $x_0 \notin \overline{A}$. Since $\overline{A} = A \cup A'$, we have $x_0 \notin A$ and $x_0 \notin A'$.

Since $x_0 \notin A'$, there exists a $\delta > 0$ such that

$$B(x_0, \delta) \cap (A \setminus \{x_0\}) = \emptyset,$$

and since $x_0 \notin A$, we have $A \setminus \{x_0\} = A$, and therefore,

$$B(x_0, \delta) \cap A = \emptyset.$$

Suppose there exists a point $z_0 \in B(x_0, \delta) \cap A'$. Let $\epsilon = \delta - \|x_0 - z_0\|$.

If $y \in B(z_0, \epsilon)$, then

$$
\begin{aligned}
\|y - x_0\| &= \|y - z_0 + z_0 - x_0\| \\
&\leq \|y - z_0\| + \|z_0 - x_0\| \\
&< \epsilon + \|z_0 - x_0\| \\
&= \delta,
\end{aligned}
$$

so that $y \in B(x_0, \delta)$.

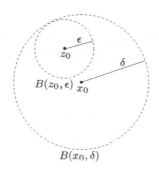

Therefore,

$$B(z_0, \epsilon) \subset B(x_0, \delta).$$

Since $B(x_0, \delta) \cap A = \emptyset$, this implies that

$$B(z_0, \epsilon) \cap A = \emptyset,$$

but this contradicts the fact that $z_0 \in A'$. Therefore, $B(x_0, \delta) \cap A' = \emptyset$ also.

From the above,

$$B(x_0, \delta) \cap (A \cup A') = (B(x_0, \delta) \cap A) \cup (B(x_0, \delta) \cap A') = \emptyset,$$

and

$$B(x_0, \delta) \cap \overline{A} = \emptyset.$$

Hence,

$$B(x_0, \delta) \subset \mathbb{R}^n \setminus \overline{A}$$

so that x_0 is an interior point of $\mathbb{R}^n \setminus \overline{A}$.

Since $x_0 \in \mathbb{R}^n \setminus \overline{A}$ was arbitrary, it follows that $\mathbb{R}^n \setminus \overline{A}$ is open, and so $\overline{A}$ is closed.

$\square$

For the closure of a set, we have a theorem analogous to the result for the interior of a set A.

Theorem 2.3.11. *For every set* $A \subset \mathbb{R}^n$, *the closure of A is a closed set containing A and is the smallest closed set containing A, in fact,*

$$\overline{A} = \bigcap \{F \subset \mathbb{R}^n : F \text{ is closed, and } A \subset F\}.$$

Proof. Let $B \subset \mathbb{R}^n$ be defined to be the intersection of all closed sets containing A, that is,

$$B = \bigcap \{F \subset \mathbb{R}^n : F \text{ is closed, and } A \subset F\}.$$

Since $\overline{A}$ is a closed set and $A \subseteq \overline{A}$, it follows that $\overline{A}$ is one of the sets F in the definition of B, so that $B \subseteq \overline{A}$.

Now, A is contained in each closed subset F containing A in the definition of B, so that

$$A \subseteq \bigcap \{F \subset \mathbb{R}^n : F \text{ is closed, and } A \subset F\} = B,$$

Therefore, any accumulation point of A must also be an accumulation point of B, so that $\overline{A} \subseteq \overline{B}$, and since B is closed, it contains all of its accumulation points, so $\overline{A} \subseteq \overline{B} = B$ and $\overline{A} \subseteq B$.

Therefore, $\overline{A} = B = \bigcap \{F \subset \mathbb{R}^n : F \text{ is closed, and } A \subset F\}$.

$\square$

Note. Since the closure of A is the intersection of all closed sets containing A, the closure of A is sometimes defined as the *smallest closed set that contains A.*

2.3.4 Problems

1. Identify which of the following subsets of $\mathbb{R}^2$ are closed. For those sets that are not closed, identify an accumulation point that does not belong to the set. In each case, draw a diagram of the set.
   (a) $A = \{x \in \mathbb{R}^2 : \|x\| \geq 1\}$.
   (b) $B = \{x \in \mathbb{R}^2 : \|x\| > 1\}$.
   (c) $C = \{x \in \mathbb{R}^2 : \|x\| \leq 1, x \neq \overline{0}\}$.
   (d) $D = \{x \in \mathbb{R}^2 : x = (\alpha, \beta), \beta = \alpha + 1\}$.
   (e) $E = \left\{(\alpha, \beta) \in \mathbb{R}^2 : \beta = \dfrac{\alpha^2 - 1}{\alpha - 1}\right\}$.
   (f) $F = \{(\alpha, \beta) \in \mathbb{R}^2 : \beta = \alpha^2 + 1\}$.

(g) $G = \{(\alpha, \beta) \in \mathbb{R}^2 : \beta > \alpha^2 + 1\}$.

(h) $H = \{(\alpha, \beta) \in \mathbb{R}^2 : \beta \geq \alpha^2 + 1\}$.

(i) $I = \{(\alpha, \beta) \in \mathbb{R}^2 : \beta = \alpha, \ \alpha = 1/i, \ i = 1, 2, \ldots\}$.

(j) $J = \{(\alpha, 0) \in \mathbb{R}^2 : -1 \leq \alpha < 1\}$.

(k) $K = \{(\alpha, 0) \in \mathbb{R}^2 : -1 < \alpha \leq 1\}$.

(l) $L = J \cup K$, J and K as above.

(m) $M = J \cap K$, J and K as above.

2. Prove that a set that has no accumulation points is closed.

3. Show that a finite set has no accumulation points. *Hint.* Prove the contrapositive: a set that has an accumulation point must be infinite.

4. Give an example of an infinite set that does not have any accumulation points.

5. In each of the following, identify the boundary of the set.

(a) $A = \{(x, y) \in \mathbb{R}^2 : y \geq 1/x, \ 0 < x < \infty\}$.

(b) The closed unit ball in $\mathbb{R}^3$.

(c) The open unit ball in $\mathbb{R}^3$.

(d) The unit sphere in $\mathbb{R}^3$.

(e) The set $B - \{x \in \mathbb{R}^2 : \|x\| > 1\}$.

6. Identify which of the sets in Problem 1 are open. If a set is not open, identify a point of the set which is not an interior point.

7. Identify which of the sets in Problem 1 are bounded.

8. (a) What is the interior of the subset A of Problem 1?

(b) What is the interior of the subset B of Problem 1?

* 9. Identify the boundary and interior of the set

$$\{x \in \mathbb{R} : -1 \leq x \leq 1, \ x \text{ rational}\}.$$

10. Prove that the union of any finite family of closed sets is closed.

11. (a) Give an example in $\mathbb{R}$ of an infinite family of closed sets whose union is not a closed set.

(b) Give an example in $\mathbb{R}$ of an infinite family of open sets whose intersection is not an open set.

12. Give an example of a set that is neither open nor closed.

13. Which two sets in $\mathbb{R}^n$ are both open and closed?

14. Prove that a subset $G \subset \mathbb{R}^n$ is open if and only if the complement of G is closed.

15. Explain why the empty set is open.

16. In $\mathbb{R}^n$, show that if the sequence $\{x_k\}_{k \geq 1}$ converges to x_0 in the ℓ_2 norm, then it also converges to x_0 in

(a) the ℓ_1 norm, and

(b) the ℓ_∞ norm.

17. In $\mathbb{R}$, does the following sequence converge?

$$x_k = \begin{cases} 1 & \text{if } k = n! \text{ for some } n \geq 1, \\ \dfrac{1}{k} & \text{otherwise.} \end{cases}$$

18. In $\mathbb{R}^n$, let $S = B(\overline{0}, 1) \setminus \{\overline{0}\}$.

    (a) Using convergent sequences, show that $\overline{0}$ is not an interior point of S.
    (b) What are the interior, closure, and boundary of the set S.

19. Show that the closure operation is idempotent. (An *idempotent operator* is one that when applied twice has the same effect as being applied once.)

20. Give an example to show that the interior of A and the interior of $\overline{A}$ may be different.

* 21. Give an example to show that the closure of A and the closure of the interior of A may be different.

22. Show that every open subset in $\mathbb{R}^n$ is the union of a family of balls.

* 23. Show that in $\mathbb{R}^n$, if p is an interior point of a subset A, then every ray with endpoint p has the property that an initial segment of the ray is contained in A. Show that the converse need not hold. (In $\mathbb{R}^2$, Schatz's apple is a simple counterexample.)

2.4 COMPACT SETS IN $\mathbb{R}$

We start this section with a discussion of bounded subsets of the real line and the order properties of $\mathbb{R}$. We introduce a notion of compactness and then give some applications. It should be mentioned at the start that many of the results about compactness are specific to the usual topology on $\mathbb{R}$, that is, the norm topology generated by the absolute value.

Consider the following subset of $\mathbb{R}$:

$$S = \left\{ \frac{1}{k} \in \mathbb{R} : k = 1, 2, 3, \ldots \right\}.$$

The real number 0 is smaller than all of the numbers in S, and no number larger than 0 has that property. We describe this situation as saying that 0 is the *infimum* of S, and we write

$$\inf \left\{ \frac{1}{k} : k = 1, 2, 3, \ldots \right\} = 0.$$

More generally, we have the following:*

---

* The existence of a greatest lower bound or a least upper bound must be taken as an axiom for the set of real numbers, called the *Least Upper Bound Property* or *Order Completeness Axiom* for $\mathbb{R}$: Every nonempty subset of $\mathbb{R}$ that is bounded above has a least upper bound (equivalently, every nonempty subset of $\mathbb{R}$ that is bounded below has a greatest lower bound).

- A nonempty set of real numbers S is said to be ***bounded below*** if there is a number α such that $\alpha \leq s$ for all $s \in S$. The number α is called a ***lower bound*** for S.

  If S has a lower bound, then it has infinitely many lower bounds. Among all the lower bounds for S, there is a greatest one, and it is called the ***greatest lower bound*** of S or the ***infimum*** of S.

  Thus, a real number α is the ***greatest lower bound*** for a nonempty set $S \subset \mathbb{R}$ if and only if

  (i) α is a lower bound for S, and
  (ii) if $\alpha < \gamma$, then there is an $s \in S$ such that $\alpha \leq s < \gamma$.

  The infimum of a nonempty set $S \subset \mathbb{R}$ is denoted by $\inf(S)$ or g.l.b.(S).

- A nonempty set of real numbers S is said to be ***bounded above*** if there is a number β such that $s \leq \beta$ for all $s \in S$. The number β is called an ***upper bound*** for S.

  If S has an upper bound, then it has infinitely many upper bounds. Among all the upper bounds for S there is a smallest one, and it is called the ***least upper bound*** of S or the ***supremum*** of S.

  Thus, a real number β is the ***least upper bound*** for a nonempty set $S \subset \mathbb{R}$ if and only if

  (i) β is an upper bound for S, and
  (ii) if $\gamma < \beta$, then there is an $s \in S$ such that $\gamma < s \leq \beta$.

  The supremum of a nonempty set $S \subset \mathbb{R}$ is denoted by $\sup(S)$ or l.u.b.(S).

- A nonempty set of real numbers S is said to be ***bounded*** if it is bounded above and bounded below.

Note. If the set S is not bounded below, we may write $\inf(S) = -\infty$, and when S is not bounded above, we may write $\sup(S) = +\infty$. Except for these two cases, for every nonempty set $S \subset \mathbb{R}$, $\sup(S)$ and $\inf(S)$ are always real numbers.

In the definition of $\inf(S)$, statement (ii) is equivalent to saying that there cannot be a lower bound for S that is strictly greater than α. Thus, a nonempty set of real numbers can have only one infimum (if there were two, one would be smaller than the other, which leads to a contradiction). A similar argument applies to the supremum of S. Thus, we have the following theorem:

Theorem 2.4.1. *A nonempty set of real numbers has at most one infimum and at most one supremum.*

Note. The infimum or supremum of a nonempty set of real numbers may or may not belong to the set. For example, if

$$S = \left\{ \frac{1}{k} : k = 1, 2, 3, \dots \right\},$$

then $\inf(S) = 0$, which *does not* belong to S; while $\sup(S) = 1$, which *does* belong to S. However, if a set is finite, then its infimum and supremum both belong to the set, and we use the words *minimum* and *maximum* to describe that situation.

In general, whether S is finite or not, to say that α is the **minimum** of S, means that $\alpha = \inf(S)$ and that $\alpha \in S$. Similarly, to say that β is the **maximum** of S, means that $\beta = \sup(S)$ and that $\beta \in S$.

Sometimes, the crux of a problem is to establish that a supremum is actually a maximum or that an infimum is actually a minimum. Several applications are given later that show how the notion of compactness is used to accomplish this.

Theorem 2.4.2. *Let S be a nonempty set of real numbers.*

(a) *If S is closed and bounded below, then S has a minimum.*
(b) *If S is closed and bounded above, then S has a maximum.*
(c) *If S is closed and bounded, then S has both a maximum and a minimum.*

Proof. Let S be a nonempty subset of $\mathbb{R}$.

(a) Suppose that S is closed and bounded below and let $\alpha = \inf(S)$.
Let $\delta > 0$ and let

$$B(\alpha, \delta) = \{\, x \in \mathbb{R} : |x - \alpha| < \delta \,\}$$

be the open ball centered at α with radius δ.
Since $\alpha + \delta$ is not a lower bound for S, then there is a point $x \in S$ such that

$$\alpha \leq x < \alpha + \delta$$

as in the figure below.

Thus, for every $\delta > 0$, the open ball $B(\alpha, \delta)$ contains a point of S different from α. Therefore, α is an accumulation point of S, and since S is closed, this implies that $\alpha \in S$ and α is the minimum for S.
(b) Suppose that S is closed and bounded above and let $\beta = \sup(S)$. An argument similar to that in part (a) shows that β is the maximum for S.

(c) If the set S is closed and bounded, then it is bounded above and bounded below, so that $\alpha = \inf(S) \in S$ and $\beta = \sup(S) \in S$. Thus, in this case, S has both a maximum and a minimum.

□

Example 2.4.3. *If A and B are nonempty subsets of $\mathbb{R}$ and $A \subset B$, then*

$$\inf(B) \leq \inf(A) \quad and \quad \sup(A) \leq \sup(B).$$

Solution. Let $\alpha = \inf(B)$. We have to show that α is a lower bound for A. If $x \in A$, since $A \subset B$, we also have $x \in B$. Therefore, $\alpha \leq x$ for each $x \in A$ and α is a lower bound for A. However, since the infimum of A is the greatest lower bound for A, then we have $\alpha \leq \inf(A)$. A similar argument shows that $\sup(A) \leq \sup(B)$.

□

Example 2.4.4. *Give an example of two sets A and B such that $A \subsetneq B$ but*

$$\inf(A) = \inf(B) \quad and \quad \sup(A) = \sup(B).$$

Solution. Let $A = (0, 1)$ and $B = [0, 1]$, then $A \subsetneq B$, but $\inf(A) = \inf(B) = 0$ and $\sup(A) = \sup(B) = 1$.

⊔

Theorem 2.4.5. *If S is a nonempty subset of $\mathbb{R}$, then*

$$\inf(S) = \inf(\overline{S}) \quad and \quad \sup(S) = \sup(\overline{S}).$$

Proof. Since $S \subset \overline{S}$, then $\inf(\overline{S}) \leq \inf(S)$, and we only have to show that $\inf(S) \leq \inf(\overline{S})$.

Let $\sigma = \inf(S)$, we will show that σ is a lower bound for $\overline{S}$. From the definition of σ, no point of S can be smaller than σ. Now suppose that ξ is an accumulation point of S and $\xi < \sigma$, this implies that there is a $\delta > 0$ such that $\xi + \delta < \sigma$ and $B(\xi, \delta)$ contains a point of S.

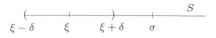

However, this contradicts the fact that σ is the infimum of S. Thus, no accumulation point of S can be smaller than σ, so that $\sigma \le \xi$ for all $\xi \in \overline{S}$, and σ is a lower bound for $\overline{S}$. Therefore, $\sigma \le \inf(\overline{S})$.

A similar argument shows that $\sup(S) = \sup(\overline{S})$ and is left as an exercise.

$\square$

Here is the definition of compactness for subsets of the real line with the usual topology.

A subset K of $\mathbb{R}$ is said to be **compact** if every infinite subset of K has an accumulation point that belongs to K.

Note. For $K \subset \mathbb{R}$ to be compact, the definition requires two things:

(i) Every infinite subset of K has at least one accumulation point p.
(ii) The accumulation point p must belong to K.

There are several equivalent definitions of compactness, in fact the usual definition is given in terms of coverings by open sets: A family of sets is said to **cover** K if the union of the family contains K.

We will show later that a subset of $\mathbb{R}$ is **compact** if and only if *every* family of open sets that covers K has a finite subfamily that also covers K. For $\mathbb{R}$ with the norm topology, this definition and the one we gave are equivalent. However, our definition is a useful one for the context in which we want to use it. Other characterizations of compactness will be discussed later.

Example 2.4.6. *Any finite set (even the empty set) is compact.*

Solution. The condition for compactness cannot be violated, since such sets do not contain infinite subsets.

$\square$

Example 2.4.7. *The set $\mathbb{R}$ is not compact.*

Solution. The infinite subset $\mathbb{N} = \{1, 2, 3, \dots\}$ of $\mathbb{R}$ has no accumulation points; that is, condition (i) in the definition of compactness fails.

$\square$

Note. Condition (ii) in the definition of compactness can be replaced by a much stronger one, namely:

Theorem 2.4.8. *A subset K of $\mathbb{R}$ is compact if and only if*

(i) *Every infinite subset of K has at least one accumulation point p.*
(iii) *All accumulation points of K belong to K.*

Proof. Suppose that $K \subset \mathbb{R}$ is compact, so that both (i) and (ii) in the definition are satisfied. If p is an accumulation point of K, then for each positive integer $k \geq 1$, we can choose a point $s_k \in B\left(p, \frac{1}{k}\right) \cap K$ with $s_k \neq p$. The set $\{ s_1, s_2, s_3, \dots \}$ is an infinite subset of K, and this set has only one accumulation point, namely, p. Clearly, p is also an accumulation point of K, and since (i) holds, $p \in K$. Therefore, (iii) holds.

Conversely, if $K \subset \mathbb{R}$ satisfies (i) and (iii), then clearly it satisfies (i) and (ii), and so K is compact.

□

Example 2.4.9. *The set $I = (0, 1]$ in $\mathbb{R}$ is not compact.*

Solution. The infinite subset of I given by

$$A = \left\{ \frac{1}{k} : k \in \mathbb{N} \right\}$$

has exactly one accumulation point that is not in the set I, and so condition (ii) in the definition of compactness fails.

□

2.4.1 Basic Properties of Compact Sets

The following is just a restatement of condition (iii) in the theorem above giving an equivalent definition of compactness.

Theorem 2.4.10. *A compact subset K of $\mathbb{R}$ must be closed.*

Proof. If K is compact, then by (iii), it contains all of its accumulation points; thus, K is closed.

□

Colloquially, "compact" means "packed together, not diffuse or dispersed," and the mathematical meaning is close to the colloquial meaning.

Theorem 2.4.11. *A compact subset K of $\mathbb{R}$ must be bounded.*

Proof. If K is not bounded, then for each positive integer $i \geq 1$, we can find points $k_i \in K \setminus B(0, i)$. Thus, the infinite subset $\{ k_1, k_2, k_3, \ldots \}$ of K has no accumulation points, showing that K cannot be compact.

$\square$

The open unit ball in $\mathbb{R}$, that is, the open interval $(-1, 1)$, is an example of a bounded set that is not compact. It fails to be compact because it fails to be closed. To some eyes, it would be much more significant if it contained an infinite subset that failed to have an accumulation point. There is a well-known theorem that says that bounded subsets cannot be this badly noncompact.

Theorem 2.4.12. *(Bolzano–Weierstrass)*
A bounded infinite subset of $\mathbb{R}$ has an accumulation point.

Proof. If you have ever taken a computing science course, you will recognize that the following proof is really the "divide-and-conquer" algorithm.*

Let A be a bounded infinite subset of $\mathbb{R}$, so that there is an interval $[a_1, b_1]$ that contains A. If p_1 is the midpoint of $[a_1, b_1]$, then either $[a_1, p_1]$ or $[p_1, b_1]$ contains infinitely many points of A. Choose whichever interval it is, and relabel it as $[a_2, b_2]$.

Now repeat the process: Let p_2 be the midpoint of $[a_2, b_2]$. Since $[a_2, b_2] \cap A$ contains infinitely many points, at least one of the intervals $[a_2, p_2]$ or $[p_2, b_2]$ contains infinitely many points of A. Choose one that does, and relabel it as $[a_3, b_3]$. It is obvious that we can continue in this manner, choosing a nested sequence of intervals

$$[a_1, b_1] \supset [a_2, b_2] \supset [a_3, b_3] \supset \cdots$$

such that at each stage $[a_i, b_i]$ contains infinitely many points of A.

Now, we claim that the intersection of all of these intervals is not empty. To see why, let
$$q = \sup\{ a_i : i = 1, 2, 3, \ldots \},$$

this supremum exists because the set $\{ a_i : i = 1, 2, 3, \ldots \}$ is bounded above by b_1, and by definition of q, we have $a_i \leq q$ for all $i = 1, 2, 3, \ldots$ We claim

---

* Of course, neither Bolzano nor Weierstrass learned this in a computing science course. It may be the case that Computer Scientists learned the divide-and-conquer algorithm from the proof of the Bolzano–Weierstrass Theorem.

also that $q \le b_i$ for all $i = 1, 2, 3, \ldots$ If $q > b_i$ for some $i \ge 1$, then for all $m > i$, we would have

$$q - a_m > q - b_m \ge q - b_i > 0,$$

so that $a_m < q$ for all $m > i$, showing that q cannot be the least upper bound of $\{a_{i+1}, a_{i+2}, \ldots\}$. Therefore, $q \in [a_i, b_i]$ for all $i \ge 1$.

Now, there are infinitely many points of A in each $[a_i, b_i]$, and since

$$|b_i - a_i| \le \frac{|b_1 - a_1|}{2^i}$$

for all $i \ge 1$, and since $\lim_{i \to \infty} |b_i - a_i| = 0$, this means that there are points of A arbitrarily close to q. Thus, q is an accumulation point of A, finishing the proof.

$\square$

A consequence of the Bolzano–Weierstrass theorem is that a bounded infinite subset of $\mathbb{R}$ must be compact whenever it is closed. Of course, a bounded finite set is always compact, closed, and bounded. These two facts give us the following theorem.

Theorem 2.4.13. *(Heine–Borel)*
A subset of $\mathbb{R}$ is compact if and only if it is closed and bounded.

Corollary 2.4.14. *A closed subset of a bounded set is compact.*

Corollary 2.4.15. *A closed subset of a compact set is compact.*

Corollary 2.4.16. *If K is a compact set and S is closed, then $K \cap S$ is compact.*

Note. The Heine–Borel theorem is often considered as an alternative characterization of compactness in $\mathbb{R}$. It should be emphasized that it is very specific to $\mathbb{R}$ and does not characterize compactness in other topological settings.

In the proof of the Bolzano–Weierstrass theorem, it was shown that a particular family of nested closed intervals had nonempty intersection. This has a very useful generalization.

Theorem 2.4.17. *(Nested Sets Theorem)*
If $K_1, K_2, K_3, \ldots$ is a family of nonempty compact subsets of $\mathbb{R}$ such that

$$K_1 \supseteq K_2 \supseteq K_3 \supseteq \cdots,$$

then $\bigcap_{i=1}^{\infty} K_i \ne \emptyset$.

Proof. For each $i \geq 1$, choose a point x_i from K_i. The set

$$S = \{\, x_1, x_2, x_3, \ldots \,\}$$

is either finite or infinite, since some of the points x_i may be repeated.

If S is finite, say $S = \{\, x_1, x_2, x_3, \ldots, x_m \,\}$, then at least one point x_{i_0} occurs infinitely often. If there exists a set K_i such that $x_{i_0} \notin K_i$, then since the sets are nested, $x_{i_0} \notin K_j$ for any $j \geq i$, and so is in only finitely many of the sets. This contradiction shows that x_{i_0} is in every K_i, and in this case, $\bigcap\limits_{m=1}^{\infty} K_m \neq \emptyset$.

If S is infinite, then it is an infinite subset of the compact set K_1; hence, S has an accumulation point, say p, which belongs to K_1.

In fact, p is an accumulation point of $\{\, x_2, x_3, \ldots \,\}$, and since

$$\{\, x_2, x_3, \ldots \,\} \subset K_2,$$

then p is an accumulation point of K_2.

In general, for each $m \geq 1$, the point p is an accumulation point of the set $\{\, x_m, x_{m+1}, \ldots \,\}$ which is a subset of K_m, and therefore, p is an accumulation point of K_m. Since K_m is closed, then $p \in K_m$ for all $m \geq 1$, that is, $p \in \bigcap\limits_{m=1}^{\infty} K_m$ and $\bigcap\limits_{m=1}^{\infty} K_m \neq \emptyset$.

$\square$

For completeness, we give a proof that our definition of compactness is equivalent to the usual definition in terms of coverings of a set by open sets.

Theorem 2.4.18. *A subset $K \subset \mathbb{R}$ is compact if and only if every family of open sets that covers K has a finite subfamily that also covers K.*

Proof. From what we have done previously, we need only show that K is closed and bounded (and, hence, compact) if and only if every family of open sets that covers K has a finite subfamily that also covers K.

Let K be a closed and bounded subset of $\mathbb{R}$ and let $\mathcal{U} = \{\, U_\alpha : \alpha \in I \,\}$ be a family of open subsets of $\mathbb{R}$ such that $K \subset \bigcup\limits_{\alpha \in I} U_\alpha$.

- Assume first that $K = [a, b]$, and let

  $F = \{\, x \in [a, b] : [a, x] \text{ can be covered by a finite number of sets from } \mathcal{U} \,\}$.

Since $a \in F$, the set F is nonempty, and since F is bounded above by b, then the supremum of F exists. Let $\gamma = \sup(F)$. Since $\gamma \in [a, b]$, there is a set $U_{\alpha_0} \in \mathcal{U}$ such that $\gamma \in U_{\alpha_0}$, and since U_{α_0} is open, there exists a $\delta > 0$ such that the interval $(\gamma - \delta, \gamma + \delta)$ is contained in U_{α_0}. Now, $\gamma - \delta$ is not an upper bound for F; hence, there exists an $x \in F$ with $x > \gamma - \delta$. Since $x \in F$, there is a finite collection

$$\{ U_{\alpha_1}, U_{\alpha_2}, \ldots, U_{\alpha_m} \}$$

of sets from $\mathcal{U}$ that covers the interval $[a, x]$. Therefore, the finite collection

$$\{ U_{\alpha_0}, U_{\alpha_1}, U_{\alpha_2}, \ldots, U_{\alpha_m} \}$$

covers the interval $[a, \gamma + \delta)$. Now note that we must have $\gamma = b$ since if $\gamma < b$, then there is a point x in $(\gamma, \gamma + \delta) \cap [a, b]$ such that $[a, x]$ is covered by $\{ U_{\alpha_0}, U_{\alpha_1}, U_{\alpha_2}, \ldots, U_{\alpha_m} \}$. That is, γ is not an upper bound for F. Therefore, the closed and bounded interval $[a, b]$ can be covered by a finite number of sets from $\mathcal{U}$.

- Now assume that K is any closed and bounded subset of $\mathbb{R}$, and that $\mathcal{U}$ is a family of open sets with $K \subset \bigcup \{ U_\alpha : \alpha \in I \}$. Since K is bounded, it is contained in some closed and bounded interval $[a, b]$, and since K is closed, then $U_0 = \mathbb{R} \setminus K$ is open. Let $\mathcal{V} = \mathcal{U} \cup \{ U_0 \}$, since $\mathcal{U}$ covers K, then $\mathcal{V}$ covers $[a, b]$. By the case considered above, there is a finite subcollection of $\mathcal{V}$ that covers $[a, b]$ and hence K. If the set U_0 occurs in the finite subcover, we remove it. What remains is a finite subcover of $\mathcal{U}$ that covers K.

Therefore, every covering of K by open sets has a finite subcovering that also covers K.

Conversely, suppose that every covering of K by open sets has a finite subcover that also covers K. Let $\mathcal{U}$ be the family of open balls of the form $B(k, 1)$ where $k \in K$. This is an open cover of the set K and so has a finite subcover. If $\{ k_1, k_2, \ldots, k_m \}$ are the centers of members of this subcover, then

$$K \subset \bigcup_{i=1}^{m} B(k_i, 1),$$

and since the union is a bounded set, then K is bounded.

Now we will show that if every covering of K by open sets has a finite subcover that also covers K, then K is closed. Let $U = \mathbb{R} \setminus K$ be the complement of K. Since K is bounded, then $U \neq \emptyset$. We will show that no point $q \in U$ is an accumulation point of K. For each point $p \in K$, consider the open ball

$$B(p, \delta(p)) = \{ x \in \mathbb{R} : |x - p| < \delta(p) \}$$

where $\delta(p) = \frac{1}{2}|p - q| > 0$ since $p \neq q$. The family $\mathcal{U} = \{B(p, \delta(p)) : p \in K\}$ is an open cover of K and, hence, has a finite subcover. Thus, there are points $\{p_1, p_2, \ldots, p_m\}$ of K such that

$$K \subset \bigcup_{i=1}^{m} B(p_i, \delta(p_i)).$$

Now let $r = \min\{\delta(p_i) : i = 1, 2, \ldots, \}$ and consider the neighborhood $B(q, r)$ of q. We have

$$B(q, r) \cap K \subset B(q, r) \cap \left[\bigcup_{i=1}^{m} B(p_i, \delta(p_i))\right] = \emptyset,$$

since $s \in B(q, r)$ implies that

$$2\delta(p_i) = |q - p_i| \leq |q - s| + |s - p_i| < r + |s - p_i| \leq \delta(p_i) + |s - p_i|,$$

which in turn implies that $|s - p_i| > \delta(p_i)$ for all $i = 1, 2, \ldots, m$. Therefore, $s \notin B(p_i, \delta(p_i))$ for any $i = 1, 2, \ldots, m$, so that $B(q, r) \subset \mathbb{R} \setminus K$. Thus, $\mathbb{R} \setminus K$ is open, and so K is closed.

$\square$

2.4.2 Sequences and Compact Sets in $\mathbb{R}$

In this section, we give an equivalent characterization of compactness in terms of sequences. Again, we emphasize that these results are specific to $\mathbb{R}$.

We say that a sequence of real numbers $\{a_k\}_{k \geq 1}$ is

- **increasing** if $a_k \leq a_{k+1}$ for all $k \geq 1$,
- **decreasing** if $a_k \geq a_{k+1}$ for all $k \geq 1$,
- **bounded above** if there exists an $M \in \mathbb{R}$ such that $a_k \leq M$ for all $k \geq 1$,
- **bounded below** if there exists an $m \in \mathbb{R}$ such that $m \leq a_k$ for all $k \geq 1$,
- A sequence is said to be **monotone** if it is either increasing or decreasing.

The next result says that bounded, monotone sequences converge.

Theorem 2.4.19. *If the sequence $\{a_k\}_{k \geq 1}$ is increasing and bounded above, or if it is decreasing and bounded below, then the sequence $\{a_k\}_{k \geq 1}$ converges.*

Proof. Let $\{a_k\}_{k \geq 1}$ be a sequence of real numbers that is bounded above, and let $A = \{a_k : k \in \mathbb{N}\}$ be the set containing the terms of the sequence.

The sequence is bounded above, so that A is a nonempty set that is bounded above, and so $\alpha = \sup(A)$ exists. We claim that $\alpha = \lim_{k \to \infty} a_k$. In fact, if $\epsilon > 0$, then there exists an integer k_0 such that $\alpha - a_{k_0} < \epsilon$ since $\alpha = \sup(A)$. Hence, if $k > k_0$, then $a_k \geq a_{k_0}$ and

$$0 \leq \alpha - a_k \leq \alpha - a_{k_0} < \epsilon.$$

Since $\epsilon > 0$ was arbitrary, then $\lim_{k \to \infty} a_k = \alpha$.

A similar argument shows that if $\{a_k\}_{k \geq 1}$ is decreasing and bounded below, then the sequence converges to the infimum of the set $A = \{ a_k : k \in \mathbb{N} \}$ and is left as an exercise.

$\square$

A *subsequence* of the sequence of real numbers $\{a_k\}_{k \geq 1}$ is a sequence of the form

$$a_{k_1}, a_{k_2}, a_{k_3}, \ldots,$$

where each $k_j \in \mathbb{N}$ and $k_1 < k_2 < k_3 < \cdots$ and is denoted by $\{a_{k_j}\}_{j > 1}$.

Theorem 2.4.20. *Any sequence $\{a_k\}_{k \geq 1}$ of real numbers contains a subsequence that is either increasing or decreasing.*

Proof. Call an integer $k \geq 1$ a *peak point* for the sequence $\{a_i\}_{i \geq 1}$ if $a_j < a_k$ for all $j > k$.

(i) If the sequence has infinitely many peak points, say $k_1, k_2, k_3, \ldots$, then

$$k_1 < k_2 < k_3 < \cdots,$$

and in this case,

$$a_{k_1} > a_{k_2} > a_{k_3} > \cdots,$$

so that $\{a_{k_j}\}$ is the decreasing subsequence.

(ii) If the sequence has only finitely many peak points, let k_1 be greater than all the peak points. Since k_1 is not a peak point, there exists an integer $k_2 > k_1$ such that $a_{k_2} \geq a_{k_1}$. Since k_2 is not a peak point, there exists an integer $k_3 > k_2$ such that $a_{k_3} > a_{k_2}$. Continuing in this manner, we get an increasing subsequence

$$a_{k_1} \leq a_{k_2} \leq a_{k_3} \leq \cdots.$$

$\square$

As corollary to this theorem, we have the sequential form of the Bolzano–Weierstrass theorem.

Corollary 2.4.21. *(Bolzano–Weierstrass)*
Every bounded sequence of real numbers has a convergent subsequence.

Summarizing the results on compactness in $\mathbb{R}$, we have the following theorem.

Theorem 2.4.22. *If K is a subset of $\mathbb{R}$, then the following are equivalent.*

(i) *Every family of open sets that covers K has a finite subfamily that also covers K.*
(ii) *K is closed and bounded.*
(iii) *Every infinite subset of K has an accumulation point in K.*
(iv) *Every sequence in K has a subsequence that converges to a point of K.*

Thus, $K \subset \mathbb{R}$ is **compact** *if and only if any one of the above conditions are true.*

2.4.3 Completeness

A sequence of real numbers $\{a_k\}_{k \geq 1}$ is a **Cauchy sequence** if given any $\epsilon > 0$, there exists an integer $k_0 = k_0(\epsilon)$ such that

$$|a_k - a_m| < \epsilon$$

whenever $k, m > k_0$.

We will show that a sequence of real numbers converges if and only if it is a Cauchy sequence. Thus, to determine whether or not a sequence converges, we do not have to know what the limit is; we only have to show that it is a Cauchy sequence.

Lemma 2.4.23. *If $\{a_k\}_{k \geq 1}$ is a Cauchy sequence, then the sequence $\{a_k\}_{k \geq 1}$ is bounded.*

Proof. Since the sequence $\{a_k\}_{k \geq 1}$ is a Cauchy sequence, if we take $\epsilon = 1$, then there exists an integer k_0 such that

$$|a_k - a_m| < 1$$

for all $k, m > k_0$.

Thus,

$$|a_k - a_{k_0+1}| < 1$$

for all $k > k_0$, and so

$$|a_k| < 1 + |a_{k_0+1}|$$

for all $k > k_0$.

Now we let

$$M = \max\{ |a_1|, |a_2|, \ldots, |a_{k_0}|, 1 + |a_{k_0+1}| \},$$

so that $|a_k| \leq M$ for all $k \geq 1$, and the sequence $\{a_k\}_{k\geq 1}$ is bounded.

$\square$

Lemma 2.4.24. *If $\{a_k\}_{k\geq 1}$ is a Cauchy sequence and the subsequence $\{a_{k_i}\}$ of $\{a_k\}_{k\geq 1}$ converges, say $\lim\limits_{i\to\infty} a_{k_i} = \gamma$, then the sequence $\{a_k\}$ also converges to γ.*

Proof. Suppose that $\{a_k\}_{k\geq 1}$ is a Cauchy sequence and the subsequence $\{a_{k_i}\}_{i\geq 1}$ of $\{a_k\}_{k\geq 1}$ converges. If $\lim\limits_{i\to\infty} a_{k_i} = \gamma$, then given $\epsilon > 0$, there exists an integer i_0 such that

$$|a_{k_i} - \gamma| < \frac{\epsilon}{2}$$

for all $i \geq i_0$.

Since the sequence $\{a_k\}_{k\geq 1}$ is a Cauchy sequence, there exists an integer k_0 such that

$$|a_k - a_m| < \frac{\epsilon}{2}$$

for all $k, m > k_0$.

Now, $|a_k - \gamma| = |a_k - a_{k_i} + a_{k_i} - \gamma|$, and from the triangle inequality,

$$|a_k - \gamma| \leq |a_k - a_{k_i}| + |a_{k_i} - \gamma|. \tag{$*$}$$

Thus, if $k, k_i > k_0$, then

$$|a_k - a_{k_i}| < \frac{\epsilon}{2},$$

and if $i \geq i_0$, then

$$|a_{k_i} - \gamma| < \frac{\epsilon}{2}.$$

Now choose $i \geq i_0$ so that $k_i \geq \max\{ k_0, k_{i_0} \}$, from $(*)$, we have

$$|a_k - \gamma| < \frac{\epsilon}{2} + \frac{\epsilon}{2} = \epsilon,$$

for all $k \geq k_0$. Therefore, $\lim\limits_{k\to\infty} a_k = \gamma$.

$\square$

Theorem 2.4.25. *A sequence $\{a_k\}_{k\geq 1}$ of real numbers converges if and only if it is a Cauchy sequence.*

Proof. If $\{a_k\}_{k\geq 1}$ is convergent, say $\lim\limits_{k\to\infty} a_k = \alpha$, choose k_0 such that

$$|a_k - \alpha| < \frac{\epsilon}{2}$$

for all $k \geq k_0$, then

$$|a_k - a_m| \leq |a_k - \alpha| + |\alpha - a_m| < \frac{\epsilon}{2} + \frac{\epsilon}{2} = \epsilon$$

for all $k, m > k_0$. Therefore, $\{a_k\}_{k\geq 1}$ is a Cauchy sequence.

Conversely, suppose that $\{a_k\}_{k\geq 1}$ is a Cauchy sequence. From the first lemma, we know that the sequence is bounded, and therefore, $\{a_k\}_{k\geq 1}$ has a convergent subsequence $\{a_{k_i}\}_{i\geq 1}$, say $\lim\limits_{i\to\infty} a_{k_i} = \alpha$. From the second lemma, since the sequence $\{a_k\}_{k\geq 1}$ is a Cauchy sequence, then the sequence converges and $\lim\limits_{k\to\infty} a_k = \alpha$.

$\square$

2.5 COMPACT SETS IN $\mathbb{R}^n$

The definition of compactness in $\mathbb{R}^n$ is the same as the definition we gave for subsets of $\mathbb{R}$.

A subset K of $\mathbb{R}^n$ is **compact** if every infinite subset of K has an accumulation point that belongs to K.

Note. As before, for $K \subset \mathbb{R}^n$ to be compact, the definition requires two things:

(i) Every infinite subset of K has at least one accumulation point p.
(ii) The accumulation point p must belong to K.

And as before, we have

Note. Condition (ii) in the definition of compactness can be replaced by a much stronger one, namely:

Theorem 2.5.1. *A subset K of $\mathbb{R}^n$ is compact if and only if*

(i) *Every infinite subset of K has at least one accumulation point p.*
(iii) *All accumulation points of K belong to K.*

The proof of the following theorem is word for word the same as the proof given for subsets of $\mathbb{R}$.

Theorem 2.5.2. *A compact subset K of $\mathbb{R}^n$ must be closed and bounded.*

In the proof of the Bolzano–Weierstrass theorem, we need to replace the intervals by closed **cells** of the form

$$I = \{ (x_1, x_2, \ldots, x_n) \in \mathbb{R}^n : a_i \leq x_i \leq b_i \text{ for } i = 1, 2, \ldots, n \},$$

where $a = (a_1, a_2, \ldots, a_n)$ and $b = (b_1, b_2, \ldots, b_n)$, and we let $\ell(I)$ be the longest side of I, that is,

$$\ell(I) = \sup\{b_1 - a_1, b_2 - a_2, \ldots, b_n - a_n\}.$$

Theorem 2.5.3. *(Bolzano–Weierstrass)*
A bounded infinite subset of $\mathbb{R}^n$ has an accumulation point.

Proof. Let A be a bounded infinite subset of $\mathbb{R}^n$ and let I_1 be a closed cell containing A. Divide I_1 into 2^n closed cells by bisecting each of its sides. Since I_1 contains infinitely many points of A, at least one cell I_2 in this subdivision will contain infinitely many points of A. Next divide I_2 into 2^n closed cells by bisecting its sides, so that at least one cell I_3 in this subdivision must contain infinitely many points of A. Continuing in this manner, we obtain a nested sequence $\{I_k\}_{k \geq 1}$ of nonempty closed cells each of which contains infinitely many points of A.

Now note that the longest side of the kth cell satisfies

$$0 < \ell(I_k) = \frac{\ell(I_1)}{2^{k-1}}$$

and the sequence of real numbers $\{\ell(I_k)\}_{k \geq 1}$ converges to 0. Thus, for each i with $1 < i \leq n$, the closed and bounded *coordinate* intervals $[a_{k,i}, b_{k,i}]$ of I_k form a nested sequence of nonempty compact subsets of $\mathbb{R}$, and by the nested sets theorem for $\mathbb{R}$, there is an $p_i \in \mathbb{R}$ such that

$$p_i \in \bigcap_{k=1}^{\infty} [a_{k,i}, b_{k,i}].$$

If we let $p = (p_1, p_2, \ldots, p_n) \in \mathbb{R}^n$, then it is easy to see that each ball $B(p, \epsilon)$ contains points of A different from p. Thus, p is an accumulation point of A.

$\square$

As a consequence of the Bolzano–Weierstrass theorem, a bounded infinite subset of $\mathbb{R}^n$ must be compact whenever it is closed, so we have the following theorem.

Theorem 2.5.4. *(Heine–Borel)*
A subset of $\mathbb{R}^n$ is compact if and only if it is closed and bounded.

And the corresponding corollaries for subsets of $\mathbb{R}^n$:

Corollary 2.5.5. *A closed subset of a bounded set is compact.*

Corollary 2.5.6. *A closed subset of a compact set is compact.*

Corollary 2.5.7. *If K is a compact set and S is closed, then $K \cap S$ is compact.*

The Nested Sets theorem in $\mathbb{R}$ is also valid for $\mathbb{R}^n$, and the proof follows word for word the proof given for subsets of $\mathbb{R}$ and is left as an exercise.

Theorem 2.5.8. *(Nested Sets Theorem)**
If K_1, K_2, K_3, ... is a family of nonempty compact subsets of $\mathbb{R}^n$ such that

$$K_1 \supseteq K_2 \supseteq K_3 \supseteq \cdots,$$

then $\bigcap\limits_{i=1}^{\infty} K_i \neq \emptyset$.

Again, we give a proof that our definition of compactness in $\mathbb{R}^n$ is equivalent to the usual definition in terms of coverings of a set by open sets.

Theorem 2.5.9. *A subset $K \subset \mathbb{R}^n$ is compact if and only if every family of open sets that covers K has a finite subfamily that also covers K.*

Proof. From what we have done previously, we need only show that K is closed and bounded (and hence compact) if and only if every family of open sets that covers K has a finite subfamily that also covers K.

Suppose that every covering of K by open sets has a finite subcover that also covers K. Let $\mathcal{U}$ be the family of open balls of the form $B(k, 1)$ where $k \in K$. This is an open cover of the compact set K and so has a finite subcover. If $\{k_1, k_2, \ldots, k_m\}$ are the centers of members of this subcover, then

$$K \subset \bigcup_{i=1}^{m} B(k_i, 1),$$

and since the union is a bounded set, then K is bounded.

Suppose now that every covering of K by open sets has a finite subcover that also covers K. Let $U = \mathbb{R}^n \setminus K$ be the complement of K, since K is bounded, then $U \neq \emptyset$. We will show that no point $q \in U$ is an accumulation point of K.

---

* This is also called the ***Cantor Intersection theorem***.

Let $q \in U$ be arbitrary, and for each point $p \in K$, consider the open ball

$$B(p, \delta(p)) = \{ x \in \mathbb{R}^n \; : \; \|x - p\| < \delta(p) \},$$

where $\delta(p) = \frac{1}{2}\|p - q\| > 0$ since $p \neq q$. The family $\mathcal{U} = \{B(p, \delta(p)) : p \in K\}$ is an open cover of K and, hence, has a finite subcover. Thus, there are points $\{ p_1, p_2, \ldots, p_m \}$ of K such that

$$K \subset \bigcup_{i=1}^{m} B(p_i, \delta(p_i)).$$

Now let $r = \min\{\delta(p_i) \; : \; i = 1, 2, \ldots\}$ and consider the neighborhood $B(q, r)$ of q. If $1 \leq i \leq m$ and $s \in B(q, r)$, then

$$\begin{aligned}
2\delta(p_i) = \|q - p_i\| \\
\leq \|q - s\| + \|s - p_i\| \\
< r + \|s - p_i\| \\
\leq \delta(p_i) + \|s - p_i\|,
\end{aligned}$$

so that $\|s - p_i\| > \delta(p_i)$ for all $i = 1, 2, \ldots, m$, and hence, $s \notin B(p_i, \delta(p_i))$ for any $i = 1, 2, \ldots, m$. Therefore,

$$B(q, r) \cap K \subset B(q, r) \cap \left[\bigcup_{i=1}^{m} B(p_i, \delta(p_i)) \right] = \bigcup_{i=1}^{m} B(q, r) \cap B(p_i, \delta(p_i)) = \emptyset.$$

Hence, $B(q, r) \subset \mathbb{R}^n \setminus K$, and so $\mathbb{R}^n \setminus K$ is open, and K is closed.

Conversely, let K be a closed and bounded subset of $\mathbb{R}^n$ and let

$$\mathcal{U} = \{ U_\alpha \; : \; \alpha \in I \}$$

be a family of open sets of $\mathbb{R}^n$ such that $K \subset \bigcup\{U_\alpha : \alpha \in I\}$. Assume that K is not contained in the union of any finite number of the sets in $\mathcal{U}$. We will show that this assumption leads to a contradiction.

Since K is bounded, there is an $M > 0$ such that $K \subseteq I_1$ where I_1 is the closed cell in $\mathbb{R}^n$ given by

$$I_1 = \{ (x_1, x_2, \ldots, x_n) \in \mathbb{R}^n \; : \; |x_k| \leq M \text{ for } k = 1, 2, \ldots, n \}.$$

Bisecting the sides of I_1, we obtain 2^n closed cells contained in I_1 containing points of K that are *not* contained in the union of any finite number of sets from $\mathcal{U}$. Let I_2 be any one of the closed cells in this subdivision of I_1 such that the nonempty set $K \cap I_2$ is not contained in the union of any finite number of sets from $\mathcal{U}$.

Continuing, we bisect the sides of I_2 to get 2^n closed cells contained in I_2. Let I_3 be any one of the closed cells in the subdivision of I_2 such that the nonempty set $K \cap I_3$ is not contained in the union of any finite number of sets from $\mathcal{U}$.

We continue this process and obtain a nested sequence $\{I_k\}_{k \geq 1}$ of closed cells such that the length $\ell(I_k)$ of the longest side of I_k satisfies

$$0 < \ell(I_k) \leq \frac{\ell(I_1)}{2^{k-1}} \leq \frac{2M}{2^{k-1}}$$

for $k \geq 1$.

As in the proof of the Bolzano–Weierstrass theorem for $\mathbb{R}^n$, since the sequence of real numbers $\{\ell(I_k)\}_{k \geq 1}$ converges to 0, then for each $i \geq 1$, the closed and bounded coordinate intervals $[a_{k,i}, b_{k,i}]$ of I_k form a nested sequence of compact subsets of $\mathbb{R}$, and by the Nested Sets Theorem for $\mathbb{R}$, there is a $p_i \in \mathbb{R}$ such that

$$p_i \in \bigcap_{k=1}^{\infty} [a_{k,i}, b_{k,i}].$$

If we let $p = (p_1, p_2, \ldots, p_n) \in \mathbb{R}^n$, then $p \in \bigcap_{k=1}^{\infty} I_k$. Since each I_k contains points from K, this implies that the point p is an accumulation point of K, and since K is closed, that $p \in K$.

Now, since $\mathcal{U}$ is a covering of K, there exists an α such that $p \in U_\alpha$, and so there exists a $\delta > 0$ such that $B(p, \delta) \subset U_\alpha$. Since $\lim_{k \to \infty} \ell(I_k) = 0$, we can choose k so large that $0 < \ell(I_k) < \delta/2$ and all the points of I_k are contained in the set U_α. However, this contradicts the construction of I_k as a set such that $K \cap I_k$ is *not* contained in the union of a finite number of sets from $\mathcal{U}$. Therefore, K is covered by a finite subfamily of $\mathcal{U}$.

$\square$

2.5.1 Sequences and Compact Sets in $\mathbb{R}^n$

In this section, we give an equivalent characterization of compactness in terms of sequences. Again, we emphasize that these results are specific to $\mathbb{R}^n$.

First, the sequential form of the Bolzano–Weierstrass theorem.

Theorem 2.5.10. *(Bolzano–Weierstrass)*
Every bounded sequence of points from $\mathbb{R}^n$ has a convergent subsequence.

Proof. Let $\{x_k\}_{k \geq 1}$ be a bounded sequence in $\mathbb{R}^n$. If there are only a finite number of distinct points in this sequence, then at least one of these points, say x_0, must occur infinitely often. Define a subsequence of the sequence $\{x_k\}_{k \geq 1}$ by selecting this element each time it appears. This is a convergent subsequence of the original sequence.

On the other hand, if the sequence $\{x_k\}_{k \geq 1}$ contains an infinite number of distinct points in $\mathbb{R}^n$, then the Bolzano–Weierstrass theorem implies that there is at least one accumulation point, say x_0, of this infinite set of distinct points from the sequence. Let x_{k_1} be an element of the sequence such that

$$\|x_{k_1} - x_0\| < 1.$$

Now consider the open ball $B(x_0, \frac{1}{2})$, since x_0 is an accumulation point of the set $S_1 = \{x_k : k \geq 1\}$, it is also an accumulation point of the set $S_2 = \{x_k : k > k_1\}$ obtained by deleting a finite number of points from S_1. Therefore, there is a point x_{k_2} of S_2 belonging to $B(x_0, \frac{1}{2})$, note that since $x_{k_2} \in S_2$, then $k_2 > k_1$.

Now consider the open ball $B(x_0, \frac{1}{3})$, and let $S_3 = \{x_k : k > k_2\}$, since x_0 is an accumulation point of S_3, there is a point $x_{k_3} \in S_3$ such that $x_{k_3} \in B(x_0, \frac{1}{3})$, note that $k_3 > k_2$ since $x_{k_3} \in S_3$.

Continuing in this manner, we obtain a subsequence $\{x_{k_i}\}_{i \geq 1}$ of the sequence $\{x_k\}_{k \geq 1}$ such that

$$\|x_{k_i} - x_0\| < \frac{1}{i},$$

so that $\lim_{i \to \infty} x_{k_i} = x_0$.

What we have actually shown is that if $\{x_k\}_{k \geq 1}$ is a bounded sequence and x_0 is a accumulation point of the *set* $S = \{x_k : k \geq 1\}$, then there is a subsequence $\{x_{k_i}\}_{i \geq 1}$ of the sequence $\{x_k\}_{k \geq 1}$ that converges to x_0.

$\square$

Next, a characterization of compactness in $\mathbb{R}^n$ in terms of sequences.

Theorem 2.5.11. *A subset K of $\mathbb{R}^n$ is compact if and only if every sequence in K has a subsequence that converges to a point of K.*

Proof. Suppose that K is a compact subset of $\mathbb{R}^n$, and let $\{x_k\}_{k \geq 1}$ be a sequence of points with $x_k \in K$ for all $k \geq 1$. Since K is bounded, then the sequence is bounded, and by the Bolzano–Weierstrass theorem, the sequence has a convergent subsequence $\{x_{k_i}\}_{i \geq 1}$, with $\lim_{i \to \infty} x_{k_i} = x_0$, say. Since K is closed and $\{x_{k_i}\}_{i \geq 1}$ is a sequence of points from K converging to x_0, then $x_0 \in K$.

Conversely, suppose that whenever $\{x_k\}_{k \geq 1}$ is a sequence of points with $x_k \in K$ for all $k \geq 1$, there exists a convergent subsequence $\{x_{k_i}\}_{i \geq 1}$ with $x_0 = \lim\limits_{i \to \infty} x_{k_i} \in K$.

We show first that K is bounded. If K is not bounded, then for every positive integer k, we can find a point $x_k \in K$ such that $\|x_k\| \geq k$. Now, by hypothesis, there exists a convergent subsequence $\{x_{k_i}\}_{i \geq 1}$ of this sequence such that $x_0 = \lim\limits_{i \to \infty} x_{k_i} \in K$, and by the triangle inequality,

$$\|x_{k_i}\| \leq \|x_{k_i} - x_0\| + \|x_0\|$$

for all $i \geq 1$.

Choose the integer i_0 such that $\|x_{k_i} - x_0\| < 1$ for all $i \geq i_0$, so that

$$\|x_{k_i}\| \leq \max\{\|x_{k_1}\|, \ldots, \|x_{k_{i_0-1}}\|, 1 + \|x_0\|\}$$

for all $i \geq 1$.

However, from the way the sequence $\{x_k\}_{k \geq 1}$ was constructed, we have

$$\|x_{k_i}\| \geq k_i \to +\infty$$

as $i \to \infty$, which is a contradiction. Therefore, K is bounded.

Next, we show that K is closed. Let a be an accumulation point of K, for each integer $k \geq 1$, there exists a point x_k in $B\left(a, \frac{1}{k}\right) \setminus \{a\}$. Since

$$\|x_k - a\| < \frac{1}{k}$$

for all $k \geq 1$, it is clear that the sequence $\{x_k\}_{k \geq 1}$ converges to a. Now, by hypothesis, since the sequence $\{x_k\}_{k \geq 1}$ is in K, it has a subsequence that converges to some $x_0 \in K$. However, the sequence $\{x_k\}$ converges to a, so that every subsequence of $\{x_k\}_{k \geq 1}$ also converges to a. Therefore, since limits are unique, then $x_0 = a$ and $a \in K$. Hence, K is closed. Therefore, K is closed and bounded and so is compact.

$$\square$$

Summarizing the results on compactness in $\mathbb{R}^n$, we have the following theorem.

Theorem 2.5.12. *If K is a subset of $\mathbb{R}^n$, then the following are equivalent.*

(i) *Every family of open sets that covers K has a finite subfamily that also covers K.*
(ii) *K is closed and bounded.*
(iii) *Every infinite subset of K has an accumulation point in K.*
(iv) *Every sequence in K has a subsequence that converges to a point of K.*

Thus, $K \subset \mathbb{R}$ is **compact** *if and only if any one of the conditions above is true.*

2.5.2 Completeness

A sequence of points $\{x_k\}_{k \geq 1}$ in $\mathbb{R}^n$ is a ***Cauchy sequence*** if given any $\epsilon > 0$, there exists an integer $k_0 = k_0(\epsilon)$ such that

$$\|x_k - x_m\| < \epsilon$$

whenever $k, m > k_0$.

We have the following lemmas, and the proofs follow exactly the proofs given for Cauchy sequences in $\mathbb{R}$.

Lemma 2.5.13. *If $\{x_k\}_{k \geq 1}$ is a Cauchy sequence in $\mathbb{R}^n$, then the sequence $\{x_k\}_{k \geq 1}$ is bounded.*

Lemma 2.5.14. *If $\{x_k\}_{k \geq 1}$ is a Cauchy sequence in $\mathbb{R}^n$ and the subsequence $\{x_{k_i}\}_{i \geq 1}$ of $\{x_k\}_{k \geq 1}$ converges, say $\lim_{i \to \infty} x_{k_i} = x_0$, then the sequence $\{x_k\}_{k \geq 1}$ also converges to x_0.*

Now we want to show that the normed linear space $\mathbb{R}^n$ is complete; that is, a sequence converges if and only if it is a Cauchy sequence. We no longer have an order relation on $\mathbb{R}^n$ as we did for $\mathbb{R}$, but we can still use the order completeness of $\mathbb{R}$, but in an indirect way.

Given a nonempty bounded subset A of $\mathbb{R}^n$, the ***diameter*** of A, denoted by $\mathrm{diam}(A)$ is defined to be

$$\mathrm{diam}(A) = \sup\{\, \|x - y\| \, : \, x \in A, \, y \in A \,\}.$$

If A is not bounded, we define $\mathrm{diam}(A) = \infty$.

Theorem 2.5.15. *A nonempty set A in $\mathbb{R}^n$ and its closure $\overline{A}$ have the same diameter, that is, $\mathrm{diam}\left(\overline{A}\right) = \mathrm{diam}(\mathrm{A})$.*

Proof. Since $A \subseteq \overline{A}$, this implies that

$$\sup\{\, \|x - y\| \, : \, x \in A, \ y \in A \,\} \leq \sup\{\, \|x - y\| \, : \, x \in \overline{A}, \ y \in \overline{A} \,\}$$

so that $\operatorname{diam}(A) \leq \operatorname{diam}\left(\overline{A}\right)$.

Now suppose that $x' \in \overline{A}$ and $y' \in \overline{A}$, and let $\delta > 0$ be arbitrary. There exist $x \in A$ and $y \in A$ such that

$$\|x - x'\| < \frac{\delta}{2} \quad \text{and} \quad \|y - y'\| < \frac{\delta}{2},$$

so that

$$\|x' - y'\| \leq \|x' - x\| + \|x - y\| + \|y - y'\| \leq \operatorname{diam}(A) + \delta$$

for all $x', y' \in \overline{A}$.

Thus,

$$\operatorname{diam}\left(\overline{A}\right) \leq \operatorname{diam}(A) + \delta,$$

and since $\delta > 0$ was arbitrary, then $\operatorname{diam}\left(\overline{A}\right) \leq \operatorname{diam}(A)$.

Therefore, $\operatorname{diam}\left(\overline{A}\right) = \operatorname{diam}(A)$.

$\square$

Theorem 2.5.16. *A sequence $\{x_k\}_{k \geq 1}$ in $\mathbb{R}^n$ converges if and only if it is a Cauchy sequence.*

Proof. Let $\{x_k\}_{k \geq 1}$ be a Cauchy sequence in $\mathbb{R}^n$, and let

$$A_k = \{\, x_k, \ x_{k+1}, \ x_{k+2}, \ \dots \,\}$$

for $k = 1, 2, 3, \dots$ We have

$$A_1 \supset A_k \supset A_{k+1} \quad \text{and} \quad \overline{A}_1 \supset \overline{A}_k \supset \overline{A}_{k+1}$$

for $k = 1, 2, 3, \dots$ Since A_1 is bounded, then each $\overline{A}_k$ is a closed and bounded set and is therefore compact.

From the Nested Sets Theorem, $\bigcap_{k=1}^{\infty} \overline{A}_k$ is a nonempty compact set. Let $x_0 \in \mathbb{R}^n$ be such that $x_0 \in \bigcap_{k=1}^{\infty} \overline{A}_k$. The sequence $\{x_k\}_{k \geq 1}$ is a Cauchy sequence, so given any $\epsilon > 0$, there exists an integer k_0 such that

$$\|x_k - x_j\| < \epsilon$$

whenever $k, j \geq k_0$. However, $k \geq k_0$ and $j \geq k_0$ imply that $x_k \in A_{k_0}$ and $x_j \in A_{k_0}$, so that $\text{diam}(A_{k_0}) \leq \epsilon$.

Now, $x_0 \in \overline{A}_{k_0}$ and $x_k \in A_{k_0} \subset \overline{A}_{k_0}$ for all $k \geq k_0$.

Therefore, since $\text{diam}\left(\overline{A}_{k_0}\right) = \text{diam}(A_{k_0})$, we have

$$\|x_k - x_0\| \leq \text{diam}\left(\overline{A}_{k_0}\right) = \text{diam}\left(A_{k_0}\right) \leq \epsilon$$

whenever $k \geq k_0$, that is, $\{x_k\}_{k \geq 1}$ converges to x_0.

Conversely, it is easily seen that any convergent sequence in $\mathbb{R}^n$ is a Cauchy sequence.

$\square$

2.6 APPLICATIONS OF COMPACTNESS

2.6.1 Continuous Functions

A function $f : \mathbb{R}^n \longrightarrow \mathbb{R}^m$ is said to be **continuous** at a point x_0 in the domain of f if given any $\epsilon > 0$, there exists a $\delta > 0$ such that

$$\|f(x) - f(x_0)\| < \epsilon$$

whenever $x \in \text{domain}(f)$ and $\|x - x_0\| < \delta$. Equivalently, f is continuous at x_0 if and only if $\lim_{x \to x_0} f(x) = f(x_0)$.

We will show that the notion of continuity on $\mathbb{R}^n$ does not depend on the norm that is being used (see Theorem 2.6.5).

Continuity of a function f can be described using sequences as in the following theorem.

Theorem 2.6.1. *If x_0 is a point in the domain of the function $f : \mathbb{R}^n \longrightarrow \mathbb{R}^m$, then f is continuous at x_0 if and only if whenever $\{x_k\}_{k \geq 1}$ is a sequence of points in the domain of f such that $\lim_{k \to \infty} x_k = x_0$, then $\lim_{k \to \infty} f(x_k) = f(x_0)$.*

Proof. Suppose that f is continuous at x_0, and let $\{x_k\}_{k \geq 1}$ be a sequence of points in $\text{domain}(f)$, which converges to x_0. Let $\epsilon > 0$, and let $\delta = \delta(\epsilon) > 0$ be such that $\|f(x) - f(x_0)\| < \epsilon$ whenever $x \in \text{domain}(f)$ and $\|x - x_0\| < \delta$. Since $\|x_k - x_0\| \to 0$ as $k \to \infty$, there is a positive integer $k_0 = k_0(\delta)$ such that $\|x_k - x_0\| < \delta$ for all $k \geq k_0$, and since each $x_k \in \text{domain}(f)$, then

$$\|f(x_k) - f(x_0)\| < \epsilon$$

for all $k \geq k_0$; therefore, $\lim_{k \to \infty} f(x_k) = f(x_0)$.

Conversely, suppose that x_0 is in domain(f) and f is not continuous at x_0, then there exists an $\epsilon > 0$ such that given any $\delta > 0$, there is an element $x_\delta \in$ domain(f) such that $\|x_\delta - x_0\| < \delta$, but $\|f(x_\delta) - f(x_0)\| \geq \epsilon$.

Thus, for each positive integer k, there is an element $x_k \in$ domain(f) such that

$$\|x_k - x_0\| < \frac{1}{k}$$

but

$$\|f(x_k) - f(x_0)\| \geq \epsilon.$$

Thus, $\lim_{k \to \infty} x_k = x_0$, but the sequence $\{f(x_k)\}_{k \geq 1}$ does not converge to $f(x_0)$, since

$$\|f(x_k) - f(x_0)\| \geq \epsilon$$

for all $k \geq 1$. Therefore, if $\lim_{k \to \infty} f(x_k) = f(x_0)$ whenever $x_k \in$ domain(f) and $\lim_{k \to \infty} x_k = x_0$, then f is continuous at x_0.

$\square$

Theorem 2.6.2. *If $f : \mathbb{R}^n \longrightarrow \mathbb{R}^m$ is continuous at each point of a compact subset $K \subset$ domain(f), then*

$$f(K) = \{y \in \mathbb{R}^m : y = f(x) \text{ for some } x \in K\}$$

is a compact subset of $\mathbb{R}^m$.

Proof. Let $K \subseteq$ domain(f) be compact and suppose that f is continuous at each point of K, in order to show that $f(K)$ is compact, we need only show that it is closed and bounded.

If f is not bounded on K, then for each $k \geq 1$, there is a point $x_k \in K$ such that $\|f(x_k)\| \geq k$. Now, since K is compact, the sequence $\{x_k\}_{k \geq 1}$ is bounded, and by the Bolzano–Weierstrass theorem for sequences, it has a convergent subsequence, say $\{x_{k_i}\}_{i \geq 1}$, which converges to x_0. Since K is closed, then $x_0 \in K$, and since f is continuous at x_0, then $\lim_{i \to \infty} f(x_{k_i}) = f(x_0)$. However, a convergent sequence is bounded, and this contradicts the assumption that $\|f(x_k)\| \geq k$ for all $k \geq 1$. Therefore, $f(K)$ is bounded.

In order to show that $f(K)$ is closed, let $\{y_k\}_{k \geq 1}$ be a sequence in $f(K)$ such that $y_k \to y_0$ as $k \to \infty$. For each $k \geq 1$, $y_k \in f(K)$, so that $y_k = f(x_k)$, where $x_k \in K$. Since K is compact, the sequence $\{x_k\}_{k \geq 1}$ has a convergent subsequence, say $\{x_{k_i}\}_{i \geq 1}$ with $x_{k_i} \to x_0$ as $i \to \infty$, and $x_0 \in K$, since K

is closed. Now, f is continuous at x_0, so that $\lim_{i \to \infty} f(x_{k_i}) = f(x_0)$, and since limits are unique, then $y_0 = f(x_0)$, that is, $y_0 \in f(K)$, and so $f(K)$ is closed.

$\square$

As with real-valued continuous functions, we have the following theorem.

Theorem 2.6.3. *If f is a continuous real-valued function and $K \subset \mathbb{R}^n$ is a compact subset of* domain(f), *then f attains its maximum and minimum values on K, that is, there exist $x_{\min}$ and $x_{\max}$ in K such that*

$$f(x_{\max}) = \sup\{ f(x) : x \in K \} \quad and \quad f(x_{\min}) = \inf\{ f(x) : x \in K \}.$$

Proof. Since K is compact, from the previous theorem, $f(K)$ is a compact subset of $\mathbb{R}$ and, hence, is bounded above and below. If we let

$$m = \inf\{f(x) : x \in K\} \quad \text{and} \quad M = \sup\{f(x) : x \in K\},$$

from Theorem 2.4.2, the set $f(K)$ has both a maximum and a minimum, so there exist points $x_{\max}$ and $x_{\min}$ such that $M = f(x_{\max})$ and $m = f(x_{\min})$, that is,

$$f(x_{\max}) = \sup\{ f(x) : x \in K \} \quad \text{and} \quad f(x_{\min}) = \inf\{ f(x) : x \in K \}.$$

$\square$

2.6.2 Equivalent Norms on $\mathbb{R}^n$

All of the preceding material on compactness is valid if we use the Euclidean norm on $\mathbb{R}^n$. Therefore, our use of the Heine–Borel theorem in the proof of Theorem 2.6.5 is justified. After proving this theorem, that is, that all norms on $\mathbb{R}^n$ are equivalent to $\| \cdot \|_2$, the preceding material (as well as all that follows) is valid for **any** norm on $\mathbb{R}^n$.

Two norms p and q on $\mathbb{R}^n$ are said to be **equivalent** if there are positive scalars m and M such that

$$m \cdot p(x) \le q(x) \le M \cdot p(x)$$

for all $x \in \mathbb{R}^n$.

Note that we are using the functional notation p and q for the norms, rather than the usual notation such as $\| \cdot \|$ and $\| \cdot \|'$.

Theorem 2.6.4. *Equivalence is a transitive relation on the set of all norms on $\mathbb{R}^n$, that is, if p, q, and r are norms on $\mathbb{R}^n$ such that p is equivalent to q and q is equivalent to r, then p is equivalent to r.*

Proof. Suppose there exist positive scalars m, M, k, and K such that

$$m \cdot p(x) \leq q(x) \leq M \cdot p(x) \qquad \text{and} \qquad k \cdot q(x) \leq r(x) \leq K \cdot q(x),$$

for all $x \in \mathbb{R}^n$, then

$$mk \cdot p(x) \leq r(x) \leq MK \cdot p(x)$$

for all $x \in \mathbb{R}^n$, so that p is equivalent to r.

$\square$

Theorem 2.6.5. *(Equivalent Norms)*
If p is any norm on $\mathbb{R}^n$, then p is equivalent to the Euclidean norm $\| \cdot \|_2$.

Proof. We assume that $\mathbb{R}^n$ is equipped with the Euclidean norm $\| \cdot \|_2$, and let $\{e_1, e_2, \ldots, e_n\}$ be the standard basis for $\mathbb{R}^n$.

Let $x \in \mathbb{R}^n$, with

$$x = x_1 e_1 + x_2 e_2 + \cdots + x_n e_n,$$

since p is a norm on $\mathbb{R}^n$, we have

$$p(x) \leq |x_1| \, p(e_1) + |x_2| \, p(e_2) + \cdots + |x_n| \, p(e_n),$$

and from the Cauchy–Schwarz inequality,

$$p(x) \leq \left(p(e_1)^2 + p(e_2)^2 + \cdots + p(e_n)^2 \right)^{\frac{1}{2}} \|x\|_2 \leq M \|x\|_2,$$

where $M = \left(p(e_1)^2 + p(e_2)^2 + \cdots + p(e_n)^2 \right)^{\frac{1}{2}}$.

Therefore,

$$p(x) \leq M \|x\|_2$$

for all $x \in \mathbb{R}^n$.

Now, from the triangle inequality, we have

$$|p(x) - p(y)| \leq p(x - y) \leq M \|x - y\|_2$$

for all $x, y \in \mathbb{R}^n$, so that the norm p is a continuous function on $\mathbb{R}^n$ with respect to the Euclidean norm. If we let

$$S = \{x \in \mathbb{R}^n \, : \, \|x\|_2 = 1\},$$

from the Heine–Borel theorem, S is a compact subset of $\mathbb{R}^n$, and therefore, p attains its minimum at a point $x_0 \in S$, so that

$$p(x_0) \leq p(x)$$

for all $x \in S$, and $p(x_0) > 0$ since $\|x_0\|_2 = 1$ implies that $x_0 \neq \overline{0}$.

Now, if $x \in \mathbb{R}^n$ with $x \neq \overline{0}$, then $\dfrac{x}{\|x\|_2} \in S$, so that

$$p(x_0) \leq p\left(\frac{x}{\|x\|_2}\right) = \frac{1}{\|x\|_2}\, p(x),$$

so that

$$m\,\|x\|_2 \leq p(x)$$

for all $x \in \mathbb{R}^n$, where $m = p(x_0) > 0$.

Therefore, p and $\|\ \|_2$ are equivalent norms on $\mathbb{R}^n$.

$\square$

2.6.3 Distance Between Sets in $\mathbb{R}^n$

The notion of infimum of a nonempty set of real numbers allows us to define precisely the notion of distance between two subsets of $\mathbb{R}^n$.

If x_0 is a point in $\mathbb{R}^n$ and A is a nonempty subset of $\mathbb{R}^n$, then the *distance* between x_0 and A, denoted by $\mathrm{dist}(x_0, A)$, is defined to be

$$\mathrm{dist}(x_0, A) = \inf\{\,\|x_0 - x\| \,:\, x \in A\,\}.$$

Also, if A and B are nonempty subsets of $\mathbb{R}^n$, then the *distance* between A and B, denoted by $\mathrm{dist}(A, B)$, is defined to be

$$\mathrm{dist}(A, B) = \inf\{\,\|x - y\| \,:\, x \in A,\, y \in B\,\}.$$

Note that if $x_0 \in A$ and $y_0 \in B$, then $\mathrm{dist}(A, B) \leq \|x_0 - y_0\| < \infty$.

Clearly, the distance between a point and a set is zero if the point is in the set, but we also have the following.

Theorem 2.6.6. *If $x_0 \in \mathbb{R}^n$ and A is a nonempty subset of $\mathbb{R}^n$, with $x_0 \notin A$, then $\mathrm{dist}(x_0, A) = 0$ if and only if x_0 is an accumulation point of A.*

Proof. If $x_0 \notin A$ and $\mathrm{dist}(x_0, A) = 0$, then we have

$$\inf\{\|x_0 - x\| \,:\, x \in A\} = 0.$$

Hence, given any $\epsilon > 0$, the real number ϵ is not a lower bound for the nonempty set of real numbers $\{\,\|x_0 - x\| \,:\, x \in A\,\}$, and so there exists an $x \in A$ such that

$$0 < \|x_0 - x\| < \epsilon.$$

Therefore, any open ball $B(x_0, \epsilon)$ contains a point of A different from x_0, and thus x_0 is an accumulation point of A.

Conversely, suppose $x_0 \notin A$ and x_0 is an accumulation point of A, and suppose $\text{dist}(x_0, A) = \gamma$, that is,

$$\gamma = \inf\{\, \|x_0 - x\| \, : \, x \in A\,\}.$$

If $\gamma > 0$, then $B\left(x_0, \frac{\gamma}{2}\right)$ misses A, which implies that x_0 is not an accumulation point of A, which is a contradiction. Therefore, $\gamma = 0$.

$\square$

Example 2.6.7. *If $A = B(\overline{0}, 1)$ is the open unit ball in $\mathbb{R}^n$ and $x_0 \in \mathbb{R}^n$ with $\|x_0\| = 1$, then*

$$\text{dist}(x_0, A) = \inf\{\, \|x_0 - x\| \, : \, x \in A\,\} = 0$$

since x_0 is an accumulation point of A, but $x_0 \notin A$.

Corollary 2.6.8. *If $x_0 \in \mathbb{R}^n$, A is a closed subset of $\mathbb{R}^n$, and $\text{dist}(x_0, A) = 0$, then $x_0 \in A$.*

The next theorem is the classic illustration of the power of the nested sets theorem and shows that if A is closed and $x_0 \notin A$, then there is a point $y_0 \in A$ such that

$$\|x_0 - y_0\| = \text{dist}(x_0, A),$$

that is, the distance is actually attained for some point in A.

Theorem 2.6.9. *If $A \subset \mathbb{R}^n$ is closed and $x_0 \in \mathbb{R}^n$, then there exists a point $y_0 \in A$ such that*

$$\|x_0 - y_0\| = \text{dist}(x_0, A).$$

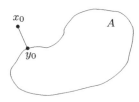

Proof. Suppose that $\text{dist}(x_0, A) = r_0 > 0$, then for every positive integer $k \geq 1$, we can find a point $y_k \in A$ such that

$$\|x_0 - y_k\| < r_0 + \frac{1}{k},$$

in other words, the set $A_k = A \cap \overline{B}\left(x_0, r_0 + \frac{1}{k}\right)$ is nonempty. Also, each of the sets A_k is compact, and they are nested, that is, $A_{k+1} \subset A_k$ for all $k \geq 1$. Thus, the Nested Sets Theorem assures us that $\bigcap\limits_{k=1}^{\infty} A_k \neq \emptyset$.

Let $y_0 \in \bigcap\limits_{k=1}^{\infty} A_k$, then $\|x_0 - y_0\| \geq r_0$ from the definition of dist(x_0, A). Also,

$$\|x_0 - y_0\| < r_0 + \frac{1}{k}$$

for all $k \geq 1$ so that $\|x_0 - y_0\| \leq r_0$.

$\square$

Example 2.6.10. *Show that if L is a straight line and A is a closed set, then it is possible that* dist$(A, L) = 0$ *without A and L having any points in common, that is, $A \cap L = \emptyset$.*

Solution.

Let A be the set

$$A = \{\, (x, y) \in \mathbb{R}^2 \;:\; y \geq \frac{1}{x}, \; 0 < x < \infty \,\}$$

and let

$$L = \{\, (x, y) \in \mathbb{R}^2 \;:\; y = 0 \,\},$$

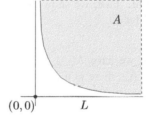

as shown in the figure on the right.

The boundary of A is the set

$$\text{bdy}(A) = \{\, (x, y) \in \mathbb{R}^2 \;:\; y = \frac{1}{x}, \; 0 < x < \infty \,\},$$

and since bdy$(A) \subset A$, then A is closed.

Also, if $p = (x, 1/x) \in \text{bdy}(A)$ and $q = (x, 0) \in L$, then

$$\text{dist}(A, L) \leq \|p - q\| = \sqrt{(x - x)^2 + (1/x)^2} = \frac{1}{x}.$$

Given $\epsilon > 0$, we can choose x so large that

$$0 < \frac{1}{x} < \epsilon,$$

so dist$(A, L) < \epsilon$, and since ϵ is arbitrary, then dist$(A, L) = 0$.

Now note that if $(x, y) \in A$, then $y > 0$ for all $0 < x < \infty$, so that $(x, y) \notin L$, and $A \cap L = \emptyset$, even though $\mathrm{dist}(A, L) = 0$.

$\square$

This cannot happen if one of the sets is compact, as the next theorem shows. The proof is left as an exercise, see Problem 2.6.5 at the end of this section.

Theorem 2.6.11. *Let A and K be nonempty subsets of $\mathbb{R}^n$, with A closed and K compact. If $\mathrm{dist}(A, K) = 0$, then $A \cap K \neq \emptyset$.*

Compact sets sometimes act like points, for example, in Theorem 2.6.9, if we replace the singleton $\{x_0\}$ (which is compact) with a compact set B, then the theorem is still true. First, a lemma.

Lemma 2.6.12. *If K is a compact subset of $\mathbb{R}^n$ and $\delta > 0$, then the set*

$$K_\delta = \{\, x \in \mathbb{R}^n \,:\, \mathrm{dist}(x, K) \leq \delta \,\}$$

is compact.

Proof. First, we show that K_δ is bounded. Let $y_0 \in K$ be a fixed point in K, and let

$$M = \sup\{\, \|y - y_0\| \,:\, y \in K \,\},$$

then we have $M < \infty$ since K is bounded. From Theorem 2.6.9, for each $x \in K_\delta$, there is a point $y \in K$ such that $\|x - y\| = \mathrm{dist}(x, K)$. Thus,

$$\|x - y_0\| \leq \|x - y\| + \|y - y_0\| \leq \delta + M$$

for all $x \in K_\delta$, and K_δ is bounded.

Next, we show that K_δ is closed. If x_0 is an accumulation point of K_δ, then given any $\epsilon > 0$, there exists a point $x \in K_\delta$ such that $\|x_0 - x\| < \epsilon$, and for this particular x, from Theorem 2.6.9, there is a point $y \in K$ such that

$$\|x - y\| = \mathrm{dist}(x, K) \leq \delta.$$

Thus,

$$\mathrm{dist}(x_0, K) \leq \|x_0 - y\| \leq \|x_0 - x\| + \|x - y\| \leq \epsilon + \delta,$$

for all $\epsilon > 0$, so that $\mathrm{dist}(x_0, K) \leq \delta$. Therefore, $x_0 \in K_\delta$, and K_δ contains all of its accumulation points, that is, K_δ is closed.

$\square$

And we have the following theorem.

Theorem 2.6.13. *If A is a nonempty closed subset of $\mathbb{R}^n$ and B is a nonempty compact subset of $\mathbb{R}^n$, then there exist points $a_0 \in A$ and $b_0 \in B$ such that*

$$\|a_0 - b_0\| = \mathrm{dist}(A, B) = \inf\{\, \|x - y\| \,:\, x \in A,\ y \in B \,\}.$$

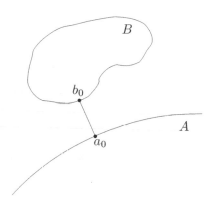

Proof. Since $\mathrm{dist}(A, B) = \inf\{\, \|a - b\| \,:\, a \in A,\ b \in B \,\}$, for each $k \geq 1$, the real number

$$\mathrm{dist}(A, B) + \frac{1}{k}$$

is no longer a lower bound for the set $\{\, \|a - b\| \,:\, a \in A,\ b \in B \,\}$, so there exist points $a_k \in A$ and $b_k \in B$ such that

$$\|a_k - b_k\| \leq \mathrm{dist}(A, B) + \frac{1}{k}$$

for all $k \geq 1$.

Now, the set $S = \{\, b_1,\, b_2,\, b_3,\, \dots \,\}$ is either finite or infinite.

If the set S is finite, then there is a point $b_0 \in B$ such that $b_k = b_0$ for infinitely many values of k.

If the set S is infinite, then since B is compact, it has an accumulation point, call it b_0, and $b_0 \in B$.

In either case, we have

$$\mathrm{dist}(b_0, A) \leq \|b_0 - a_k\| \leq \|b_0 - b_k\| + \|b_k - a_k\| \leq \|b_0 - b_k\| + \mathrm{dist}(A, B) + \frac{1}{k}$$

for all $k \geq 1$. Given $\epsilon > 0$, choose k such that

$$\frac{1}{k} < \frac{\epsilon}{2} \quad \text{and} \quad \|b_0 - b_k\| < \frac{\epsilon}{2}.$$

Note that this is possible in *both* cases, since either $b_k = b_0$ for infinitely many k or b_0 is an accumulation point of B. Therefore,

$$\text{dist}(b_0, A) \leq \frac{\epsilon}{2} + \text{dist}(A, B) + \frac{\epsilon}{2} = \text{dist}(A, B) + \epsilon.$$

Since $\epsilon > 0$ was arbitrary, then $\text{dist}(b_0, A) \leq \text{dist}(A, B)$.

On the other hand, since $b_0 \in B$, we have

$$\text{dist}(A, B) \leq \|b_0 - a\|$$

for all $a \in A$, so that $\text{dist}(A, B)$ is a lower bound for the set $\{\|b_0 - a\| \ : \ a \in A\}$, and so

$$\text{dist}(A, B) \leq \text{dist}(b_0, A).$$

Therefore, $\text{dist}(b_0, A) = \text{dist}(A, B)$, and since A is closed, then from theorem 2.6.9, the distance $\text{dist}(b_0, A)$ is attained, and there exists an $a_0 \in A$ such that

$$\text{dist}(b_0, A) = \|a_0 - b_0\| = \text{dist}(A, B).$$

$\square$

Note. The previous theorem says that if A and B are disjoint, with A closed and B compact, then there exist $a_0 \in A$ and $b_0 \in B$ such that $\text{dist}(A, B) = \|a_0 - b_0\| > 0$.

Example 2.6.14. *If A is a nonempty subset of $\mathbb{R}^n$ and for each $x \in \mathbb{R}^n$, we define*

$$f(x) = \text{dist}(x, A) = \inf\{\|x - a\| \ : \ a \in A\},$$

show that f is continuous at each point $x \in \mathbb{R}^n$.

Solution. Let x and y be points in $\mathbb{R}^n$ and let $\epsilon > 0$. Since $\text{dist}(y, A) + \epsilon$ is no longer a lower bound for the set of real numbers $\{\|y - a\| \ : \ a \in A\}$, there exists a point $a_0 \in A$ such that

$$\|y - a_0\| \leq \text{dist}(y, A) + \epsilon.$$

Therefore,

$$\begin{aligned}
f(x) - f(y) &= \inf\{\|x - a\| \ : \ a \in A\} - \inf\{\|y - a\| \ : \ a \in A\} \\
&\leq \|x - a_0\| - \|y - a_0\| + \epsilon \\
&\leq \|x - y\| + \epsilon,
\end{aligned}$$

and since $\epsilon > 0$ was arbitrary, then $f(x) - f(y) \le \|x - y\|$. Interchanging x and y, we see that $f(y) - f(x) \le \|y - x\| = \|x - y\|$ also. Thus,

$$|f(x) - f(y)| \le \|x - y\|$$

for all x, $y \in \mathbb{R}^n$, and f is continuous everywhere.

$\square$

2.6.4 Support Hyperplanes for Compact Sets in $\mathbb{R}^n$

This section introduces a generalization of the notion of a tangent line.

If A is a subset of $\mathbb{R}^2$, a *support line* for A is a straight line ℓ that intersects A and such that A is contained in one of the closed halfspaces determined by ℓ.

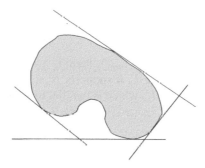

In the figure above, support lines to the set A are illustrated.

More generally, if A is a subset of $\mathbb{R}^n$, a *supporting hyperplane* for A is a hyperplane H that intersects A and such that A is contained in one of the closed halfspaces determined by H.

Thus, in $\mathbb{R}^2$, a supporting hyperplane is a line, while in $\mathbb{R}^3$, a supporting hyperplane is a plane.

Example 2.6.15. *(Parallel Supporting Hyperplanes)*
Let K be a nonempty compact subset of $\mathbb{R}^2$ and let L be a straight line in $\mathbb{R}^2$, then there are two lines parallel to L that support K.

Solution. We consider one of the closed halfspaces H determined by L and consider the intersection of K with parallel translates of H.

The hyperplane L can be written as

$$L = \{(x, y) \in \mathbb{R}^2 : f(x, y) = \delta\}$$

for some $\delta \in \mathbb{R}$, where f is a nonzero linear functional given by

$$f(x, y) = ax + by$$

for $(x, y) \in \mathbb{R}^2$.

For each $(x, y) \in \mathbb{R}^2$, we have

$$|f(x + h, y + k) - f(x, y)| = |a(x + h - x) + b(y + k - y)|$$
$$\leq |a| \cdot |h| + |b| \cdot |k|,$$

so that f is continuous at (x, y). Since K is compact, then $f(K)$ is compact, and so f attains its maximum and minimum on K. Thus, there exist points $z_{\max}$ and $z_{\min}$ in K such that

$$f(z_{\max}) = \max\{f(x, y) : (x, y) \in K\}$$

and

$$f(z_{\min}) = \min\{f(x, y) : (x, y) \in K\}.$$

If we let $\alpha = f(z_{\max})$ and $\beta = f(z_{\min})$, then $f^{-1}(\alpha)$ is a supporting hyperplane for K, which is parallel to L, it supports K at $z_{\max}$. Similarly, $f^{-1}(\beta)$ is a supporting hyperplane for K, which is parallel to L, it supports K at $z_{\min}$.

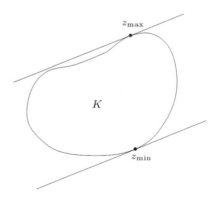

Note that if $\beta < \alpha$, then the support lines are distinct, but if $\alpha = \beta$, then

$$f(x) = \alpha = \beta$$

for all $x \in K$, that is, K lies in a hyperplane parallel to L. We leave it as an exercise to show that the same proof works for $\mathbb{R}^n$ where $n > 2$.

$\square$

The next theorem is the "thought experiment" of Section 2.1 brought to mathematical life. The proof of the theorem uses the notion of infimum to establish the existence of the limiting line that we discussed in Section 2.1. Compactness is then used to verify that the limiting line is at distance zero from the set K, so that Theorem 2.6.11 applies.

Theorem 2.6.16. *If K is a nonempty compact subset of $\mathbb{R}^2$ and p is a point in the boundary of some halfspace that contains K, then there is a support line for K that passes through p.*

Proof. The boundary of the halfspace is a line L_0 and p is a point on this line. If L_0 touches K, we are finished, since then L_0 would be the support line.

Otherwise, since K is not empty, there is at least one line L_1 passing through the point p and some point k_1 of K that makes a positive angle α with L_0.*

Let S be the set of all positive angles between L_0 and the lines that pass through p and the points of K. The set S is a nonempty set of real numbers, which is bounded below by the number 0. Consequently, S has an infimum, say

$$\inf(S) = \beta.$$

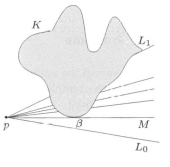

Let M be the line through p that makes a positive angle β with L_0, this will be our candidate for the limiting line.

Note that K is contained in one of the halfspaces determined by M, since if not, then there would be points from K on both sides of M, and this would violate the definition of β—it would no longer be an infimum. We want to show that $\text{dist}(M, K) = 0$. Since M is a closed set and K is a nonempty compact set, then Theorem 2.6.11 would guarantee that M and K have points in common, and so this would finish the proof.

We will use the fact that K is bounded to show that the line M is at a distance 0 from K. Since K is bounded, there is some small ball of finite radius centered at p that contains K. If $\text{dist}(M, K) > \epsilon > 0$ for some $\epsilon > 0$, then there would be two lines parallel to M at distance ϵ from M with the property that no point of K is on or between these parallel lines as in the following figure.

---

* The positive angle from L_0 to L_1 is the one that is in the counterclockwise direction from L_0.

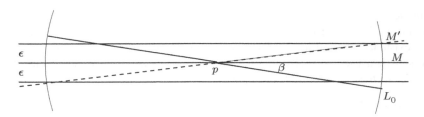

However, this implies that there is a line M' through p making an angle greater than β with L_0 that misses K, and this contradicts the fact that β is the infimum of the set of positive angles between L_0 and lines that pass through p and points of K. Consequently, we must conclude that the distance from M to K is zero, and so this limiting line M is a support line for K passing through p, completing the proof.

$\square$

2.6.5 Problems

1. Prove that the union of finitely many closed sets is a closed set.
2. Give an example of a family of infinitely many closed sets whose union is not closed.
3. Explain why a translate of a closed set is a closed set.
4. Explain why a closed subset of a compact set is compact.
5. Show that if x_0 is an accumulation point of $A \subset \mathbb{R}^n$, then every open ball centered at x_0 contains infinitely many points of A.
6. If A and B are subsets of $\mathbb{R}^n$, show that

    (a) $\text{int}(A \cap B) = \text{int}(A) \cap \text{int}(B)$.
    (b) $\text{int}(A) \cup \text{int}(B) \subseteq \text{int}(A \cup B)$, and give an example to show that equality many not hold.

7. If A and B are subsets of $\mathbb{R}^n$, show that

    (a) $\overline{A \cup B} = \overline{A} \cup \overline{B}$.
    (b) $\overline{A \cap B} \subseteq \overline{A} \cap \overline{B}$, and give an example to show that equality many not hold.

8. Let A be a nonempty subset of $\mathbb{R}^n$.

    (a) Show that $\text{bdy}(A) = \overline{A} \setminus \text{int}(A)$.
    (b) Show that $\text{bdy}(A)$ is a closed set.
    (c) Show that A' is a closed set, where A' is the set of all accumulation points of A.

9. Let the sets F_k of $\mathbb{R}^n$ be defined as follows:

$$F_1 = \{ x_1, x_2, x_3, \ldots \}$$
$$F_2 = \{ x_2, x_3, x_4, \ldots \}$$
$$F_3 = \{ x_3, x_4, x_5, \ldots \}$$

$$\vdots$$

Show that if x_0 is an accumulation point of F_1, then x_0 is an accumulation point of F_k for every $k = 1, 2, 3, \ldots$

10. Prove Theorem 2.6.11: Let A and K be nonempty subsets of $\mathbb{R}^n$, with A closed and K compact. If $\operatorname{dist}(A, K) = 0$, then $A \cap K \neq \emptyset$.

11. Let A and B be nonempty subsets of $\mathbb{R}$. Show that

$$\sup(A + B) = \sup(A) + \sup(B),$$

where $A + B = \{ a + b \in \mathbb{R} : a \in A \text{ and } b \in B \}$.

12. The **Cartesian product** of $\mathbb{R}^n$ with itself is

$$\mathbb{R}^n \times \mathbb{R}^n = \{ (x, y) : x \in \mathbb{R}^n \text{ and } y \in \mathbb{R}^n \}.$$

If A and B are subsets of $\mathbb{R}^n$, the Cartesian product of A and B is

$$A \times B = \{ (a, b) : a \in A, b \in B \},$$

so that $A \times B$ is a subset of $\mathbb{R}^n \times \mathbb{R}^n$.

(a) Show that for $x \in \mathbb{R}^n$ and $y \in \mathbb{R}^n$, the function $\|\cdot\| : \mathbb{R}^n \times \mathbb{R}^n \longrightarrow \mathbb{R}$

$$\|(x, y)\| = \left(\|x\|^2 + \|y\|^2 \right)^{\frac{1}{2}}$$

defines a norm on $\mathbb{R}^n \times \mathbb{R}^n$.

(b) Show that if A and B are nonempty compact subsets of $\mathbb{R}^n$, then $A \times B$ is a compact subset of $\mathbb{R}^n \times \mathbb{R}^n$.

13. Let H be a hyperplane in $\mathbb{R}^n$:

(a) Show that H is a proper subset of $\mathbb{R}^n$.

(b) Show that H does not contain an open ball in $\mathbb{R}^n$.

(c) Show that $\mathbb{R}^n$ is not the union of a finite number of hyperplanes.

(d) Show that $\mathbb{R}^n$ is not the union of countably many hyperplanes.

14. Show that a linear functional $f : \mathbb{R}^n \longrightarrow \mathbb{R}$ is continuous at each point $x \in \mathbb{R}^n$.

15. Let K be a nonempty compact subset of $\mathbb{R}^2$ and let L be a line in the plane, show that some translate of L supports K.

16. The **graph** of a real-valued function $f : \mathbb{R} \longrightarrow \mathbb{R}$ is the set

$$\operatorname{graph}(f) = \{ (x, y) : y = f(x), -\infty < x < \infty \}.$$

Show that if f is continuous, then graph(f) is a closed subset of $\mathbb{R}^2$, and use this result to show that if the function f is continuous on a closed and bounded interval $[a, b]$, then f has a maximum and a minimum on $[a, b]$.

17. Give an example of a function $f : [0, 1] \longrightarrow \mathbb{R}$ whose graph is not closed.

18. Give an example of a function $f : [0, 1] \longrightarrow \mathbb{R}$ that has neither a maximum nor a minimum.

19. More generally, the **graph** of a function $f : \mathbb{R}^n \longrightarrow \mathbb{R}^m$ is the set

$$\text{graph}(f) = \{\, (x, y) \in \mathbb{R}^n \times \mathbb{R}^m \; : \; y = f(x), \;\; \text{for some } x \in \mathbb{R}^n \,\}.$$

Show that if f is continuous, then the graph(f) is a closed subset of $\mathbb{R}^n \times \mathbb{R}^m$,

20. Give a proof of Theorem 2.6.16 that uses the nested sets theorem.

21. A point x_0 of $A \subset \mathbb{R}^n$ is said to be **farthest** from the point p if

$$\|x_0 - p\| = \sup\{\, \|x - p\| \; : \; x \in A \,\}.$$

This definition does not require that the farthest point be unique.

(a) In the Euclidean norm, if $A = \overline{B}(\overline{0}, 1)$, explain how to find the point or points that are farthest from the given point $p \neq \overline{0}$.

(b) Using the Euclidean norm, identify all points of $\mathbb{R}^2$ that admit more than one farthest point in the set $\overline{B}(\overline{0}, 1)$.

(c) Using the Euclidean norm, identify all points of $\mathbb{R}^2$ that admit more than one farthest point in the set $\{\, (x, y) \in \mathbb{R}^2 \; : \; |x| + |y| \leq 1 \,\}$.

(d) Show that if K is a nonempty compact subset of $\mathbb{R}^2$ and if p is any point in $\mathbb{R}^2$, then some point of K is farthest from p.

22. Show that

(a) In $\mathbb{R}^n$, there is a nested family of closed sets F_1, F_2, F_3, $\ldots$ with empty intersection.

(b) In $\mathbb{R}^n$, there is a nested family of bounded sets B_1, B_2, B_3, $\ldots$ with empty intersection.

23. Let K be a nonempty compact subset of $\mathbb{R}^n$ and let H be a hyperplane in $\mathbb{R}^n$, show that there are two hyperplanes parallel to H that support K.

24. A family of sets $\mathcal{F}$ is said to have the **finite intersection property** if every finite subcollection of $\mathcal{F}$ has a nonempty intersection.

(a) Using the usual definition of compactness in terms of coverings by open sets, prove that a subset K of $\mathbb{R}^n$ is compact if and only if whenever $\mathcal{F}$ is a family of closed subsets of K with the finite intersection property, then $\mathcal{F}$ has a nonempty intersection.

(b) Using part (a), prove the nested sets theorem in $\mathbb{R}^n$.

25. Let K be a nonempty compact subset of $\mathbb{R}^n$ and let U be an open set containing K. Show that there exists an open set V such that

$$K \subset V \subset \overline{V} \subset U.$$

26. If A and B are nonempty subsets of $\mathbb{R}^n$, then the **Minkowski sum** of A and B, denoted by $A + B$, is defined to be

$$A + B = \{a + b \in \mathbb{R}^n \, : \, a \in A \text{ and } b \in B\}.$$

   (a) Show that if one of A or B is open, then $A + B$ is open.
   (b) Show that if A is compact and B is closed, then $A + B$ is closed.
   (c) Give an example of two closed subsets A and B of $\mathbb{R}^n$ for which $A + B$ is not closed.

27. If K is a compact subset of $\mathbb{R}^n$ and $\delta > 0$, let

$$K_\delta = \{x \subset \mathbb{R}^n \, : \, \text{dist}(x, K) \leq \delta\}.$$

   Show that $K_\delta = K + \overline{B}(\overline{0}, \delta)$.

∗28. Let ℓ_1 be the set of all real-valued sequences $x = \{x_k\}_{k \geq 1}$ such that

$$\|x\|_1 = \sum_{k=1}^{\infty} |x_k| < \infty,$$

   that is, the set of all **absolutely summable sequences**.
   Let ℓ_2 be the set of all real-valued sequences $x = \{x_k\}_{k \geq 1}$ such that

$$\|x\|_2 = \left(\sum_{k=1}^{\infty} x_k^2\right)^{\frac{1}{2}} < \infty,$$

   that is, the set of all **square summable sequences**.

   (a) Show that with the usual notions of addition and scalar multiplication, ℓ_1 and ℓ_2 are real linear spaces, that is, vector spaces over the field $\mathbb{R}$ of real numbers.
   Note. You only need to show that they are closed under addition and scalar multiplication since they are in fact subsets of the larger vector space of functions $f : \mathbb{N} \longrightarrow \mathbb{R}$ with pointwise addition and scalar multiplication.
   (b) Show that $\| \cdot \|_1$ is a norm on ℓ_1.
   (c) Show that $\| \cdot \|_2$ is a norm on ℓ_2.
   (d) Show that $\ell_1 \subsetneqq \ell_2$.
   Hint. If $a = \{a_k\}_{k \geq 1} \in \ell_1$, then $\sum\limits_{k=1}^{\infty} |a_k| < \infty$, so that $|a_k| \to 0$ as $k \to \infty$, and there exists an integer k_0 such that $|a_k| < 1$ for all $k \geq k_0$. Therefore $|a_k|^2 \leq |a_k|$ for all $k \geq k_0$, ...

∗29. In ℓ_2, let $\{e_k\}_{k\geq 1}$ be the unit coordinate vectors where e_k has a 1 in the kth coordinate and 0's everywhere else, that is,

$$e_1 = (1, 0, 0, 0, \ldots)$$
$$e_2 = (0, 1, 0, 0, \ldots)$$
$$e_3 = (0, 0, 1, 0, \ldots)$$
$$\vdots$$

and let
$$A = \{\, e_k \,:\, k = 1, 2, 3, \ldots \}.$$

(a) Show that $\|e_k\|_2 = 1$ for each k.
(b) Show that $\|e_k - e_m\|_2 = \sqrt{2}$ if $k \neq m$.
(c) Show that $A \subset \ell_2$ and $\mathrm{diam}(A) = \sqrt{2} < \infty$.
(d) Show that A has no accumulation points; hence, A is a closed subset of ℓ_2.
(e) Show that $\mathcal{U} = \{\, B(e_k, 1) \,:\, k = 1, 2, \ldots \}$ is an open cover for A that has no finite subcover; hence, A is not compact.
(f) Let $\overline{B}(\overline{0}, 1)$ be the closed unit ball in ℓ_2, that is,

$$\overline{B}(\overline{0}, 1) = \{\, x \in \ell_2 \,:\, \|x\|_2 \leq 1 \,\}.$$

Show that $\overline{B}(\overline{0}, 1)$ is not compact.
Hint. Note that the set A above is a closed subset of the ball.

3 CONVEXITY

3.1 INTRODUCTION

Recall from Chapter 1 that the part of a straight line between two distinct points p and q in $\mathbb{R}^n$ is called a *straight line segment*, or simply a *line segment*.

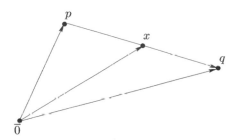

The notation used was the same as that for intervals of the real line, and it is common to adopt the terminology of $\mathbb{R}$, although (p, q) is not really an open set in $\mathbb{R}^n$ unless $n = 1$. The actual definitions were

- $[p, q]$ is the *closed line segment* joining p and q

$$[p, q] = \{\, x \in \mathbb{R}^n \ : \ x = (1 - \mu)p + \mu q \ : \ 0 \leq \mu \leq 1 \,\}.$$

- (p, q) is the *open line segment* joining p and q

$$(p, q) = \{\, x \in \mathbb{R}^n \ : \ x = (1 - \mu)p + \mu q \ : \ 0 < \mu < 1 \,\}.$$

Geometry of Convex Sets, First Edition. I. E. Leonard and J. E. Lewis.
© 2016 John Wiley & Sons, Inc. Published 2016 by John Wiley & Sons, Inc.

- $[p, q)$ and $(p, q]$ are the **half-open line segments** joining p and q

$$[p, q) = \{\, x \in \mathbb{R}^n \;:\; x = (1 - \mu)p + \mu q \;:\; 0 \leq \mu < 1 \,\},$$
$$(p, q] = \{\, x \in \mathbb{R}^n \;:\; x = (1 - \mu)p + \mu q \;:\; 0 < \mu \leq 1 \,\}.$$

We noted that as the scalar μ increased from $\mu = 0$ to $\mu = 1$, the point x moved along the line segment from $x = p$ to $x = q$.

Also, we defined a subset C of $\mathbb{R}^n$ to be **convex** if it contained all of the straight line segments determined by pairs of points of the set C. It followed from the definition that the empty set, a single point, a subspace, or indeed any translate of a subspace, are all convex sets.

In $\mathbb{R}^1$, the convex sets are easy to identify. They are the empty set, single points, intervals, halflines, and the entire real line. In $\mathbb{R}^2$, the supply of convex sets is much more varied (see the figure below).

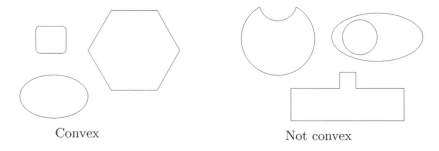

Convex　　　　　　　　　　　　Not convex

A convex set is one that has no holes, hollows, pits, or dimples. One may describe convexity in terms of visibility: a set is convex if from each point in the set it is possible to see every other point without having the line of sight pass outside the set. The strong visual component of convexity is one of the appealing aspects of the subject.

We saw earlier that points, lines, and hyperplanes have properties that are dimension free. The same is true about more general convex sets. As a consequence, we can gain some understanding of n-dimensional convex sets by studying two-dimensional and three- dimensional convex sets. Because of this, convexity forms an important part of n-dimensional geometry.

Although we will confine our study of convex sets to geometric applications, it should be understood that convexity is not without value in applied mathematics. In fact, much of the recent interest in convexity is due to its applicability to areas such as linear and nonlinear programming and optimization theory. For example, using linear programming to find an optimal solution to even the simplest problems, leads to the realization that the "feasibility region" is a convex set. In the areas where convexity is applied, an understanding of the geometry of convex sets is almost indispensable.

Having stressed the usefulness of convexity in applications, perhaps a word of caution might be in order: we should not devalue the importance of convexity in "pure mathematics" simply because it happens to be applicable.

We have already mentioned that points, lines, and hyperplanes in $\mathbb{R}^n$ are convex sets. Other basic n-dimensional sets that we have studied are the open and closed balls in $\mathbb{R}^n$. We have shown earlier that the sets $B(x, \rho)$ and $\overline{B}(x, \rho)$ are convex; we will end the introduction by showing that a hyperplane and a halfspace are also convex.

Theorem 3.1.1. *Let f be a nonzero linear functional on $\mathbb{R}^n$ and let β be a scalar. The hyperplane*

$$H_\beta = f^{-1}(\beta) = \{ x \in \mathbb{R}^n : f(x) = \beta \}$$

and the open halfspace

$$H = \{ x \subset \mathbb{R}^n : f(x) < \beta \}$$

are convex.

Proof. Let x and y be in H_β and let $0 < \lambda < 1$, we have to show that the convex combination $\lambda x + (1 - \lambda)y$ is also in H_β. Since f is linear and $x, y \in H_\beta$, we have

$$f(\lambda x + (1 - \lambda)y) = \lambda f(x) + (1 - \lambda)f(y) = \lambda \beta + (1 - \lambda)\beta = \beta,$$

so that $\lambda x + (1 - \lambda)y \in H_\beta$.

Similarly, let x and y be in H and let $0 < \lambda < 1$, since $\lambda > 0$ and $1 - \lambda > 0$, we have

$$f(\lambda x + (1 - \lambda)y) = \lambda f(x) + (1 - \lambda)f(y) < \lambda \beta + (1 - \lambda)\beta = \beta,$$

so that $\lambda x + (1 - \lambda)y \in H$.

$\square$

3.2 BASIC PROPERTIES OF CONVEX SETS

Recall that if X is any set in $\mathbb{R}^n$, then αX is the set obtained by multiplying each member of X by the scalar α, that is,

$$\alpha X = \{ \alpha x \in \mathbb{R}^n : x \in X \}.$$

Recall also that when $\alpha > 0$, the set αX, and any translate $\alpha X + x_0$, is called a *positive homothet* of X.

Theorem 3.2.1. *If $[p, q]$ is a straight line segment in $\mathbb{R}^n$, then*

(a) $[p, q] + x_0 = [p + x_0, q + x_0]$ *for every $x_0 \in \mathbb{R}^n$, and*
(b) $\alpha[p, q] = [\alpha p, \alpha q]$ *for every real number $\alpha \neq 0$.*

Proof.

(a) We have $x \in [p, q] + x_0$ if and only if there exists a λ, with $0 \leq \lambda \leq 1$, such that

$$
\begin{aligned}
x &= (1 - \lambda)p + \lambda q + x_0 \\
&= (1 - \lambda)p + \lambda q + (1 - \lambda)x_0 + \lambda x_0 \\
&= (1 - \lambda)(p + x_0) + \lambda(q + x_0),
\end{aligned}
$$

that is, if and only if $x \in [p+x_0, q+x_0]$. Thus, $[p, q]+x_0 = [p+x_0, q+x_0]$ for every $x_0 \in \mathbb{R}^n$.

(b) Also, $x \in \alpha[p, q]$ if and only if there exists a λ, with $0 \leq \lambda \leq 1$, such that

$$
\begin{aligned}
x &= \alpha\left[(1 - \lambda)p + \lambda q\right] \\
&= (1 - \lambda)\alpha p + \lambda\alpha q,
\end{aligned}
$$

that is, if and only if $x \in [\alpha p, \alpha q]$. Therefore, $\alpha[p, q] = [\alpha p, \alpha q]$.

$\square$

Corollary 3.2.2. *If C is a convex subset of $\mathbb{R}^n$, then $C + x_0$ and αC are convex for every $x_0 \in \mathbb{R}^n$ and every scalar $\alpha \in \mathbb{R}$.*

Proof. To show that $C + x_0$ is convex, let p and q be points in $C + x_0$, so that $p - x_0$ and $q - x_0$ are points in C. Since C is convex, the segment $[p - x_0, q - x_0]$ is also in C, and from the previous theorem,

$$
[p - x_0, q - x_0] + x_0 = [p - x_0 + x_0, q - x_0 + x_0] = [p, q].
$$

Therefore, $[p, q] \subset C + x_0$ whenever p and q are points of $C + x_0$, and so $C + x_0$ is convex.

If $\alpha = 0$, then $\alpha C = \{\overline{0}\}$ is convex whether or not C is convex, so we may assume that $\alpha \neq 0$. Let p and q be points in αC, so that $p = \alpha x$ and $q = \alpha y$, where x and y are points in C. Since C is convex, we have $[x, y] \subset C$, and, therefore, $\alpha[x, y] \subset \alpha C$. From the previous theorem, this implies that $[p, q] \subset \alpha C$, that is, αC is convex.

$\square$

The preceding corollary often eases the problem of checking for convexity. For example, to show that the ball $B(x_0, \rho)$ is convex, we note that by an appropriate translation and scalar multiplication, we may assume that the ball is the open unit ball $B(\overline{0}, 1)$. The proof is then as follows: if x and y are in $B(\overline{0}, 1)$, we need only show that if $0 \leq \lambda \leq 1$, then $(1 - \lambda)x + \lambda y$ is also in $B(\overline{0}, 1)$. Again we use the triangle inequality, but its application is a little easier:

$$
\begin{aligned}
\|(1 - \lambda)x + \lambda y\| &\leq \|(1 - \lambda)x\| + \|\lambda y\| \\
&= (1 - \lambda)\|x\| + \lambda\|y\| \\
&< (1 - \lambda) + \lambda \\
&= 1,
\end{aligned}
$$

which shows that $(1 - \lambda)x + \lambda y \in B(\overline{0}, 1)$ whenever $x, y \in B(\overline{0}, 1)$ and $0 \leq \lambda \leq 1$.

The next theorem, which was stated and proved in Chapter 1, shows that convexity is preserved under intersection. Again, note that the family $\mathcal{F}$ in the statement of the theorem may be finite or infinite.

Theorem 3.2.3. *If $\mathcal{F}$ is a nonempty family of convex subsets of $\mathbb{R}^n$, then the set*

$$
C = \bigcap \{ A : A \in \mathcal{F} \}
$$

is convex.

Proof. Let x and y be points in C. We will show that $[x, y] \subset C$. Since C is the intersection of all members of $\mathcal{F}$, it follows that $x \in A$ for each $A \in \mathcal{F}$; similarly, $y \in A$ for each $A \in \mathcal{F}$. Since each A is convex, we know that $[x, y] \subset A$, and so

$$
[x, y] \subset C = \bigcap \{ A : A \in \mathcal{F} \},
$$

and C is convex.

$\square$

Example 3.2.4. *Show that the set*

$$
S = \{ (x, y) \in \mathbb{R}^2 : y \geq 1/x, \ x > 0 \}
$$

is convex.

Solution.

The set S is the same one we used when we discussed support lines in Chapter 2. We will show later (see Theorem 3.6.19) that if we can show that S is the intersection of all closed halfspaces that contain it, from the previous theorem we can then conclude that S is convex, since closed halfspaces are convex. For now, we will use the definition to show that S is convex.

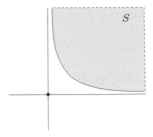

$S = \{(x, y) \in \mathbb{R}^2 : y \geq 1/x,\ 0 < x < \infty\}$

The proof follows easily from the fact that $a + 1/a \geq 2$ for all $a > 0$, with equality if and only if $a = 1$, the proof of which we leave as an exercise.

Now, if $p_1 = (x_1, y_1)$, $p_2 = (x_2, y_2) \in C$, and $0 \leq \lambda \leq 1$, then

$$(1 - \lambda)x_1 + \lambda x_2 > 0$$

since $x_1 > 0$ and $x_2 > 0$. Therefore,

$$[(1 - \lambda)x_1 + \lambda x_2]\,[(1 - \lambda)y_1 + \lambda y_2]$$
$$= (1 - \lambda)^2 x_1 y_1 + \lambda(1 - \lambda)\,[x_1 y_2 + x_2 y_1] + \lambda^2 x_2 y_2.$$

However, $x_1 y_1 \geq 1$, $x_2 y_2 \geq 1$, and

$$x_1 y_2 + x_2 y_1 \geq x_1 \frac{1}{x_2} + x_2 \frac{1}{x_1} \geq 2,$$

so that

$$[(1 - \lambda)x_1 + \lambda x_2]\,[(1 - \lambda)y_1 + \lambda y_2]$$
$$\geq (1 - \lambda)^2 + 2\lambda(1 - \lambda) + \lambda^2$$
$$= [(1 - \lambda) + \lambda]^2$$
$$= 1.$$

Therefore,

$$(1 - \lambda)y_1 + \lambda y_2 \geq \frac{1}{(1 - \lambda)x_1 + \lambda x_2},$$

so that $(1 - \lambda)p_1 + \lambda p_2 = ((1 - \lambda)x_1 + \lambda x_2, (1 - \lambda)y_1 + \lambda y_2) \in S$ and S is convex.

$\square$

One of the reasons that convex sets are so useful is that convexity is preserved by linear transformations.

A *linear transformation* is a mapping $T : \mathbb{R}^n \longrightarrow \mathbb{R}^m$ that satisfies

(i) $T(x + y) = T(x) + T(y)$ for all $x\,y \in \mathbb{R}^n$, and
(ii) $T(\alpha x) = \alpha T(x)$ for all $x \in \mathbb{R}^n$ and all $\lambda \in \mathbb{R}$.

Note that (i) says that T is *additive*, while (ii) says that T is *homogeneous*.

Often, (i) and (ii) are combined, and we say that T is *linear* provided

$$T(\alpha x + \beta y) = \alpha T(x) + \beta T(y)$$

for all x, $y \in \mathbb{R}^n$ and all scalars α and β.

In the plane, a rotation about the origin and a reflection whose axis contains the origin are special types of linear transformations. Every Euclidean motion in the plane is either a linear transformation or a linear transformation combined with a translation.

As well, any similarity with center $\overline{0}$, also known as a *positive homothety*, is a linear transformation.

Theorem 3.2.5. *The image of a straight line segment under a linear transformation is a straight line segment.*

Proof. Let T be a linear transformation, and suppose that the straight line segment is $[p, q]$. We want to show that

$$\{\, T(x) \in \mathbb{R}^m \,:\, x \in [p, q] \,\} = [T(p), T(q)].$$

Now, $x \in [p, q]$ if and only if

$$x = (1 - \lambda)p + \lambda q$$

for some $\lambda \in [0, 1]$. Since T is linear, we have

$$\begin{aligned} T(x) &= T\left((1 - \lambda)p + \lambda q\right) \\ &= (1 - \lambda)T(p) + \lambda T(q), \end{aligned}$$

so that $x \in [p, q]$ if and only if $T(x) \in [T(p), T(q)]$.

$\square$

Theorem 3.2.6. *If C is a convex subset of $\mathbb{R}^n$, and if T is a linear transformation with domain $\mathbb{R}^n$, then $T(C)$ is also convex.*

Proof. Let u and v be two points in $T(C)$. We want to show that the segment $[u, v]$ is a subset of $T(C)$.

Now, since u and v are in $T(C)$, we must have $u = T(p)$ and $v = T(q)$ for some p and q in C. By convexity of C, we have $[p, q] \subset C$, so that $T([p, q]) \subseteq T(C)$. All that remains is to show that $T([p, q])$ is the straight line segment $[u, v]$. But this is what the previous theorem says.

$\square$

Example 3.2.7. *Let A and B be subsets of $\mathbb{R}^n$, and let*

$$A + B = \{\, a + b \, : \, a \in A, \ b \in B \,\}$$

be the **Minkowski sum** *of A and B.*

(a) *Show that if A and B are convex, then $A + B$ is also convex.*
(b) *In $\mathbb{R}^2$, let*

$$A = \{\, (x, y) \in \mathbb{R}^2 \, : \, 0 \leq x < 1, \ y = 0 \,\}$$

and

$$B = \{\, (x, y) \in \mathbb{R}^2 \, : \, x = 0, \ 0 \leq y < 1 \,\}.$$

Sketch the Minkowski sum $A + B$.

Solution.

(a) If $x, y \in A + B$ and $0 < \lambda < 1$, then

$$(1 - \lambda)x + \lambda y = (1 - \lambda)(a_1 + b_1) + \lambda(a_2 + b_2),$$

where $a_1, a_2 \in A$ and $b_1, b_2 \in B$, and so

$$(1 - \lambda)x + \lambda y = [(1 - \lambda)a_1 + \lambda a_2] + [(1 - \lambda)b_1 + \lambda b_2] \in A + B,$$

since A and B are convex. Therefore, $A + B$ is also convex.
(b) The Minkowski sum of A and B is shown in the figure below.

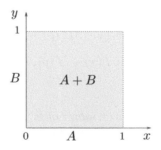

$\square$

Example 3.2.8. *Let A be a convex subset of $\mathbb{R}^n$ and let $\alpha > 0$, $\beta > 0$.*

(a) *Show that*
$$\alpha A + \beta A = (\alpha + \beta) A.$$

(b) *Give an example to show that convexity of A is necessary.*

Solution.

(a) If $x \in \alpha A + \beta A$, then there exist a_1, $a_2 \in A$ such that

$$x = \alpha a_1 + \beta a_2 = (\alpha + \beta) \left(\frac{\alpha}{\alpha + \beta} a_1 + \frac{\beta}{\alpha + \beta} a_2 \right),$$

so that $x \in (\alpha + \beta) A$ since A is convex. Therefore,

$$\alpha A + \beta A \subset (\alpha + \beta) A.$$

If $x \in (\alpha + \beta) A$, then there exists an $a \in A$ such that

$$x = (\alpha + \beta) a = \alpha a + \beta a,$$

so that $x \in \alpha A + \beta A$, and

$$(\alpha + \beta) A \subset \alpha A + \beta A.$$

Note that convexity of A is not needed for this inclusion. Thus, if A is convex, then $\alpha A + \beta A = (\alpha + \beta) A$.

(b) Let
$$X = \{ (x, 0) \in \mathbb{R}^2 : x \geq 0 \}$$

and
$$Y = \{ (0, y) \in \mathbb{R}^2 : y \geq 0 \}.$$

Now let $A = X \cup Y$, then

$$A + A = \{ (x, y) \in \mathbb{R}^2 : x \geq 0, \, y \geq 0 \},$$

but $2A = A$, so that

$$1 \cdot A + 1 \cdot A \neq (1 + 1) A = 2A.$$

Therefore, convexity of A is needed to guarantee equality in part (a).

□

3.2.1 Problems

1. Let A and B be subsets of $\mathbb{R}^n$ with B convex.

   (a) Show that the set

   $$C = \{\, x \in \mathbb{R}^n \ : \ x + A \subset B \,\}$$

   is a convex subset of $\mathbb{R}^n$.

   (b) Show that the set

   $$C = \{\, x \in \mathbb{R}^n \ : \ A \subset x + B \,\}$$

   is a convex subset of $\mathbb{R}^n$.

2. Let A and B be convex subsets of $\mathbb{R}^n$. Show that the set

   $$C = \{\, x \in \mathbb{R}^n \ : \ A \cap (x + B) \neq \emptyset \,\}$$

   is convex.

3. Let A and B be nonempty subsets of $\mathbb{R}^n$ and let $A + B$ be the Minkowski sum of A and B.

   (a) Show that

   $$A + B = \bigcup_{a \in A} (a + B) = \bigcup_{b \in B} (b + A).$$

   (b) Let A and B be the following intervals in $\mathbb{R}^2$:

   $$A = [\,(-1, 0), (1, 0)\,] \qquad \text{and} \qquad B = [\,(0, -1), (0, 1)\,].$$

   Show that

   $$A + B = \{\, (x, y) \in \mathbb{R}^2 \ : \ -1 \leq x \leq 1, \ -1 \leq y \leq 1 \,\},$$

   and use part (a) to sketch $A + B$.

4. Given a nonzero point $p \in \mathbb{R}^n$ and a scalar $\alpha \in \mathbb{R}$, show that the open halfspace

   $$H = \{\, x \in \mathbb{R}^n \ : \ \langle p, x \rangle > \alpha \,\}$$

   is a convex set.

5. Let C be a convex subset of $\mathbb{R}^n$ and let

   $$H_\alpha = \{\, x \in \mathbb{R}^n \ : \ \langle p, x \rangle = \alpha \,\}$$

   be a hyperplane such that $C \cap H_\alpha = \emptyset$. Show that C is contained in one of the open halfspaces determined by H_α.

6. Show that a translate of a straight line segment is a straight line segment.

7. Earlier, we defined what was meant by a ***midpoint convex*** set, namely, one that contains the midpoints of every pair of points of the set. A convex set must be midpoint convex, but there are midpoint convex sets that are not convex. Find one.

8. Let $A = \{0, 1\}$ be the two-element subset of the real line containing only the numbers 0 and 1. Let A_1 be the set obtained by adding to A the midpoints of all points in A. Let A_2 be the set obtained by adding to A_1 all the midpoints of A_1. Continue this process indefinitely. At each stage, A_{k+1} is the set obtained by adding to A_k all the midpoints of A_k, ... Describe the set B obtained. Is B midpoint convex?

9. An *affine transformation* is a linear transformation followed by a translation, that is, $A : \mathbb{R}^n \longrightarrow \mathbb{R}^m$ is *affine* if and only if

$$A(x) = T(x) + x_0,$$

where $T : \mathbb{R}^n \longrightarrow \mathbb{R}^m$ is linear and $x_0 \in \mathbb{R}^n$. Show that convexity is preserved by affine transformations; that is, show that if $C \subset \mathbb{R}^n$ is convex, then $A(C)$ is a convex subset of $\mathbb{R}^m$.

10. Show that an affine transformation A is a linear transformation if and only if $A(\overline{0}) = \overline{0}$.

11. Show that a mapping $A : \mathbb{R}^n \longrightarrow \mathbb{R}^m$ is an affine transformation if and only if

$$A(\lambda x + \mu y) = \lambda A(x) + \mu A(y)$$

for all x, $y \in \mathbb{R}^n$ and scalars λ, μ such that $\lambda + \mu = 1$.

12. Show that a translation followed by a linear transformation is also an affine transformation. Note the order is reversed from that of the definition.

13. Give an example of a translation in $\mathbb{R}^2$ and a linear transformation from $\mathbb{R}^2$ to $\mathbb{R}^2$ that do not commute with each other.

14. Show that the composition of two affine transformations is again an affine transformation.

15. Given a set of points $S = \{v_1, v_2, \ldots, v_k\}$ in $\mathbb{R}^n$, a point $x \in \mathbb{R}^n$ is an *affine combination* of S if there exist scalars $\lambda_1, \lambda_2, \ldots, \lambda_k$ such that $\lambda_1 + \lambda_2 + \cdots + \lambda_k = 1$ and $x = \lambda_1 v_1 + \lambda_2 v_2 + \cdots + \lambda_k v_k$.
Let z be a point inside the ball $B(x_0, \delta)$ with $z \neq x_0$, so that

$$0 < \|z - x_0\| = \kappa < \delta.$$

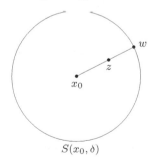

$S(x_0, \delta)$

Let w be the point where the ray from x_0 through z intersects the sphere $S(x_0, \delta)$. Express w as an affine combination of x_0 and z and verify that $\|w - x_0\| = \delta$.

16. An $n \times n$ matrix $A = (a_{ij})$ is called **doubly stochastic** if $a_{ij} \geq 0$ for all $1 \leq i, j \leq n$, and the sum of the entries in any row or any column is 1. Let Γ_n denote the set of all doubly stochastic $n \times n$ matrices. Show that Γ_n is a convex subset of the real linear space of all $n \times n$ matrices.

3.3 CONVEX HULLS

The notion of a *linear combination* is something that you are familiar with. Certain types of linear combinations are singled out for special attention, namely, *affine combinations* and *convex combinations*. The definitions, including that of a linear combination, are as follows.

Let $S = \{x_1, x_2, \ldots, x_k\}$ be a subset of $\mathbb{R}^n$, and let $\lambda_1, \lambda_2, \ldots, \lambda_k$ be scalars, the point

$$x = \lambda_1 x_1 + \lambda_2 x_2 + \cdots + \lambda_k x_k$$

is called

- a *linear combination*, no matter what values the scalars λ_i take,
- an *affine combination*, provided $\sum_{i=1}^{k} \lambda_i = 1$, and
- a *convex combination*, provided $\sum_{i=1}^{k} \lambda_i = 1$ and $\lambda_i \geq 0$ for $1 \leq i \leq k$.

These definitions are for finite subsets S of $\mathbb{R}^n$. They are extended to an infinite subset S by taking finite subsets of the points in S.

For example, if S is any subset of $\mathbb{R}^n$, then by a *convex combination* of points from S, we mean a convex combination of some finite subset of S. We get all convex combinations by taking all possible finite subsets of S

Note. It is important to understand that two different convex combinations may use different finite subsets of S, and the finite subsets do not have to contain the same number of elements.

Let S be a subset of $\mathbb{R}^n$.

- The collection of all linear combinations of points from S is usually called the *linear span* of S or simply the *span* of S and is denoted by span(S).

- The set of all affine combinations of points from S is called the **_affine hull_** of S and is denoted by aff(S).
- The set of all convex combinations of points from S is called the **_convex hull_** of S and is denoted by conv(S).

Formally, the definition of the **_convex hull_** of a set S is as follows.

A point x is in conv(S) if and only if there exist $x_1, x_2, \ldots, x_k \in S$ and scalars $\lambda_1, \lambda_2, \ldots, \lambda_k \in \mathbb{R}$ such that

$$x = \sum_{i=1}^{k} \lambda_i x_i = \lambda_1 x_1 + \lambda_2 x_2 + \cdots + \lambda_k x_k,$$

where $\lambda_i \geq 0$, for $i = 1, 2, \ldots, k$, and $\sum_{i=1}^{k} \lambda_i = 1$.

For any set $S \subset \mathbb{R}^n$, the span of S, the affine hull of S, and the convex hull of S all contain S. We recall from linear algebra that span(S) is the linear subspace generated by S, which is typically much larger than S. We would also expect aff(S) and conv(S) to be larger than S.

Example 3.3.1. *If $S = \{p, q\}$, where p and q are distinct points in $\mathbb{R}^n$, then*

- *aff(S) is the straight line through p and q.*
- *span(S) is a line through the origin containing p and q if the set S is linearly dependent or a plane through the origin containing p and q if S is linearly independent.*
- *conv(S) is the straight line segment $[p, q]$.*

Example 3.3.2. *If $S = \{a, b, c\}$, where a, b, and c are noncollinear points in $\mathbb{R}^n$, then*

- *aff(S) is a plane passing through the points a, b, and c.*
- *span(S) is either a plane through the origin containing a, b, and c if S is linearly dependent or a three-dimensional subspace if S is linearly independent.*
- *conv(S) is the triangle whose vertices are the points a, b, and c, including the points interior to the triangle.*

Example 3.3.3. *The following figure illustrates the formation of the convex hull for several different sets. In the figure, a tiny open circle indicates that the point is not in the set.*

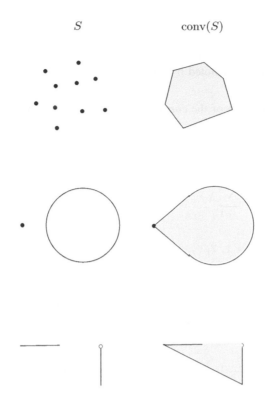

There are some cases where $\text{span}(S)$, $\text{aff}(S)$, and $\text{conv}(S)$ are no larger than S. For example, if S is a linear subspace of $\mathbb{R}^n$, then S already contains all of its linear combinations. We will prove the similar statement for convex sets.

Lemma 3.3.4. *A convex set $S \subset \mathbb{R}^n$ contains all of its convex combinations.*

Proof. Every convex combination of two points x_1 and x_2 is a point on the segment $[x_1, x_2]$. As a consequence, if S is convex, then it contains every convex combination of every two of its points.

Suppose now that x_1, x_2, and x_3 are three points of S and that z is a convex combination of these three points, that is,

$$z = \lambda_1 x_1 + \lambda_2 x_2 + \lambda_3 x_3,$$

where $\lambda_1 + \lambda_2 + \lambda_3 = 1$ and $\lambda_1 \geq 0$, $\lambda_2 \geq 0$, and $\lambda_3 \geq 0$.

If any one of λ_1, λ_2, or λ_3 is zero, then we really only have a convex combination of two points, and there would be nothing to prove. Therefore, we may assume that $\lambda_i > 0$ for $i = 1$, 2, 3. Now we rewrite z as

$$z = (\lambda_1 + \lambda_2) \left[\frac{\lambda_1}{\lambda_1 + \lambda_2} x_1 + \frac{\lambda_2}{\lambda_1 + \lambda_2} x_2 \right] + \lambda_3 x_3.$$

The expression in the square brackets is a convex combination of two points of S and is therefore a point w in S. This means that

$$z = (\lambda_1 + \lambda_2)w + \lambda_3 x_3.$$

However, this is a convex combination of the two points w and x_3 of S, which is, therefore, a point in S. This shows that the convex combination of any three points of S is a point in S. The proof may be completed by induction, using an argument similar to the one above as the inductive step, showing that S contains the convex combination of any finite number of its points.

$\square$

Lemma 3.3.5. *If S is any subset of $\mathbb{R}^n$, then any convex combination of convex combinations of S is a convex combination of S.*

Proof. Suppose that x_1 and x_2 are convex combinations of points from S. We may assume that x_1 and x_2 are both convex combinations of the same k points from S, since we could insert zero multiples of the missing points in x_1 and x_2. Therefore, we may write

$$x_1 = \sum_{i=1}^{k} \alpha_i s_i \qquad \text{and} \qquad x_2 = \sum_{i=1}^{k} \beta_i s_i,$$

where each $s_i \in S$, $\sum_{i=1}^{k} \alpha_i = \sum_{i=1}^{k} \beta_i = 1$, and $\alpha_i \geq 0$, $\beta_i \geq 0$ for $i = 1, 2, \ldots, k$.

Now, consider a convex combination of x_1 and x_2,

$$z = \lambda_1 x_1 + \lambda_2 x_2 = \lambda_1 \sum_{i=1}^{k} \alpha_i s_i + \lambda_2 \sum_{i=1}^{k} \beta_i s_i = \sum_{i=1}^{k} (\lambda_1 \alpha_i + \lambda_2 \beta_i) s_i,$$

where $\lambda_1, \lambda_2 \geq 0$, and $\lambda_1 + \lambda_2 = 1$. We also have

$$\sum_{i=1}^{k} (\lambda_1 \alpha_i + \lambda_2 \beta_i) = \lambda_1 \sum_{i=1}^{k} \alpha_i + \lambda_2 \sum_{i=1}^{k} \beta_i = \lambda_1 \cdot 1 + \lambda_2 \cdot 1 = 1,$$

and $\lambda_1 \alpha_i + \lambda_2 \beta_i \geq 0$ for $i = 1, 2, \ldots, k$. Hence, z is a convex combination of points from S. This proves the lemma for the case where we have two convex combinations of points from S, and the proof clearly extends to any finite number of convex combinations of points from S.

$\square$

Theorem 3.3.6. *If S is a convex subset of $\mathbb{R}^n$, then* conv(S) = S.

Proof. We already know that $S \subseteq \text{conv}(S)$. The reverse containment, $\text{conv}(S) \subseteq S$, is just a restatement of Lemma 3.3.4.

$\square$

Theorem 3.3.7. *Given a subset S of $\mathbb{R}^n$,* conv(S) *is the smallest convex set that contains S.*

Proof. Lemma 3.3.5 tells us that $\text{conv}(S)$ is a convex set. We want to show that every convex subset C that contains S also contains $\text{conv}(S)$.

Let C be a convex set that contains S. If C does not contain $\text{conv}(S)$, we may assume that $C \subset \text{conv}(S)$, otherwise, define C' to be $C \cap \text{conv}(S)$ and replace C with C'.

We may, therefore, assume that $S \subset C \subset \text{conv}(S)$, but then we have

$$\text{conv}(S) \subset \text{conv}(C) \subset \text{conv}\left(\text{conv}(S)\right).$$

Now, since C is convex, we have $\text{conv}(C) = C$, and since $\text{conv}(S)$ is convex, we have $\text{conv}\left(\text{conv}(S)\right) = \text{conv}(S)$, and therefore,

$$\text{conv}(S) \subset C \subset \text{conv}(S),$$

showing that $C = \text{conv}(S)$, which completes the proof.

$\square$

The convex hull has been defined in an algebraic way in order to emphasize the similarity with the notion of the span of a set of points. There is a more geometric description that is often used.

Theorem 3.3.8. *If S is a subset of $\mathbb{R}^n$, then the convex hull of S is the intersection of all convex sets that contain S, that is,*

$$\text{conv}(S) = \bigcap \{ C : S \subset C, \text{ and } C \text{ is convex} \}.$$

Proof. Let $\mathcal{F}$ be the family of all convex sets that contain S. The set $\bigcap \mathcal{F}$ is a convex set containing S. Since it is the intersection of all convex sets that contain S, it must be a subset of every convex set that contains S. In other words, $\bigcap \mathcal{F}$ is the smallest convex set that contains S and so must be the convex hull by the previous theorem.

$\square$

Example 3.3.9. *Show that if A is a bounded nonempty subset of $\mathbb{R}^n$, then* conv(A) *is also bounded and* diam(A) = diam $($conv$(A))$.

Solution. Let A be a bounded subset of $\mathbb{R}^n$, then there exists an $M > 0$ such that $\|a\| \leq M$ for all $a \in A$.

Now let $x \in \text{conv}(A)$, then there exist points $a_1, a_2, \ldots, a_k \in A$ and scalars $\lambda_1 \geq 0$, $\lambda_2 \geq 0$, $\ldots$, $\lambda_k \geq 0$ such that

$$x = \lambda_1 a_1 + \lambda_2 a_2 + \cdots + \lambda_k a_k,$$

with $\lambda_1 + \lambda_2 + \cdots + \lambda_k = 1$.

Therefore, from the triangle inequality,

$$\|x\| \leq \lambda_1\|a_1\| + \lambda_2\|a_2\| + \cdots + \lambda_k\|a_k\| \leq (\lambda_1 + \lambda_2 + \cdots + \lambda_k)\, M = M,$$

so that $\|x\| \leq M$ for all $x \in \text{conv}(A)$, and conv(A) is bounded.

Now note that if A is a bounded subset of $\mathbb{R}^n$, then $A \subset \text{conv}(A)$, and therefore diam$(A) \leq$ diam$(\text{conv}(A))$.

Let $x, y \in \text{conv}(A)$, then

$$x = \lambda_1 a_1 + \lambda_2 a_2 + \cdots + \lambda_m a_m$$

for some $a_1, a_2, \ldots, a_m \in A$ with $\lambda_1, \lambda_2, \ldots, \lambda_m \geq 0$ and $\lambda_1 + \lambda_2 + \cdots + \lambda_m = 1$, and

$$y = \mu_1 b_1 + \mu_2 b_2 + \cdots + \mu_p b_p$$

for some $b_1, b_2, \ldots, b_p \in A$ with $\mu_1, \mu_2, \ldots, \mu_p \geq 0$ and $\mu_1 + \mu_2 + \cdots + \mu_p = 1$.

Therefore,

$$x = \sum_{i=1}^{m} \sum_{j=1}^{p} \lambda_i \mu_j a_i \qquad \text{and} \qquad y = \sum_{i=1}^{m} \sum_{j=1}^{p} \lambda_i \mu_j b_j,$$

and from the triangle inequality,

$$\|x - y\| = \left\| \sum_{i=1}^{m} \sum_{j=1}^{p} \lambda_i \mu_j (a_i - b_j) \right\|$$

$$\leq \sum_{i=1}^{m} \sum_{j=1}^{p} \lambda_i \mu_j \|a_i - b_j\|$$

$$\leq \sum_{i=1}^{m} \sum_{j=1}^{p} \lambda_i \mu_j \text{diam}(A)$$

$$= \text{diam}(A)$$

for all $x,\, y \in \text{conv}(A)$. Thus, $\text{diam}(\text{conv}(A)) \leq \text{diam}(A)$.

$\square$

The next theorem, proved by Carathéodory in 1907, shows that for any subset S of $\mathbb{R}^n$, every point in $\text{conv}(S)$ can be written as a convex combination of no more than $n + 1$ points from S.

Theorem 3.3.10. *(Carathéodory's Theorem)*
If C is a subset of $\mathbb{R}^n$ and $x \in \text{conv}(C)$, then there exists a set of at most $n + 1$ points $\{ x_1,\, x_2,\, \ldots,\, x_{n+1} \} \subset C$ such that

$$x = \lambda_1 x_1 + \lambda_2 x_2 + \cdots + \lambda_{n+1} x_{n+1},$$

where $0 \leq \lambda_i \leq 1$ for $i = 1, 2, \ldots, n+1$ and $\lambda_1 + \lambda_2 + \cdots + \lambda_{n+1} = 1$.

Proof. Let $x \in \text{conv}(C)$, then there exists a positive integer k, points $x_i \in C$, and $\lambda_i \geq 0$ for $i = 1, 2, \ldots, k$, with $\lambda_1 + \lambda_2 + \cdots + \lambda_k = 1$, such that

$$x = \lambda_1 x_1 + \lambda_2 x_2 + \cdots + \lambda_k x_k.$$

We will show that we can find such an expression for x with $k \leq n + 1$.

Suppose that $k > n + 1$, then the set $\{x_1, x_2, \ldots, x_k\}$ is linearly dependent, so there exist real numbers $\alpha_1, \alpha_2, \ldots, \alpha_k$, not all zero, such that

$$\alpha_1 x_1 + \alpha_2 x_2 + \cdots + \alpha_k x_k = \overline{0}.$$

However, since $k > n + 1$, then we can also require that the α_i's satisfy

$$\alpha_1 + \alpha_2 + \cdots + \alpha_k = 0.$$

This follows, since the homogeneous system of $n + 1$ linear equations in the k unknowns $\alpha_1, \alpha_2, \ldots, \alpha_k$ given by

$$\alpha_1 x_1 + \alpha_2 x_2 + \cdots + \alpha_k x_k = \overline{0}$$
$$\alpha_1 + \alpha_2 + \cdots + \alpha_k = 0$$

always has a nontrivial solution if $k > n + 1$.

Now let

$$T = \{t \in \mathbb{R} : t \cdot \alpha_i \geq -\lambda_i \text{ for } i = 1, 2, \ldots, k\},$$

then T is a closed subset of $\mathbb{R}$ and T is nonempty, since it contains 0. Also, it is a proper subset of $\mathbb{R}$, since there is at least one $\alpha_i \neq 0$.

Let t_0 be any point in the boundary of T; since T is closed, we have

$$t_0 \cdot \alpha_i \geq -\lambda_i,$$

for $i = 1, 2, \ldots, k$, and

$$x = (\lambda_1 + t_0 \cdot \alpha_1)x_1 + (\lambda_2 + t_0 \cdot \alpha_2)x_2 + \cdots + (\lambda_k + t_0 \cdot \alpha_k)x_k$$

with $\lambda_i + t_0 \cdot \alpha_i \geq 0$, for $i = 1, 2, \ldots, k$, and

$$\sum_{i=1}^{k}(\lambda_i + t_0 \cdot \alpha_i) = \sum_{i=1}^{k}\lambda_i + t_0 \cdot \sum_{i=1}^{k}\alpha_i = 1.$$

Also, for at least one integer i_0, we have $t_0 \cdot \alpha_{i_0} + \lambda_{i_0} = 0$, so we have written x as a convex combination of points of C, except that since one of the coefficients is now zero, it is a convex combination of $k - 1$ points of C.

This process can be continued until it is no longer true that $k > n + 1$, that is, until $k \leq n + 1$.

$\square$

In fact, we can do even better. If $S \subset \mathbb{R}^n$ lies in a translate of a subspace of dimension k, then every point in $\mathrm{conv}(S)$ can be written as a convex combination of no more than $k + 1$ points from S. We leave the proof of this as an exercise.

We can use Carathéodory's theorem to show that if C is an open subset of $\mathbb{R}^n$, then $\mathrm{conv}(C)$ is open; while if C is a compact subset of $\mathbb{R}^n$, then $\mathrm{conv}(C)$ is compact.

Theorem 3.3.11. *If $C \subseteq \mathbb{R}^n$ is open, then $\mathrm{conv}(C)$ is also open.*

Proof. If $C \subseteq \mathbb{R}^n$ and $x \in \mathrm{conv}(C)$, then Carathéodory's theorem implies that

$$x = \lambda_1 x_1 + \lambda_2 x_2 + \cdots + \lambda_m x_m,$$

where $m \leq n + 1$, and $x_1, x_2, \ldots, x_m \in C$, where $\lambda_1, \lambda_2, \ldots, \lambda_m \geq 0$, and $\lambda_1 + \lambda_2 + \cdots + \lambda_m = 1$.

Since C is open, there exist positive real numbers $r_1, r_2, \ldots, r_m$ such that the open ball centered at x_i with radius r_i is contained in C for $i = 1, 2, \ldots, m$, that is,

$$B(x_i, r_i) \subseteq C$$

for $i = 1, 2, \ldots, m$.

Now let $r = \min\{r_1, r_2, \ldots, r_m\}$, then $r > 0$, and if $z \in B(x, r)$, then

$$\|z - x\| \leq r \leq r_i$$

for $i = 1, 2, \ldots, m$, and if we let $z_i = x_i + z - x$ for $i = 1, 2, \ldots, m$, then $z_i \in B(x_i, r_i) \subseteq C$, and

$$z = \sum_{i=1}^{m} \lambda_i z = \sum_{i=1}^{m} \lambda_i (x - x_i + z_i)$$

$$= \sum_{i=1}^{m} \lambda_i x - \sum_{i=1}^{m} \lambda_i x_i + \sum_{i=1}^{m} \lambda_i z_i$$

$$= x - x + \sum_{i=1}^{m} \lambda_i z_i = \sum_{i=1}^{m} \lambda_i z_i$$

since $\lambda_1 + \lambda_2 + \cdots + \lambda_m = 1$. Therefore, $z \in \text{conv}(C)$, and since $z \in B(x, r)$ was arbitrary, this implies that $B(x, r) \subseteq \text{conv}(C)$, so that x is an interior point of $\text{conv}(C)$. Thus, we have shown that every point of $\text{conv}(C)$ is an interior point, and so $\text{conv}(C)$ is open.

$\square$

Theorem 3.3.12. *If $C \in \mathbb{R}^n$ is compact, then $\text{conv}(C)$ is also compact.*

Proof. Denote by B the closed and bounded subset of $\mathbb{R}^{n+1}$ defined by

$$B = \{(\lambda_1, \lambda_2, \ldots, \lambda_{n+1}) : \lambda_i \geq 0, \ 0 \leq i \leq n+1, \ \lambda_1 + \lambda_2 + \cdots + \lambda_{n+1} = 1\}.$$

Consider the mapping

$$\varphi : \left(\lambda_1, \lambda_2, \ldots, \lambda_{n+1}, x^{(1)}, x^{(2)}, \ldots, x^{(n+1)}\right) \longrightarrow \sum_{i=1}^{n+1} \lambda_i x^{(i)},$$

where $x^{(i)} = \left(x_1^{(i)}, x_2^{(i)}, \ldots, x_n^{(i)}\right) \in \mathbb{R}^n$.

This mapping is defined and continuous on the Cartesian product

$$\mathbb{R}^{n+1} \times \underbrace{\mathbb{R}^n \times \cdots \times \mathbb{R}^n}_{n+1 \text{ times}} = \mathbb{R}^{(n+1)^2}.$$

By Carathéodory's theorem, it maps $B \times \underbrace{C \times C \times \cdots \times C}_{n+1 \text{ times}}$ onto $\text{conv}(C)$, and since B and C are compact subsets of $\mathbb{R}^{n+1}$ and $\mathbb{R}^n$, respectively, then the Cartesian product $B \times \underbrace{C \times C \times \cdots \times C}_{n+1 \text{ times}}$ is a compact subset of $\mathbb{R}^{(n+1)^2}$.

Since the continuous image of a compact set is compact, then conv(C) is compact.

☐

Note, however, that the convex hull of a closed set need not be closed (unless of course it is also bounded).

Example 3.3.13. *Give an example of a closed set C such that* conv(C) *is not closed.*

Solution. Let C consist of a line in $\mathbb{R}^2$ together with a point not on the line, then C is closed since it is the union of two closed sets, but conv(C) is not closed.

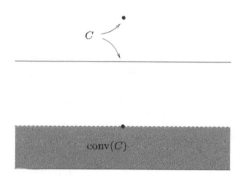

☐

3.3.1 Problems

1. Show that the set $\{(x,y) \in \mathbb{R}^2 : y \geq 1/x^2,\ x > 0\}$ is convex.
2. In $\mathbb{R}^3$, let

$$A = \{(x,y,z) \in \mathbb{R}^3 : (x-1)^2 + y^2 = 1,\ z = 0\}$$

and

$$B = \{(x,y,z) \in \mathbb{R}^3 : x = y = 0,\ -1 \leq z \leq 1\}.$$

Find and sketch $S = A \cup B$ and the convex hull of S.
3. Sketch the following subsets of $\mathbb{R}^2$ and their convex hulls.
   (a) $A = \{(1,1),\ (-1,1),\ (0,-1)\}$.
   (b) $B = \{(0,1),\ (0,4),\ (-2,-1),\ (0,0),\ (3,-2),\ (-1,2)\}$.
   (c) $C = [(1,1),(1,-1)] \cup [(-1,1),(-1,-1)]$.

    (d) $D = \overline{B}(a, 1) \cup \overline{B}(-a, 1)$, where a is the point $(1, 1)$.
    (e) $E = B(a, 1) \cup \overline{B}(-a, 1)$, where a is the point $(1, 1)$.
    (f) $F = X \cup Y$, where X is the x-axis and Y is the y-axis.
    (g) $G = \overline{B}(\overline{0}, 1) \cup \{(1, 1), (-1, -1)\}$.
    (h) $H = \{(0, 1)\} \cup X$, where X is the x-axis.
    (i) $I = \{\overline{0}\} \cup \{(x, y) : x > 0, \ y = 1/x\}$
    (j) $J = A \cup (-A)$, where A is the set in part (a).
    (k) $K = (A + x_0) \cup A$, where A is the set in part (a) and $x_0 = (3, 2)$.

4. Given a triangle in the plane with vertices p_1, p_2, and p_3, as shown in the figure below.

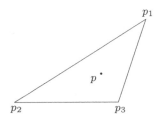

Show that the incenter of the triangle (the point of concurrency of the angle bisectors) is a convex combination of the vertices given by

$$p = \frac{1}{\alpha + \beta + \gamma}(\alpha\, p_1 + \beta\, p_2 + \gamma\, p_3),$$

where $\alpha = \|p_3 - p_2\|$, $\beta = \|p_3 - p_1\|$, $\gamma = \|p_2 - p_1\|$.
Hint. Let $a = p_3 - p_2$, $b = p_3 - p_1$, and $c = p_2 - p_1$, then the angle bisector of the angle at vertex p_1 lies on the line joining the point p_1 and the point $\dfrac{b}{\|b\|} + \dfrac{c}{\|c\|}$, ...

5. Let z be a point outside the ball $\overline{B}(x_0, \delta)$, that is,

$$\|z - x_0\| = \kappa > \delta.$$

Let w be the point where the segment $[x_0, z]$ intersects the sphere $S(x_0, \delta)$. Express w as a convex combination of x_0 and z and verify that $\|w - x_0\| = \delta$.

6. Show that if a vector $x \in \mathbb{R}^n$ has distinct representations as a convex combination of vectors from the set $\{x_0, x_1, \ldots, x_r\}$, then the set of vectors

$$\{x_1 - x_0, x_2 - x_0, \ldots, x_r - x_0\}$$

is linearly dependent.

7. Show that if A is a subset of the closed halfspace

$$H = \{x \in \mathbb{R}^n : \langle p, x \rangle \leq \alpha\},$$

and if $x \in \text{int}(A)$, then $\langle p, x \rangle < \alpha$, that is,

$$\text{int}(A) \subset H_< = \{x \in \mathbb{R}^n : \langle p, x \rangle < \alpha\}.$$

8. Let A and B be subsets of $\mathbb{R}^n$.

   (a) Show that
   $$\mathrm{conv}(A \cap B) \subseteq \mathrm{conv}(A) \cap \mathrm{conv}(B).$$

   (b) Provide an example in $\mathbb{R}^2$ of two sets A and B, with $A \cap B \neq \emptyset$, to show that the inclusion may be proper.

9. Let A and B be subsets of $\mathbb{R}^n$.

   (a) Show that
   $$\mathrm{conv}(A) \cup \mathrm{conv}(B) \subseteq \mathrm{conv}(A \cup B).$$

   (b) Provide an example in $\mathbb{R}^2$ of two sets A and B, with $A \cup B \neq \emptyset$, to show that the inclusion may be proper.

10. Determine all subsets A of $\mathbb{R}^1$ such that both A and its complement are convex. Do the same for $\mathbb{R}^2$ and $\mathbb{R}^3$. No proof is required. In $\mathbb{R}^1$, there are four different types; in $\mathbb{R}^2$, there are six different types; and in $\mathbb{R}^3$, there are eight different types.

11. Prove the following form of Carathéodory's theorem:

    Theorem. If $S \subset \mathbb{R}^n$ lies in a translate of a subspace of dimension k, then every point in $\mathrm{conv}(S)$ can be written as a convex combination of no more than $k + 1$ points from S.

12. Let $A = \{ (0,0,0),\ (1,0,0),\ (0,1,0),\ (0,0,1) \}$ and let $x_0 = \left(\frac{1}{3}, \frac{1}{4}, \frac{1}{5} \right)$.

    (a) Explain why you can write x_0 as a convex combination of points from A.

    (b) Explain why you need to use *all* of the points from A to accomplish this.

    (c) Write x_0 as a convex combination of points from A.

3.4 INTERIOR AND CLOSURE OF CONVEX SETS

We have seen that convexity is preserved by translations and linear transformations. In this section, we will show that convexity is also preserved by two topological operations: namely, the operation of taking the interior of a set and the operation of taking the closure of a set.

When checking that a point is in the closure of a set, we need to show that the point either belongs to the set or is an accumulation point of the set (it may be both). We will say that a point x is a ***closure point*** of a set S if it is a point of S or if it is an accumulation point of S. A closure point of a set S, then, is any point that is in the closure $\overline{S}$ of the set. The following lemma makes it a little easier to test for closure—a single test may be used to determine whether or not a point is in the closure of the set.

Lemma 3.4.1. *A point x is a closure point of a set S if and only if every ball centered at x contains points of S.*

Proof. Recall that if $S \subset \mathbb{R}^n$, then the derived set S' of S consists of all the accumulation points of S.

If x is a closure point of S, then either $x \in S$ or $x \in S'$. If $x \in S$, then every ball centered at x contains a point of S, namely, x. If $x \in S'$, then every open ball centered at x contains points of S different from x. In either case, every open ball centered at x contains a point of S.

Conversely, if every open ball centered at x contains a point of S and $x \notin S$, then every open ball centered at x contains points of S different from x; that is, x is in S'. Thus, either $x \in S$ or $x \in S'$; that is, x is a closure point of S.

$\square$

Theorem 3.4.2. *If $C \subset \mathbb{R}^n$ is a convex set, then the closure $\overline{C}$ is also a convex set.*

Proof. Let $\overline{C}$ be the closure of the convex set C. We want to show that if p and q are points in $\overline{C}$, then the entire segment $[p, q]$ is also in $\overline{C}$.

Let $x = (1 - \lambda)p + \lambda q$, where $0 \le \lambda \le 1$, be a typical point on the segment $[p, q]$. We claim that x is a closure point of C, that is, we claim that if δ is any positive number, the open ball $B(x, \delta)$ contains a point of C.

To see why, consider the two open balls of radius δ centered at the points p and q. Since p and q are closure points of C, each of the balls $B(p, \delta)$ and $B(q, \delta)$ must contain points of C, say y and z, respectively, as in the figure below.

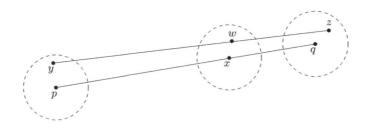

Since C is convex, the point $w = (1 - \lambda)y + \lambda z$ is in C, and as might be expected, this point is also in $B(x, \delta)$ as we now show. From the triangle inequality,

$$
\begin{aligned}
\|x - w\| &= \|x - ((1 - \lambda)y + \lambda z)\| \\
&= \|(1 - \lambda)p + \lambda q - ((1 - \lambda)y + \lambda z)\| \\
&= \|(1 - \lambda)p - (1 - \lambda)y + \lambda q - \lambda z\| \\
&\leq \|(1 - \lambda)p - (1 - \lambda)y\| + \|\lambda q - \lambda z\| \\
&= (1 - \lambda)\|p - y\| + \lambda\|q - z\| \\
&< (1 - \lambda)\delta + \lambda\delta \\
&= \delta,
\end{aligned}
$$

so that $w \in B(x, \delta)$ and x is a closure point of C. This completes the proof.

$\square$

We give an alternate proof of this result with a series of lemmas.

Lemma 3.4.3. *If $C \subset \mathbb{R}^n$ is convex and $\delta > 0$, then the set*

$$
C_\delta = \{\, z \in \mathbb{R}^n \,:\, \mathrm{dist}(z, C) < \delta \,\}
$$

is also convex.

Proof. Note that if $z \in C$, then

$$
\mathrm{dist}(z, C) = \inf\{\, \|z - c\| \,:\, c \in C \,\} = 0 < \delta,
$$

so that $z \in C_\delta$. Therefore, $C \subset C_\delta$ and $C_\delta \neq \emptyset$.

If we let $y,\, y' \in C_\delta$, then from the properties of the infimum, there are at least two points $x,\, x' \in C$ such that

$$
\|x - y\| < \delta \qquad \text{and} \qquad \|x' - y'\| < \delta.
$$

Now, if $y'' \in [y, y']$, then there exists a scalar $\lambda \in [0, 1]$ such that

$$
y'' = (1 - \lambda)y + \lambda y'.
$$

Define $x'' = (1 - \lambda)x + \lambda x'$, since C is convex, then $x'' \in C$, and from the triangle inequality,

$$
\begin{aligned}
\|x'' - y''\| &\leq \|(1 - \lambda)x - (1 - \lambda)y\| + \|\lambda x' - \lambda y'\| \\
&= (1 - \lambda)\|x - y\| + \lambda\|x' - y'\| \\
&< (1 - \lambda)\delta + \lambda\delta \\
&= \delta.
\end{aligned}
$$

Therefore,

$$
\mathrm{dist}(y'', C) = \inf\{\, \|y'' - c\| \,:\, c \in C \,\} \leq \|y'' - x''\| < \delta,
$$

so that $y'' \in C_\delta$ and C_δ is convex.

$\square$

Exercise 3.4.4. *If $C \subset \mathbb{R}^n$ and $\delta > 0$, let*

$$C_\delta = \{\, z \in \mathbb{R}^n \,:\, \mathrm{dist}(z, C) < \delta \,\}.$$

(a) *Show that C_δ is the Minkowski sum of C and the open ball $B(0, \delta)$, that is,*

$$C_\delta = C + B(0, \delta).$$

(b) *If C is convex, use Example 3.2.7 to show that C_δ is convex.*

Lemma 3.4.5. *Let $C \subset \mathbb{R}^n$ and let $\{\delta_k\}_{k \geq 1}$ be a sequence of positive real numbers decreasing to 0. If $C_{\delta_k} = \{\, z \in \mathbb{R}^n \,:\, \mathrm{dist}(z, C) < \delta_k \,\}$, then*

$$\overline{C} = \bigcap_{k=1}^{\infty} C_{\delta_k}.$$

Proof. Again, if $z \in C$, then $\mathrm{dist}(z, C) = 0 < \delta_k$ for all $k \geq 1$, so that

$$C \subset \bigcap_{k=1}^{\infty} C_{\delta_k},$$

and if $C \neq \emptyset$, then

$$\bigcap_{k=1}^{\infty} C_{\delta_k} \neq \emptyset.$$

If $z \in \bigcap_{k=1}^{\infty} C_{\delta_k}$, then $z \in C_{\delta_k}$ for all $k \geq 1$, and there exists a sequence $\{x_k\}_{k \geq 1}$ with $x_k \in C$ for all $k \geq 1$ such that $\|z - x_k\| < \delta_k$. Therefore, $\lim_{k \to \infty} x_k = z$, so that $z \in \overline{C}$, and we have $\bigcap_{k=1}^{\infty} C_{\delta_k} \subset \overline{C}$.

Conversely, if $z \in \overline{C}$, then given any $k \geq 1$, since $\delta_k > 0$, there exists an $x_k \in C$ such that $\|z - x_k\| < \delta_k$, and so

$$\mathrm{dist}(z, C_k) \leq \|z - x_k\| < \delta_k$$

for each $k \geq 1$, that is, $z \in \bigcap_{k=1}^{\infty} C_{\delta_k}$. Therefore, $\overline{C} \subset \bigcap_{k=1}^{\infty} C_{\delta_k}$.

$\square$

Corollary 3.4.6. *If $C \subset \mathbb{R}^n$ is a convex set, then the closure $\overline{C}$ is also a convex set.*

Proof. For each positive integer $k \geq 1$, let $C_k = \{z \in \mathbb{R}^n : \text{dist}(z, C) < 1/k\}$, then

$$\overline{C} = \bigcap_{k=1}^{\infty} C_k.$$

Since each C_k is convex, then the intersection is convex, that is, $\overline{C}$ is convex.

□

3.4.1 The closed Convex Hull

Often we want the smallest *closed* convex set that contains a given set A.

The **closed convex hull** of a set A is the intersection of all closed convex subsets that contain A and is denoted by $\overline{\text{conv}}(A)$, that is,

$$\overline{\text{conv}}(A) = \bigcap \{C : C \text{ is closed and convex, and } A \subset C\}.$$

The definition suggests that $\overline{\text{conv}}(A)$ is

- closed, since it is the intersection of closed sets,
- convex, since it is the intersection of convex sets, and
- is the smallest closed convex set that contains A, since it is the intersection of all such sets.

Also, the phrase "closed convex hull" of A suggests that one is taking the closure of the convex hull; that is, it suggests the following sequence of operations:

First, append to A all of the convex combinations of points from A, obtaining $\text{conv}(A)$.
Next, append to $\text{conv}(A)$ all of the accumulation points of $\text{conv}(A)$.

The result of this procedure would be denoted by $\overline{\text{conv}(A)}$. The procedure used to obtain $\overline{\text{conv}}(A)$ is very different, and it is not immediately obvious that these two different procedures result in the same set. As the next theorem shows, this is indeed the case.

Theorem 3.4.7. *If $A \subset \mathbb{R}^n$, then $\overline{\text{conv}}(A) = \overline{\text{conv}(A)}$.*

Proof. To clearly distinguish between the two sets, let us denote them by B and C as follows:

$$B = \overline{\text{conv}(A)} \quad \text{and} \quad C = \overline{\text{conv}}(A).$$

We wish to show that (i) $B \subset C$ and (ii) $C \subset B$.

(i). By definition, C is the intersection of all closed convex sets containing A. Thus, as remarked earlier, C is both closed and convex. Since $A \subset C$, we must have

$$\text{conv}(A) \subset \text{conv}(C) = C.$$

Now, taking the closures of $\text{conv}(A)$ and C, we have

$$\overline{\text{conv}(A)} \subset \overline{C}.$$

However, C is closed, so that $\overline{C} = C$. This shows that $B \subset C$.

(ii). Since $A \subset \text{conv}(A)$, then $A \subset \overline{\text{conv}(A)}$. Thus, $\overline{\text{conv}(A)}$ is a closed convex set containing A, and so contains the intersection of *all* closed convex sets that contain A. However, this says that

$$\overline{\text{conv}}(A) \subset \overline{\text{conv}(A)},$$

that is, $C \subset B$.

$\square$

3.4.2 Accessibility Lemma

The next lemma is often referred to as the *accessibility lemma* and will be useful when we discuss supporting hyperplanes for convex sets.

Lemma 3.4.8. *(Accessibility Lemma)*
Suppose that $C \subset \mathbb{R}^n$ is a convex set with nonempty interior. If $x \in \text{int}(C)$ and $y \in C$, then the half open line segment from x to y

$$[x, y) = \{ (1 - \lambda)x + \lambda y : 0 \leq \lambda < 1 \}$$

consists entirely of interior points of C, that is, $[x, y) \subset \text{int}(C)$.

Proof. Since x is an interior point of C, there is a positive number δ such that $B(x, \delta) \subset C$. Now, if $z \in (x, y)$, then

$$z = \lambda x + (1 - \lambda)y$$

for some λ with $0 < \lambda < 1$. We will show that z is an interior point of C by proving that $B(z, \lambda \delta) \subset C$.

Let $v \in B(z, \lambda \delta)$, we want to show that $v \in C$.

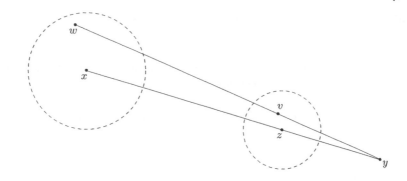

Note that if we set

$$w = \frac{1}{\lambda}\left(v - (1 - \lambda)y\right),$$

then

$$v = \lambda w + (1 - \lambda)y.$$

Thus, if we can show that $w \in C$, then by convexity, we will have $v \in C$.

To see that $w \in C$, note that

$$\|w - x\| = \|\tfrac{1}{\lambda}\left(v - (1-\lambda)y\right) - x\| = \|\tfrac{1}{\lambda}\left(v - (1-\lambda)y - \lambda x\right)\| = \|\tfrac{1}{\lambda}(v - z)\| < \delta,$$

which shows that $w \in B(x, \delta)$ and so $w \in C$. Thus, $v \in C$, so that $B(z, \lambda\delta)$ is contained in C, and z is an interior point of C. This completes the proof. $\square$

Corollary 3.4.9. *If $C \subset \mathbb{R}^n$ is a convex set, then the interior* $\mathrm{int}(C)$ *is also a convex set.*

Proof. If $\mathrm{int}(C)$ is not convex, then there exists points x, $y \in \mathrm{int}(C)$ such that some point $z \in (x, y)$ would fail to be in $\mathrm{int}(C)$. However, this contradicts the previous lemma.

$\square$

The following lemma is a generalization of Lemma 3.4.8 and is sometimes referred to as the accessibility lemma. It shows that the same conclusion holds not only if y is any point of C but also if y is any point of the closure of C.

Lemma 3.4.10. *Suppose that $C \subset \mathbb{R}^n$ is a convex set with nonempty interior. If $x \in \mathrm{int}(C)$ and $y \in \overline{C}$, then the half open line segment from x to y*

$$[x, y) = \{(1 - \lambda)x + \lambda y : 0 \leq \lambda < 1\}$$

consists entirely of interior points of C, that is, $[x, y) \subset \mathrm{int}(C)$.

Proof. Let p be a point of the open segment (x, y), say

$$p = (1 - \lambda)x + \lambda y,$$

where $0 < \lambda < 1$. Since $x \in \text{int}(C)$, there is a positive number δ such that the open ball $B(x, \delta)$ is contained in C.

Also, since y is a closure point of C, there must be some point z of C such that

$$\|y - z\| < \frac{1 - \lambda}{\lambda} \delta.$$

Now consider the point

$$w = \frac{1}{1 - \lambda} p - \frac{\lambda}{1 - \lambda} z,$$

and note that p is in the segment (w, z). We claim that w is in the ball $B(x, \delta)$. We have

$$\|w - x\| = \left\| \frac{1}{1 - \lambda} p - \frac{\lambda}{1 - \lambda} z - x \right\|$$

$$= \left\| \frac{1}{1 - \lambda} ((1 - \lambda)x + \lambda y) - \frac{\lambda}{1 - \lambda} z - x \right\|$$

$$= \left\| \frac{\lambda}{1 - \lambda} (y - z) \right\|$$

$$< \frac{\lambda}{1 - \lambda} \left(\frac{1 - \lambda}{\lambda} \right) \delta$$

$$= \delta,$$

which shows that w is in $B(x, \delta)$ as claimed. Thus, $w \in \text{int}(C)$ and $z \in C$, so Lemma 3.4.8 implies that every point of the segment (w, z) is an interior point of C. Therefore, p is an interior point of C, and since $p \in (x, y)$ was arbitrary, then $[x, y) \subset \text{int}(C)$.

□

3.4.3 Regularity of Convex Sets

There are examples of subsets S of $\mathbb{R}^n$ for which the interior of S and the interior of the closure of S are very different, that is, for which

$$\text{int}(S) \neq \text{int}\left(\overline{S}\right).$$

There are also subsets S of $\mathbb{R}^n$ that have the property that the closure of S is different from the closure of the interior of the set even if the set S has interior points, that is, for which

$$\overline{S} \neq \overline{\text{int}(S)}.$$

It is consequence of Lemmas 3.4.8 and 3.4.10 that this "pathology" can never occur for convex sets S with nonempty interior.

Theorem 3.4.11. *If A is a convex subset of $\mathbb{R}^n$ and* $\text{int}(A) \neq \emptyset$, *then*

$$\text{int}\left(\overline{A}\right) \subset A \subset \overline{\text{int}(A)}.$$

Proof. We want to show first that $\text{int}\left(\overline{A}\right) \subset A$. Since $\text{int}(A) \neq \emptyset$, we may choose a point $x \in \text{int}(A)$. Let z be a point of $\text{int}\left(\overline{A}\right)$ and consider the ray with endpoint x passing through z. Since z is an interior point of $\overline{A}$, there must be a point w in the ray beyond z that is also in the interior of $\overline{A}$ as in the figure below.

In particular, this means that w is in the closure of A, and since x is in the interior of A, Lemma 3.4.10 shows that every point of the segment (w, x) is in the interior of A. Thus, $z \in \text{int}(A) \subset A$, and since $z \in \text{int}\left(\overline{A}\right)$ is arbitrary, we have $\text{int}\left(\overline{A}\right) \subset A$.

Next we want to show that $A \subset \overline{\text{int}(A)}$. Let $x \in A$, since $\text{int}(A) \neq \emptyset$, we may choose a point $y \in \text{int}(A)$. By Lemma 3.4.8, the segment $[y, x) \subset \text{int}(A)$. Therefore, every open ball $B(x, \delta)$ contains points of $\text{int}(A)$ different from x, so that $x \in \overline{\text{int}(A)}$. Thus, since x is arbitrary, $A \subset \overline{\text{int}(A)}$.

$\square$

Example 3.4.12. *Show that convexity is necessary in Theorem 3.4.11*

Solution. In the following both cases, $\text{int}(A) \neq \emptyset$, but A is not convex.

(a) In $\mathbb{R}^1$, let A be given by

$$A = \{ q \in \mathbb{R}^1 : q \text{ is rational}, \ 0 \leq q \leq 1 \} \cup (1, 2).$$

Since the rational numbers q with $0 \leq q \leq 1$ are dense in the interval $[0, 1]$, then

$$\overline{A} = [0, 1] \cup [1, 2] = [0, 2]$$

and

$$\text{int}\left(\overline{A}\right) = (0, 2),$$

so that $\text{int}\left(\overline{A}\right) \not\subset A$.

(b) In $\mathbb{R}^1$, let
$$A = (0,1) \cup \{2\},$$

then
$$\text{int}(A) = (0,1) \quad \text{and} \quad \overline{\text{int}(A)} = [0,1],$$

so that $A \not\subset \overline{\text{int}(A)}$.

$\square$

Corollary 3.4.13. *If $A \subset \mathbb{R}^n$ is convex and $\text{int}(A) \neq \emptyset$, then*

(a) $\overline{\text{int}(A)} = \overline{A}$, *and*
(b) $\text{int}\left(\overline{A}\right) = \text{int}(A)$.

Proof. For (a), note that $\text{int}(A) \subset A \subset \overline{A}$, so that $\overline{\text{int}(A)} \subset \overline{A}$. From the previous theorem, since A is convex and $\text{int}(A) \neq \emptyset$, we have $A \subset \overline{\text{int}(A)}$, so that $\overline{A} \subset \overline{\text{int}(A)}$. Therefore, $\overline{\text{int}(A)} = \overline{A}$.

For (b), again we note that $\text{int}(A) \subset A \subset \overline{A}$, so that $\text{int}(A) \subset \text{int}\left(\overline{A}\right)$. From the previous theorem, since A is convex and $\text{int}(A) \neq \emptyset$, we have $\text{int}\left(\overline{A}\right) \subset A$, so that $\text{int}(\overline{A}) \subset \text{int}(A)$. Therefore, $\text{int}\left(\overline{A}\right) = \text{int}(A)$.

$\square$

Corollary 3.4.14. *If C is a convex set such that $\text{int}(C) \neq \emptyset$, then*

(a) C *is closed if and only if* $\overline{\text{int}(C)} = C$, *and*
(b) C *is open if and only if* $\text{int}\left(\overline{C}\right) = C$.

Example 3.4.15. *Show that a supporting hyperplane for the convex set $C \subseteq \mathbb{R}^n$ at a boundary point $x_0 \in C$ can contain no interior points of C.*
Hint. Let $x_0 \in C$ be a boundary point of the convex set C, and let
$$H = \{x \in \mathbb{R}^n : \langle p, x \rangle = \alpha\},$$

where $p \neq \overline{0}$, be a supporting hyperplane for C at x_0, such that $\langle p, x \rangle \leq \alpha$ for all $x \in C$ and $\langle p, x_0 \rangle = \alpha$. Show that if $x \in \text{int}(C)$, then $\langle p, x \rangle < \alpha$.

Solution. Following the hint, let x_0 be a boundary point of the convex set C, and let
$$H = \{x \in \mathbb{R}^n : \langle p, x \rangle = \alpha\},$$

where $p \neq \overline{0}$, be a supporting hyperplane for C at x_0, so that
$$\langle p, x \rangle \leq \alpha$$

for all $x \in C$ and $\langle p, x_0 \rangle = \alpha$.

If $x \in \text{int}(C)$, since $\text{int}(C) \subset C$, then $\langle p, x \rangle \leq \alpha$.

Suppose that $\langle p, x \rangle = \alpha$, since $x \in \text{int}(C)$, then there exists an $\epsilon > 0$ such that $B(x, \epsilon) \subset C$.

Now let

$$y = x + \eta \frac{p}{\|p\|^2} \qquad \text{and} \qquad z = x - \eta \frac{p}{\|p\|^2},$$

and choose η such that $0 < \eta < \epsilon \|p\|$, then $y, z \in B(x, \epsilon)$, so that y and z are in C.

However,

$$\langle p, y \rangle = \langle p, x \rangle + \eta = \alpha + \eta > \alpha$$

and

$$\langle p, z \rangle = \langle p, x \rangle - \eta = \alpha - \eta < \alpha,$$

which contradicts the fact that $\langle p, u \rangle \leq \alpha$ for all $u \in C$. Therefore, $\langle p, x \rangle < \alpha$ for all $x \in \text{int}(C)$, and H does not contain any interior points of C.

$\square$

Example 3.4.16. *Let G be an open subset of $\mathbb{R}^n$ with the Euclidean norm. Show that G is convex if and only if $G + G = 2G$.*

Solution. Suppose that G is an open convex subset of $\mathbb{R}^n$, and let $a \in 2G$, then $a = 2x = x + x$, where $x \in G$, so that $a \in G + G$ and $2G \subseteq G + G$.

Now, let $a \in G + G$, then $a = x + y$, where $x, y \in G$, since G is convex, then

$$z = \frac{1}{2}x + \frac{1}{2}y \in G,$$

so that

$$a = 2z \in 2G,$$

and $G + G \subseteq 2G$. Therefore, $G + G = 2G$ (we did not use the fact that G is open).

Conversely, suppose now that G is an open subset of $\mathbb{R}^n$ such that $G + G = 2G$, and let $x, y \in G$, then $x + y \in 2G$, so there exists a point $z \in G$ such that $x + y = 2z$, that is,

$$z = \frac{1}{2}x + \frac{1}{2}y \in G$$

whenever $x, y \in G$ and so G is midpoint convex.

Now, an easy induction argument shows that

$$z = (1 - \lambda)x + \lambda y \in G$$

whenever x, $y \in G$ and $0 \leq \lambda \leq 1$ is a ***dyadic rational***, that is, a real number of the form $\dfrac{k}{2^m}$ for some integers $0 \leq k \leq 2^m$.

Using the fact that the dyadic rationals $\dfrac{k}{2^m}$, for $0 \leq k \leq 2^m$, are dense* in the interval $[0, 1]$, and the fact that G is an open subset of $\mathbb{R}^n$ such that $G + G = 2G$, we will show that

$$z = (1 - \lambda)x + \lambda y \in G$$

for all $0 \leq \lambda \leq 1$, that is, that G is convex.

Suppose that x, $y \in G$ and that $z = (1 - \lambda)x + \lambda y$, where $0 < \lambda < 1$. If $\|x - z\| = \|y - z\|$, then z is the midpoint of the segment $[x, y]$, and since G is midpoint convex, this implies that $z \in G$.

On the other hand, suppose that $\rho = \|x - z\| < \|y - z\|$, and let

$$\{x, y_1\} = \overline{B}(z, \rho) \cap [x, y],$$

that is, y_1 is the point on the line segment $[x, y]$ between z and y where the closed ball $\overline{B}(z, \rho)$ intersects the segment. Since $x \in G$ and G is open, there is an $0 < \epsilon < \rho$ such that $B(x, \epsilon) \subset G$.

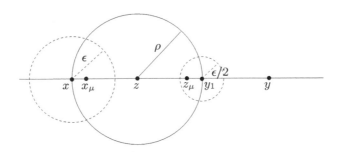

Now, since the points of the form $z_\mu = (1 - \mu)x + \mu y$, where μ is a dyadic rational in $[0, 1]$ are dense in the line segment $[x, y]$, given the $\epsilon > 0$ above, there exists a point $z_\mu \in G$, with z_μ between z and y_1, such that

$$r = \|y_1 - z_\mu\| < \frac{\epsilon}{2}.$$

If we let x_μ be the point on the line segment $[x, z]$ such that

$$\|x - x_\mu\| = r,$$

---

* If D is the set of dyadic rationals in $[0, 1]$, then D is ***dense*** in $[0, 1]$ if and only if $\overline{D} = [0, 1]$.

then x_μ, $z_\mu \in G$, and z is the midpoint of the segment $[x_\mu, z_\mu]$, and since G is midpoint convex, then $z \in G$.

We have shown that whenever x, $y \in G$ and $0 < \lambda < 1$, this implies that the point $z = (1 - \lambda)x + \lambda y \in G$, and, therefore, G is convex. It is left as an exercise to show that the dyadic rationals, that is, the real numbers of the form

$$q_{k,m} = \frac{k}{2^m}, \quad 0 \le k \le 2^m$$

are dense in $[0, 1]$.

□

3.4.4 Problems

1. If A and B are disjoint open subsets of $\mathbb{R}^n$ and $A \cup B = \mathbb{R}^n$, show that either $A = \emptyset$ or $B = \emptyset$.
2. Let A and B be convex subsets of $\mathbb{R}^n$ with $A \cap B = \emptyset$ and $A \cup B = \mathbb{R}^n$. Show that if A is closed, then A is a halfspace.
3. Let C be a convex subset of $\mathbb{R}^n$.
   (a) Show that bdy $(\bar{C}) = \mathrm{bdy}(C)$.
   (b) Find an example to show that convexity is necessary in part (a).
4. (a) Let C be a convex subset of $\mathbb{R}^n$ with $\mathrm{int}(C) \ne \emptyset$. Show that if C is compact, then the boundary of C is not convex.
   (b) Give an example of a convex subset C of $\mathbb{R}^n$ with $\mathrm{int}(C) \ne \emptyset$ such that $\mathrm{bdy}(C)$ is convex.
5. Show that if a and b are two points in the boundary of the closed convex subset C of $\mathbb{R}^n$, then either $[a, b] \subset \mathrm{bdy}(C)$ or $(a, b) \subset \mathrm{int}(C)$.
6. Suppose that a convex set $C \subset \mathbb{R}^n$ intersects each line in $\mathbb{R}^n$ in a closed set. Show that C is closed. Can convexity be omitted?
7. Let D be the set of all dyadic rationals in the interval $[0, 1]$, show that $\bar{D} = [0, 1]$.
8. Let $f : \mathbb{R}^n \to \mathbb{R}$ be a nonzero linear functional.
   (a) Show that f is continuous at each point $x \in \mathbb{R}^n$.
   (b) Show that if f is bounded above on a closed convex set $C \subset \mathbb{R}^n$ and $\mathrm{bdy}(C) \ne \emptyset$, then $\sup\limits_{x \in C} f(x) = \sup\limits_{x \in \mathrm{bdy}(C)} f(x)$.
9. Let $\| \cdot \|$ be a norm on $\mathbb{R}^n$, and let x, y, and z be points on a line *in that order*, that is,
   $$y = (1 - \lambda)x + \lambda z$$
   for some scalar with $0 < \lambda < 1$.
   Show that
   $$\frac{\|y\| - \|x\|}{\|y - x\|} \le \frac{\|z\| - \|x\|}{\|z - x\|} \le \frac{\|z\| - \|y\|}{\|z - y\|}.$$

10. A function $f : \mathbb{R} \longrightarrow \mathbb{R}$ is said to be **convex** if and only if

$$f\left(\mu x + (1 - \mu)y\right) \leq \mu f(x) + (1 - \mu)f(y)$$

for all $0 \leq \mu \leq 1$ and all x, $y \in \mathbb{R}$.

(a) If x, y, and z are real numbers with $x < y < z$, find the value of $0 < \lambda < 1$ such that

$$y = (1 - \lambda)x + \lambda z.$$

(b) If $f : \mathbb{R} \longrightarrow \mathbb{R}$ is convex and $x < y < z$, show that

$$\frac{f(y) - f(x)}{y - x} \leq \frac{f(z) - f(x)}{z - x} \leq \frac{f(z) - f(y)}{z - y}.$$

3.5 AFFINE HULLS

3.5.1 Flats or Affine Subspaces

If a hyperplane in $\mathbb{R}^n$ contains the points p and q, then it contains the entire line through p and q. In general, any subset of $\mathbb{R}^n$ with this property is called a *flat* or an *affine subspace*.

A set $A \subset \mathbb{R}^n$ is a **flat** or an **affine subspace** if and only if whenever p and q are points in A, then for all scalars λ, the point

$$x_\lambda = (1 - \lambda)p + \lambda q$$

is also in A, that is, the line determined by p and q is a subset of A.

• In $\mathbb{R}^2$, the nonempty flats are points, straight lines, and the entire space $\mathbb{R}^2$.

• In $\mathbb{R}^3$, the nonempty flats are points, straight lines, planes, and the entire space $\mathbb{R}^3$.

• In $\mathbb{R}^4$, the nonempty flats are points, straight lines, planes, hyperplanes, and the entire space $\mathbb{R}^4$.

• In $\mathbb{R}^n$, the points and the entire space are sometimes described as being the *trivial* flats, while any other type of flats is said to be *nontrivial*.

In this section, we discuss the basic geometric properties of flats and we show that nonempty flats are precisely translates of a subspace. If $A \subset \mathbb{R}^n$ is a flat, the unique subspace V such that $A = V + a$ for some $a \in \mathbb{R}^n$ is called the **parallel subspace** of A.

Theorem 3.5.1. *Let A be a nonempty flat and let $a \in A$, then $A - a$ is a subspace. Conversely, if V is a subspace and $a \in \mathbb{R}^n$, then $A = V + a$ is a flat.*

Proof. Let A be a nonempty flat, and note that if $a \in A$, then $0 = a - a \in A - a$.

(i) Suppose that $x \in A - a$ and $\lambda \in \mathbb{R}$, then $x = b - a$ for some $b \in A$, so that

$$\lambda x = \lambda b - \lambda a = (1 - \lambda)a + \lambda b - a.$$

The point $(1 - \lambda)a + \lambda b$ is on the line joining a and b and since A is a flat, then

$$(1 - \lambda)a + \lambda b \in A$$

so that $\lambda x \in A - a$ and $A - a$ is closed under scalar multiplication.

(ii) Suppose that $x, y \in A - a$, then $x + a \in A$ and $y + a \in A$, and since A is a flat, then for any $\mu \in \mathbb{R}$, we have

$$(1 - \mu)(x + a) + \mu(y + a) = (1 - \mu)x + \mu y + a \in A,$$

that is, $(1 - \mu)x + \mu y \in A - a$ for all $\mu \in \mathbb{R}$. Taking $\mu = \frac{1}{2}$, we have $\frac{1}{2}x + \frac{1}{2}y \in A - a$, and since $A - a$ is closed under scalar multiplication, then

$$x + y = 2\left(\tfrac{1}{2}x + \tfrac{1}{2}y\right) \in A - a$$

and $A - a$ is closed under addition.

Conversely, if V is a subspace and $a \in \mathbb{R}^n$, then for $x, y \in V + a$ and $\lambda \in \mathbb{R}$, we have

$$\lambda(x - a) + (1 - \lambda)(y - a) = \lambda x + (1 - \lambda)y - a \subset V,$$

so that $\lambda x + (1 - \lambda)y \subset V + a$ for all $\lambda \in \mathbb{R}$. Therefore, $V + a$ is a flat.

$\square$

Corollary 3.5.2. *Every nonempty flat is the translate of precisely one subspace of $\mathbb{R}^n$.*

Proof. If the flat A is a translate of the subspaces V and W, then $A = V + b$ and $A = W + c$ for some $b, c \in \mathbb{R}^n$, so that $W = V + a$ where $a = b - c$.

Now, since $0 \in W$, then there exists a $v \in V$ such that $0 = v + a$, that is, $v = -a$, so $-a \in V$. Since V is a subspace, then $a \in V$ also. Therefore,

$W = V + a \subseteq V$. Switching the roles of V and W, we see that $V \subseteq W$ also and so $W = V$.

$\square$

3.5.2 Properties of Flats

Recall that the **affine hull** of a set $S \subseteq \mathbb{R}^n$, denoted by $\text{aff}(S)$, was defined as the set of all affine combinations of points from S. Thus, $x \in \text{aff}(S)$ if and only if there exist points $x_1, x_2, \ldots, x_k \in S$ and scalars $\lambda_1, \lambda_2, \ldots, \lambda_k$ such that

$$x = \lambda_1 x_1 + \lambda_2 x_2 + \cdots + \lambda_k x_k,$$

where $\lambda_1 + \lambda_2 + \cdots + \lambda_k = 1$.

The following properties of flats resemble those of convex sets and convex hulls, and their proofs are left as exercises.

- If A and B are flats and λ is a scalar, then $A + B$ and λA are flats.
- If $A \subseteq B$, then $\text{aff}(A) \subseteq \text{aff}(B)$.
- If A_α is a flat for each $\alpha \in I$ and $\bigcap_{\alpha \in I} A_\alpha \neq \emptyset$, then $F = \bigcap_{\alpha \in I} A_\alpha$ is also a flat.
- $\text{aff}(A)$ is a flat that contains A, and if B is any flat containing A, then $\text{aff}(A) \subseteq B$. Thus, $\text{aff}(A)$ is the smallest flat containing A.
- The affine hull of a set A is the intersection of all flats containing A.
- If $F \subset \mathbb{R}^n$ is a flat, then F is a closed subset of $\mathbb{R}^n$.

Example 3.5.3. *Given that F is a flat and that p is a point outside F, consider the set X that is created by taking the union of all lines that pass through p and points of F. Show that if q is any point of F, then $X \cup (F + p - q)$ is a flat.*

Solution. Let q be any point of F and let L be the straight line through the origin parallel to $p - q$. We will show that $X \cup (F + p - q) = F + L$.

Let $x \in X \cup (F + p - q)$. Either $x \in X$ or $x \in (F + p - q)$, and we will consider each case separately.

The second case is obvious, since $p - q$ is a point in L, and this implies that $(F + p - q) \subset F + L$.

For the first case, if $x \in X$, then $x = (1 - \lambda)z + \lambda p$ for some $z \in F$ and $\lambda \in \mathbb{R}$. Now, since the flat F contains q as well as z, the point $(1 - \lambda)z + \lambda q$

must also be in F. Thus, a little algebra gives us

$$(1 - \lambda)z + \lambda p = ((1 - \lambda)z + \lambda q) + \lambda(p - q),$$

which shows that $x \in F + L$, and so $X \cup (F + p - q) \subset F + L$.

Therefore, in both cases, we have shown that $X \cup (F + p - q) \subset F + L$.

For the reverse containment, suppose that $y \in F + L$, then $y = z + \mu(p - q)$ for some z in F and some real number μ.

If $\mu = 1$, then $y = z + (p - q) \in F + (p - q)$ so that $y \in X \cup (F + p - q)$.

If $\mu \neq 1$, then

$$x = \frac{1}{1 - \mu} z - \frac{\mu}{1 - \mu} q$$

is a point of F, and it is easily verified that $y = (1 - \mu)x + \mu p$, which shows that $y \in X$, so that $y \in X \cup (F + p - q)$. Thus, $F + L \subset X \cup (F + p - q)$.

We have shown that $X \cup (F + p - q) = F + L$ and so is a flat (see Problem 3).

$\square$

3.5.3 Affine Basis

We proved earlier that every flat in $\mathbb{R}^n$ is a translate of a unique linear subspace. To put it another way, if $F \subset \mathbb{R}^n$ is a flat and p is a point of F, then $F - p$ is a linear subspace of $\mathbb{R}^n$ (independent of the choice of the point $p \in F$).

The previous results allow us to define the *dimension* of a flat in $\mathbb{R}^n$.

The ***dimension*** of a flat $F \subset \mathbb{R}^n$ is defined to be the dimension of the unique parallel subspace $V = F - p$ of F.

A finite set of points $B = \{x_0, x_1, \ldots, x_k\}$ is ***affinely dependent*** if there exist real numbers $\lambda_0, \lambda_1, \ldots, \lambda_k$, not all zero, such that

$$\lambda_0 x_0 + \lambda_1 x_1 + \cdots + \lambda_k x_k = \overline{0}$$

and $\lambda_0 + \lambda_1 + \cdots + \lambda_k = 0$. The set of points $B = \{x_0, x_1, \ldots, x_k\}$ is ***affinely independent*** if it is not affinely dependent.

The notions of linear independence and affine independence are not the same. For example, in $\mathbb{R}^2$, the three vertices of a nondegenerate triangle are linearly dependent but affinely independent. In general,

- In a linear subspace $V \subset \mathbb{R}^n$ of dimension k, a set of $k + 1$ or more vectors must be linearly dependent.
- In a flat $F \subset \mathbb{R}^n$ of dimension k, a set of $k + 2$ or more points must be affinely dependent.

This means that it is possible to define the dimension of a flat as *one less* than the maximum number of affinely independent points that can be found. In fact, it is this connection between dimension and affine independence that makes the notion of affine independence so useful.

Theorem 3.5.4. *A set of points $B = \{ x_0, x_1, \ldots, x_k \}$ is affinely dependent if and only if the set of vectors $S = \{ x_1 - x_0, x_2 - x_0, \ldots, x_k - x_0 \}$ is linearly dependent. Equivalently, B is affinely independent if and only if S is linearly independent.*

Proof. If the set of points $B = \{ x_0, x_1, \ldots, x_k \}$ is affinely dependent, then there exist scalars $\lambda_0, \lambda_1, \ldots, \lambda_k$, not all zero, such that $\sum_{i=0}^{k} \lambda_i x_i = \bar{0}$, where

$$\sum_{i=0}^{k} \lambda_i = 0.$$

Therefore, $\lambda_0 = -\sum_{i=1}^{k} \lambda_i$, so that $\sum_{i=1}^{k} \lambda_i (x_i - x_0) = \bar{0}$, where not all of $\lambda_1, \lambda_2, \ldots, \lambda_k$ are zero (if they were, then λ_0 would also be zero). Hence, the set of vectors $S = \{ x_1 - x_0, x_2 - x_0, \ldots, x_k - x_0 \}$ is linearly dependent.

Conversely, if the set $S = \{ x_1 - x_0, x_2 - x_0, \ldots, x_k - x_0 \}$ is linearly dependent, then there exist scalars $\lambda_1, \lambda_2, \ldots, \lambda_k$, not all zero, such that

$$\lambda_1 (x_1 - x_0) + \lambda_2 (x_2 - x_0) + \cdots + \lambda_k (x_k - x_0) = \bar{0},$$

that is,

$$\sum_{i=1}^{k} \lambda_i x_i - \left(\sum_{i=1}^{k} \lambda_i \right) x_0 = \bar{0}.$$

If we let $\lambda_0 = -\sum_{i=1}^{k} \lambda_i$, then $\sum_{i=0}^{k} \lambda_i x_i = \bar{0}$, where $\sum_{i=0}^{k} \lambda_i = 0$, and not all λ_i for $i = 0, 1, \ldots, k$ are zero (otherwise we would have $\lambda_i = 0$ for $i = 1, 2, \ldots, k$). Hence, the set of points $B = \{ x_0, x_1, \ldots, x_k \}$ is affinely dependent.

$\square$

If F is a flat with $\dim(F) = k$, then an **affine basis** for F, or simply a **basis** for F, is a set of $k + 1$ affinely independent points in F, that is, a set $B = \{ x_0, x_1, \ldots, x_k \}$ of points in F such that

(a) B is an affinely independent subset of F, and
(b) every $x \in F$ can be written as $x = \lambda_0 x_0 + \lambda_1 x_1 + \cdots + \lambda_k x_k$, where $\lambda_0 + \lambda_1 + \cdots + \lambda_k = 1$. The scalars λ_i for $i = 0, 1, \ldots, k$ are called the ***barycentric coordinates*** of x with respect to the basis B.

Theorem 3.5.5. *(Affine Basis Theorem)*

Let $F \subset \mathbb{R}^n$ be a flat, then the following statements are equivalent:

(a) *The dimension of F is k.*
(b) *There exists a subset $B = \{ x_0, x_1, \ldots, x_k \}$ of $k+1$ points in F such that B is affinely independent, and any set of $k+2$ points in F is affinely dependent.*
(c) *There exist $k+1$ points in F such that the set $B = \{ x_0, x_1, \ldots, x_k \}$ is affinely independent, and each point in F has a unique representation as an affine combination of B, that is, $x = \lambda_0 x_0 + \lambda_1 x_1 + \cdots + \lambda_k x_k$, where $\lambda_0 + \lambda_1 + \cdots + \lambda_k = 1$.*

Proof. To prove this theorem, we need only to consider the parallel subspace $V = F - a$, where $a \in F$.

$\square$

If A is a subset of $\mathbb{R}^n$, the ***dimension*** of A, denoted by $\dim(A)$, is defined to be the dimension of the smallest flat that contains A, that is, the dimension of $\text{aff}(A)$.

Note. If $A \subset \mathbb{R}^n$ and $\dim(A) = k$, then A contains $k+1$ points such that the set $B = \{ a_0, a_1, \ldots, a_k \}$ is affinely independent, and every set of $k+2$ points from A is affinely dependent. Otherwise, the dimension of $\text{aff}(A)$ is less than k. Thus, A contains an affine basis $B = \{ a_0, a_1, \ldots, a_k \}$ for $\text{aff}(A)$ and

$$\text{aff}(B) = \text{aff}(\{ a_0, a_1, \ldots, a_k \}) = \text{aff}(A).$$

In a flat F of dimension k, it is possible to find $k+1$ affinely independent points, and if $B = \{ a_0, a_1, \ldots, a_k \}$ is affinely independent, the set $\text{conv}(B)$ is called the k-dimensional ***simplex*** or k-***simplex*** spanned by B and the points $a_0, a_1, \ldots, a_k$ are called the ***vertices*** of the simplex. We note that

- a zero-dimensional simplex is a point,
- a one-dimensional simplex is a closed line segment,
- a two-dimensional simplex is a nondegenerate triangle,
- a three-dimensional simplex is a tetrahedron,

as in the figure on the following page.

0 simplex   1 simplex   2 simplex   3 simplex

A point x in an k-simplex S with vertices $x_0, x_1, \ldots, x_k$ can be written in a *unique* way as an affine combination of the vertices

$$x = \lambda_0 x_0 + \lambda_1 x_1 + \cdots + \lambda_k x_k,$$

where $\lambda_0 + \lambda_1 + \cdots + \lambda_k = 1$, since $S \subset \text{aff}(S)$. Also, since

$$S = \text{conv}(\{x_0, x_1, \ldots, x_k\}),$$

then $\lambda_i \geq 0$, for each $i = 0, 1, \ldots, k$, since S is convex. Thus, we have shown the following theorem.

Theorem 3.5.6. *A point x in an k-simplex S with vertices $x_0, x_1, \ldots, x_k$ can be written in a unique way as a convex combination of the vertices*

$$x = \lambda_0 x_0 + \lambda_1 x_1 + \cdots + \lambda_k x_k,$$

where $\lambda_0 + \lambda_1 + \cdots + \lambda_k = 1$ and $\lambda_i \geq 0$ for $i = 0, 1, \ldots, k$.

We will show that in an n-dimensional space, an n-dimensional simplex always has a nonempty interior. Since a simplex is convex, in order to show this it suffices to show that a convex set $C \subset \mathbb{R}^n$ has dimension n if and only if $\text{int}(C) \neq \emptyset$.

The following theorem seems intuitively obvious.

Theorem 3.5.7. *If $C \subset \mathbb{R}^n$ is convex, then $\text{int}(C) \neq \emptyset$ if and only if $\dim(C) = n$.*

Proof. Note that if $\text{int}(C) \neq \emptyset$ and if x_0 is an interior point of C, then there exists an $\epsilon > 0$ such that $B(x_0, \epsilon) \subset C$. Since a hyperplane has no interior points, the smallest flat that contains C is $\mathbb{R}^n$, and so $\dim(C) = n$.

Conversely, suppose that $\dim(C) = n$, then $\text{aff}(C) = \mathbb{R}^n$, and there exist points $x_0, x_1, \ldots, x_n$ in C such that the set $B = \{x_0, x_1, \ldots, x_n\}$ forms an affine basis for $\text{aff}(C) = \mathbb{R}^n$. Therefore, each $x \in \mathbb{R}^n$ can be written *uniquely* as

$$x = \lambda_0 x_0 + \lambda_1 x_1 + \cdots + \lambda_n x_n,$$

where $\lambda_0 + \lambda_1 + \cdots + \lambda_n = 1$.

Now note that this expression for x depends only on the barycentric co-ordinates of the point x, since everything else is fixed, and the mapping $f : \mathbb{R}^n \longrightarrow \mathbb{R}^{n+1}$ given by

$$f(x) = f\left(\sum_{k=0}^{n} \lambda_k x_k\right) = (\lambda_0, \lambda_1, \ldots, \lambda_n),$$

where $\lambda_0 = 1 - \sum_{k=1}^{n} \lambda_k$, is well defined and continuous.

Since C is convex, the centroid of the simplex $\text{conv}(\{x_0, x_1, \ldots, x_n\})$ is in C, that is,

$$z = \frac{1}{n+1}\,(x_0 + x_1 + \cdots + x_n)$$

is in C. Since f is continuous at z, then

$$\lim_{x \to z} f(x) = f(z) = \left(\frac{1}{n+1}, \frac{1}{n+1}, \ldots, \frac{1}{n+1}\right),$$

so given $\epsilon > 0$, there exists a $\delta = \delta(\epsilon) > 0$ such that

$$\|f(x) - f(z)\|_\infty < \epsilon$$

for all $x \in B(z, \delta)$. If we take $\epsilon = \frac{1}{2(n+1)}$, then there exists a $\delta > 0$ such that $x \in B(z, \delta)$ implies that

$$\left|\lambda_k - \frac{1}{n+1}\right| \leq \|f(x) - f(z)\|_\infty < \frac{1}{2(n+1)}$$

for all $k = 0, 1, \ldots, n$.

Thus, for each $k = 0, 1, \ldots, n$, we have

$$-\frac{1}{2(n+1)} < \lambda_k - \frac{1}{n+1} < \frac{1}{2(n+1)},$$

that is,

$$\lambda_k > \frac{1}{n+1} - \frac{1}{2(n+1)} = \frac{1}{2(n+1)} > 0.$$

Therefore, for every $x \in B(z, \delta)$, all the components λ_k of x will be strictly positive, so that $B(z, \delta) \subset C$, and so z is an interior point of C.

$\square$

And, since simplices are convex, we have the following theorem.

Corollary 3.5.8. *In a k-dimensional space, a k-simplex always has a nonempty interior.*

If $C \subset \mathbb{R}^n$ is convex, the **relative interior** of C, denoted by relint(C), is the interior of C with respect to the smallest flat that contains C, that is, with respect to aff(C). For example, if C is a triangle in $\mathbb{R}^3$, then C is contained in some plane, and the relative interior of C is the interior of the triangle with respect to the plane.

Note that since the dimension of a nonempty convex set C and the dimension of the smallest flat that contains C are one and the same, then relint(C) is never empty. Accordingly, the results that we have stated earlier about the interior of C are also valid for the relative interior of C. For example, Lemma 3.4.8 becomes the following.

Lemma 3.5.9. *(Accessibility Lemma)*

Suppose that $C \subset \mathbb{R}^n$ is a nonempty convex set. If $x \in$ relint(C) and $y \in C$, then the half open line segment from x to y

$$[x, y) = \{ (1 - \lambda)x + \lambda y : 0 \leq \lambda < 1 \}$$

consists entirely of points in the relative interior of C, that is,

$$[x, y) \subset \text{relint}(C).$$

3.5.4 Problems

1. Show that if F and G are flats, then either $F \cap G$ is a flat or else $F \cap G = \emptyset$.
2. Show that if F is a flat and x_0 is any point, then $F + x_0$ is a flat.
3. Show that if F and G are flats, then $F + G$ is also a flat.
4. Show that if F is a flat and p and q are points in F, then $F = F + p - q$.
5. Show that if the vector v is **parallel** to the flat F, that is, $v = p - q$ for some p and q in F, then $F + v \subset F$. Thus, the **parallel translate**, or simply the **translate**, $F + v$ is a subset of F.
6. Given that F is a flat and that p is point outside F. Let X be the set that is created by taking the union of all lines that pass through p and points of F. Show that if q is any point of F, then the flat $X \cup (F + p - q)$ is the smallest flat that contains F and p.
   Hint. If H is any flat containing both F and p, then H must contain both X and $F + p - q$.
7. Show that if p_1, p_2, and p_3 are points in the flat F and if λ_1, λ_2, and λ_3 are real numbers whose sum is 1, then $\lambda_1 p_1 + \lambda_2 p_2 + \lambda_3 p_3$ is also a point in F.

8. Let
$$L = \{ (\lambda, 0, 0) \in \mathbb{R}^3 : -\infty < \lambda < \infty \}$$
and
$$M = \{ (\lambda, 2\lambda, 0) \in \mathbb{R}^3 : -\infty < \lambda < \infty \}.$$
Describe the flat $L + M$ in $\mathbb{R}^3$. Draw a picture.

9. Find two flats in $\mathbb{R}^4$, neither of which is a point or a line, but whose intersection is a single point.

10. Show that F is a maximal proper affine subspace of $\mathbb{R}^n$ if and only if it is the translate of a maximal proper linear subspace of $\mathbb{R}^n$, that is, if and only if it is a hyperplane.

11. Show that if $F \subset \mathbb{R}^n$ is a flat, then F is a closed subset of $\mathbb{R}^n$.

12. Show that if $F \subsetneq \mathbb{R}^n$ is a flat, then $\mathrm{int}(F) = \emptyset$, but $\mathrm{relint}(F) \neq \emptyset$.

13. Show that if $\{ x_0, x_1, x_2, y, z, \}$ are, respectively, the three vertices, midpoint of the side opposite x_0, and the intersection of the medians of a triangle, then their barycentric coordinates are

$$(1, 0, 0), \quad (0, 1, 0), \quad (0, 0, 1), \quad (0, \tfrac{1}{2}, \tfrac{1}{2}), \quad (\tfrac{1}{3}, \tfrac{1}{3}, \tfrac{1}{3}).$$

14. The dimension of a k-simplex is k. Show that the dimension of a convex set $C \subset \mathbb{R}^n$ is the maximum of the dimensions of the simplices contained in C.

15. Let H be a hyperplane in a normed linear space V. Show the following:
    (a) If $V = \mathbb{R}^n$, then H is closed.
    (b) If V is infinite dimensional, then H need not be closed.
    (c) The closure of H is always affine.
    (d) H is either closed or dense in V.

16. Let $B = \{ (x, y) \in \mathbb{R}^2 : 0 \le x \le 1, 0 \le y \le 1 \}$ be a square in $\mathbb{R}^2$ and let $A = \{ (x, y) \in \mathbb{R}^2 : 0 \le x \le 1, y = 0 \}$ be the bottom side of the square. Show that $A \subset B$ and $\mathrm{relint}(A)$ and $\mathrm{relint}(B)$ are nonempty, but $\mathrm{relint}(A) \not\subset \mathrm{relint}(B)$.

17. Show that if A and B are subsets of $\mathbb{R}^n$ with $A \subset B$ and $\dim(A) = \dim(B)$, then $\mathrm{relint}(A) \subset \mathrm{relint}(B)$.

18. A point $a \in \mathbb{R}^n$ is a **relative boundary point** of a set $A \subset \mathbb{R}^n$ if $a \in \overline{A}$, but $a \notin \mathrm{relint}(A)$. The set of all relative boundary points of A is called the **relative boundary** of A and is denoted by $\mathrm{relbdy}(A)$.
    (a) Show that if $A \subset \mathbb{R}^n$ and $\dim(A) = n$, then $\mathrm{relbdy}(A) = \mathrm{bdy}(A)$.
    (b) If $a \in A$ and $x \in \mathrm{aff}(A)$ but $x \notin A$, define the scalar λ_0 by

$$\lambda_0 = \sup\{ \lambda \in [0, 1] : (1 - \lambda)a + \lambda x \in A \}.$$

Show that $(1 - \lambda_0)a + \lambda_0 x$ is a relative boundary point of A lying between a and x.

19. Show that the only sets $F \subset \mathbb{R}^n$, which have an empty relative boundary, are flats.

20. Given the subset A of $\mathbb{R}^n$ defined by
$$A = \{\, (x_1, x_2, \ldots, x_n) \,:\, x_1 + x_2 + \cdots + x_n \leq 1, \ x_k \geq 0, \ k = 1, 2, \ldots, n \,\},$$

(a) Show that $\dim(A) = n$ by finding $n+1$ points of A, which form an affine basis for A.

(b) Show that A is an n-simplex in $\mathbb{R}^n$.

$*$21. Let ℓ_2 be the normed linear space of all real-valued square summable sequences $x = \{x_k\}_{k \geq 1}$, with norm
$$\|x\|_2 = \left(\sum_{k=1}^{\infty} x_k^2 \right)^{1/2} < \infty,$$

(see Problems 28 and 29 in Section 2.6.1), and let
$$H = \{\, x \in \ell_2 \,:\, \sum_{k=1}^{\infty} x_k = 0 \,\}.$$

(a) Show that H is a flat.

(b) Show that H is not a closed subset of ℓ_2.

(c) Show that H is dense in ℓ_2, that is, show that $\overline{H} = \ell_2$.

3.6 SEPARATION THEOREMS

In this section, we use the Euclidean norm on $\mathbb{R}^n$. The reason for this is that the topological notions do not depend on the norm being used and that hyperplanes, being strictly linear notions, are the same in all norms. Thus, in discussing supporting or separating hyperplanes, we can use whatever norm is most convenient.

A hyperplane H is said to **support** the set $A \subset R^n$ if and only if $A \cap H \neq \emptyset$ and A is contained one of the closed halfspaces determined by H. Thus, if
$$H = \{\, x \in \mathbb{R}^n \,:\, \langle p, x \rangle = \alpha \,\},$$
where $p \in \mathbb{R}^n$ with $p \neq \overline{0}$, then H **supports** A if and only if

(i) there is a point $a_0 \in A$ such that $\langle p, a_0 \rangle = \alpha$, and

(ii) either $\langle p, x \rangle \leq \alpha$ for all $x \in A$ or $\langle p, x \rangle \geq \alpha$ for all $x \in A$.

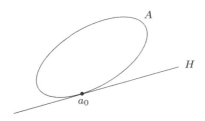

In this case, we say that H is a *supporting hyperplane* for A, and any point of $A \cap H$ is a *support point* of A.

Two subsets A and B of $\mathbb{R}^n$ are *separated* by the hyperplane

$$H = \{\, x \in \mathbb{R}^n \,:\, \langle p, x \rangle = \alpha \,\},$$

if and only if A is in one of the closed halfspaces determined by H and B is in the other, that is,

(i) $\langle p, a \rangle \le \alpha$ for all $a \in A$, and
(ii) $\langle p, b \rangle \ge \alpha$ for all $b \in B$.

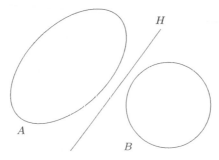

Exercise 3.6.1. *Let H be a hyperplane in $\mathbb{R}^n$ and A, $B \subset \mathbb{R}^n$.*

(a) *Show that if A and B are separated by the hyperplane H, then A and B need not be disjoint.*
(b) *Show also that it is possible that A and B can be in opposite closed halfspaces, where at least one of the sets is not contained in H and still be separated by H.*
(c) *Show, in fact, that one of the sets may even be a subset of the other, and the two sets can still be separated by H.*

To rule out these possibilities, we need a stronger notion of separation.

Two subsets A and B of $\mathbb{R}^n$ are *strictly separated* by the hyperplane

$$H = \{\, x \in \mathbb{R}^n \,:\, \langle p, x \rangle = \alpha \,\},$$

if and only if A is in one of the open halfspaces determined by H and B is in the other open halfspace determined by H, that is,

(i) $\langle p, a \rangle < \alpha$ for all $a \in A$, and
(i) $\langle p, b \rangle > \alpha$ for all $b \in B$.

This is still not a completely acceptable notion of separation.

Exercise 3.6.2. *Let H be a hyperplane in $\mathbb{R}^n$ and A, $B \subset \mathbb{R}^n$. Show that it is possible for A and B to be strictly separated by the hyperplane H and even have $\overline{A}$ and $\overline{B}$ separated by H, but not strictly separated by H.*

Thus, we need an even stronger notion.

Two subsets A and B of $\mathbb{R}^n$ are ***strongly separated*** by the hyperplane

$$H = \{\, x \in \mathbb{R}^n \,:\, \langle p, x \rangle = \alpha \,\},$$

if and only if H strictly separates A and B, with $\operatorname{dist}(A, H) > 0$ and $\operatorname{dist}(B, H) > 0$.

Example 3.6.3. *Show that in $\mathbb{R}^n$, two sets A and B are strongly separated by the hyperplane $H = \{\, x \in \mathbb{R}^n \,:\, \langle p, x \rangle = \alpha \,\}$ if and only if there exists an $\epsilon > 0$ such that*

$$A + \epsilon B(\overline{0}, 1) \quad and \quad B + \epsilon B(\overline{0}, 1)$$

are strictly separated by H.

Solution. If $A_\epsilon = A + \epsilon B(\overline{0}, 1)$ and $B_\epsilon = B + \epsilon B(\overline{0}, 1)$ are strictly separated by the hyperplane $H_\alpha = \{\, x \in \mathbb{R}^n \,:\, \langle p, x \rangle = \alpha \,\}$ where $\|p\| = 1$, then

$$\langle p, a \rangle < \alpha \quad \text{for all } a \in A_\epsilon, \quad \text{and} \quad \langle p, b \rangle > \alpha \quad \text{for all } b \in B_\epsilon. \qquad (*)$$

From the first inequality in $(*)$, if $a \in A$ and $z \in B(\overline{0}, 1)$, then

$$\langle p, a + \epsilon z \rangle = \langle p, a \rangle + \epsilon \langle p, z \rangle < \alpha.$$

Now let $z = \dfrac{\lambda p}{\|p\|}$, so that $z \in B(\overline{0}, 1)$ where $0 < \lambda < 1$, and let $\lambda \to 1^-$, then

$$\langle p, a \rangle + \epsilon \|p\| \leq \alpha,$$

and since $\|p\| = 1$, we have $\langle p, a \rangle \leq \alpha - \epsilon$ for all $a \in A$.

From the second inequality in $(*)$, since $z \in B(\overline{0}, 1)$ if and only if $-z \in B(\overline{0}, 1)$; if $b \in B$, an argument similar to that above shows that $\langle p, b \rangle \geq \alpha + \epsilon$ for all $b \in B$.

Thus, A and B are separated by the "slab" between the hyperplanes $H_{\alpha-\epsilon}$ and $H_{\alpha+\epsilon}$ parallel to H_α. Therefore,

$$\operatorname{dist}(A, H_\alpha) \geq \operatorname{dist}(H_{\alpha-\epsilon}, H_\alpha) \geq \epsilon \quad \text{and} \quad \operatorname{dist}(B, H_\alpha) \geq \operatorname{dist}(H_{\alpha+\epsilon}, H_\alpha) \geq \epsilon,$$

that is, A and B are strongly separated by H_α.

An alternate proof is as follows. If $\epsilon > 0$, then $\epsilon B(\overline{0}, 1) = B(\overline{0}, \epsilon)$, and

$$A_\epsilon = A + B(\overline{0}, \epsilon) = \{\, x \in \mathbb{R}^n \,:\, \mathrm{dist}(x, A) < \epsilon \,\},$$
$$B_\epsilon = B + B(\overline{0}, \epsilon) = \{\, x \in \mathbb{R}^n \,:\, \mathrm{dist}(x, B) < \epsilon \,\}.$$

If A_ϵ and B_ϵ are strictly separated by the hyperplane H, they lie in opposite open halfspaces determined by H and are, therefore, disjoint and neither intersects H.

Thus, for each $x \in H$, we must have

$$\mathrm{dist}(x, A) \geq \epsilon \qquad \text{and} \qquad \mathrm{dist}(x, B) \geq \epsilon,$$

so that

$$\epsilon \leq \mathrm{dist}(x, A) \leq \inf\{\, \|z - a\| \,:\, z \in H, \ a \in A \,\} = \mathrm{dist}(A, H),$$

and similarly, $\mathrm{dist}(B, H) \geq \epsilon$. Therefore, A and B are strongly separated by H.

Conversely, if the hyperplane H strongly separates A and B, then there are positive numbers ϵ_A and ϵ_B such that

$$\mathrm{dist}(A, H) > \epsilon_A \qquad \text{and} \qquad \mathrm{dist}(B, H) > \epsilon_B,$$

and if we let $\epsilon = \min\{\, \epsilon_A, \epsilon_B \,\}$, then A_ϵ and B_ϵ are strictly separated by H.

$\square$

Exercise 3.6.4. *Given a hyperplane* $H = \{\, x \in \mathbb{R}^n \,:\, \langle p, x \rangle = \alpha \,\}$, *where* $\|p\| = 1$, *and two sets* $A, B \subset \mathbb{R}^n$, *the following are equivalent:*

(i) *A and B are strongly separated by the hyperplane H,*
(ii) *there exists an $\eta > 0$ such that*

$$\langle p, a \rangle > \alpha + \eta \qquad \text{for all } a \in A,$$

and

$$\langle p, b \rangle < \alpha - \eta \qquad \text{for all } b \in B,$$

(iii) *there exists an $\epsilon > 0$ such that*

$$\inf\{\, \langle p, a \rangle \,:\, a \in A \,\} \geq \alpha + \epsilon,$$

and

$$\sup\{\, \langle p, b \rangle \,:\, b \in B \,\} \leq \alpha - \epsilon.$$

In this section, we prove the following two separation theorems.

Theorem (*Strong Separation Theorem*)

If A and B are disjoint convex subsets of $\mathbb{R}^n$ with A compact and B closed, then A and B are strongly separated by some hyperplane.

Theorem (*Separation Theorem*)

If A and B are convex subsets of $\mathbb{R}^n$ whose relative interiors are disjoint, and if $\operatorname{aff}(A \cup B) = \mathbb{R}^n$, then A and B are separated by some hyperplane.

These will be a consequence of some geometric facts that are of some use in their own right. As usual, we will assume that all references to distance and norm refer to the Euclidean norm unless otherwise stated.

Lemma 3.6.5. *If the hyperplane H supports the unit ball $\overline{B}(\bar{0}, 1)$ at a point y_0, and if z_0 is any point in the open halfspace containing $\bar{0}$, then the segment $[y_0, z_0]$ contains points of the open unit ball $B(\bar{0}, 1)$.*

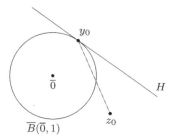

Proof. First, note that $H = \{\, x \in \mathbb{R}^n \ : \ \langle y_0, x \rangle = 1 \,\}$. If the conclusion of the lemma is not true, then the entire line through y_0 and z_0 would have to miss the open ball. Thus, $\|x\| \geq 1$ for every point x on the line. However, since $\|y_0\| = 1$, this would mean that y_0 is a point of the line that is closest to $\bar{0}$. From Theorem 1.4.5, this would imply that

$$\langle y_0, z_0 - y_0 \rangle = 0,$$

that is,

$$\langle y_0, z_0 \rangle = \langle y_0, y_0 \rangle = 1,$$

and so z_0 would be in the hyperplane H, which is a contradiction. Therefore, if z_0 is any point in the open halfspace containing $\bar{0}$, then the segment $[y_0, z_0]$ contains points of the open unit ball $B(\bar{0}, 1)$.

$\square$

Lemma 3.6.6. *If C is a closed convex subset of $\mathbb{R}^n$, and if x_0 is a point in the complement of C, with $y_0 \in C$ being the closest point to x_0, then the hyperplane H through y_0 orthogonal to the segment $[x_0, y_0]$ is a supporting hyperplane for C that separates C and x_0.*

Proof. We claim that C is contained in the closed halfspace that does not contain x_0.

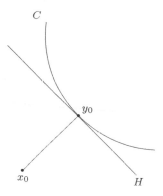

If this were not the case, then some point z of C would lie on the same side of H as x_0. Now, by convexity, the entire segment $[y_0, z]$ would be in C. However, by the previous lemma, this segment must contain points of the open ball $B(x_0, \|y_0 - x_0\|)$, and this implies that some point of C would be closer to x_0 than y_0, which is a contradiction. Therefore, H is a support hyperplane for C that separates C and x_0.

$\square$

Theorem 3.6.7. *If A and B are subsets of $\mathbb{R}^n$ with A compact and B closed, then $A + B$ is closed.*

Proof. Recall that

$$A + B = \{\, a + b \in \mathbb{R}^n \ : \ a \in A, \ b \in B \,\}$$

is the Minkowski sum of A and B (see Problem 2.6.5).

If z is an accumulation point of $A + B$, then given any positive integer m, there exists a point $z_m \in A + B$ such that

$$\|z_m - z\| < \frac{1}{m}.$$

Now since $z_m \in A + B$, we have $z_m = a_m + b_m$ for some points $a_m \in A$ and $b_m \in B$, respectively.

Since A is compact, then either the set $\{a_1, a_2, \ldots\}$ is finite, in which case one of the a_i's occurs infinitely often, or else the set must have an accumulation point. In either case, there is some point a_0 such that for every positive δ, there are infinitely many indices m for which $a_m \in B(a_0, \delta)$. Thus, a_0 is a closure point of the compact set A and, therefore, belongs to A.

Now, let $b_0 = z - a_0$. To finish the proof, it suffices to show that $b_0 \in B$. Since B is closed, we need only show that b_0 is a closure point of B

Thus, let $\epsilon > 0$ and choose $\delta > 0$ and $M > 0$ such that $\delta < \dfrac{\epsilon}{2}$ and $\dfrac{1}{M} < \dfrac{\epsilon}{2}$.

Now, if $m > M$ is an index such that $\|a_m - a_0\| < \delta$, then

$$\|b_0 - b_m\| = \|z - a_0 - (z_m - a_m)\| = \|a_m - a_0 + z - z_m\|$$
$$\leq \|a_m - a_0\| + \|z - z_m\| < \delta + \frac{1}{m}$$
$$< \delta + \frac{1}{M} < \frac{\epsilon}{2} + \frac{\epsilon}{2} = \epsilon.$$

Therefore, given any $\epsilon > 0$, we have $\|b_0 - b_m\| < \epsilon$ for all $m > M$, and so every ball centered at b_0 contains points of B. This finishes the proof.

$\square$

Remark. If neither A nor B is compact, the conclusion of the theorem may fail; that is, $A + B$ may not be closed.

Theorem 3.6.8. *If A and B are subsets of $\mathbb{R}^n$, with A compact and B closed, then there exist points $a_0 \in A$ and $b_0 \in B$ such that*

$$\|a_0 - b_0\| = \text{dist}(A, B) = \inf\{\|a - b\| : a \in A, \, b \in B\}.$$

Proof. This is actually Theorem 2.6.13, but we give a different proof here. We should also note that this theorem is true no matter what norm is being used.

Let $\delta = \text{dist}(A, B)$ be the distance from A to B. For each positive integer m, there exist points $a_m \in A$ and $b_m \in B$ such that

$$\|a_m - b_m\| < \delta + \frac{1}{m}.$$

Either the set $\{a_1, a_2, \ldots\}$ is infinite or else there are infinitely many indices for which a_m is the same point. In the former case, there must be an accumulation point a_0 of this set. In the latter case, we may assume that $a_m = a_0$ for infinitely many indices m. In either case, this means that for every $\epsilon > 0$, the ball $B(a_0, \epsilon)$ contains a_m for infinitely many indices m.

Now, we claim that $\text{dist}(a_0, B) = \text{dist}(A, B)$. To verify this, note that given any positive integer M and any positive number ϵ, we can find an index m with $m \geq M$ such that

$$a_m \in B(a_0, \epsilon) \qquad \text{and} \qquad \|a_m - b_m\| < \delta + \frac{1}{m}.$$

A simple application of the triangle inequality shows that

$$\|a_0 - b_m\| < \delta + \epsilon + \frac{1}{m},$$

and since $m \geq M$, we have

$$\|a_0 - b_m\| < \delta + \epsilon + \frac{1}{M}.$$

Since this inequality holds for all $\epsilon > 0$ and all positive integers M, it follows that

$$\text{dist}(a_0, B) \leq \inf\{\, \delta + \epsilon + 1/M \ : \ \epsilon > 0, \ M = 1, 2, 3, \dots \,\} = \delta = \text{dist}(A, B).$$

Also, since $a_0 \in A$, we must have $\text{dist}(a_0, B) \geq \text{dist}(A, B)$, and this shows that $\text{dist}(a_0, B) = \text{dist}(A, B)$ as claimed.

If $\delta = 0$, the theorem now follows from the fact that a_0 must be an accumulation point of the closed set B, and in this case, we may take $b_0 = a_0$. If $\delta > 0$, the theorem follows from Theorem 2.6.9.

$\square$

Now we are ready to prove the Strong Separation Theorem.

Theorem 3.6.9. *(Strong Separation Theorem)*
If A and B are disjoint convex subsets of $\mathbb{R}^n$ with A compact and B closed, then A and B are strongly separated by some hyperplane.

Proof. Since A is compact and B is closed and the sets are disjoint, then the sets A and B must be at a positive distance from each other, that is, $\text{dist}(A, B) > 0$.

From the previous theorem, we can find points $a_0 \in A$ and $b_0 \in B$ such that

$$\|a_0 - b_0\| = \text{dist}(A, B).$$

The parallel hyperplanes H_{a_0} and H_{b_0} through the endpoints of the segment $[a_0, b_0]$ and perpendicular to $[a_0, b_0]$ must be a positive distance apart.

By Lemma 3.6.6, since a_0 is the closest point of A to b_0, then H_{a_0} supports A at a_0 and separates A and b_0. Similarly, H_{b_0} supports B at b_0 and separates B from a_0.

This means that the open "slab" between the hyperplanes H_{a_0} and H_{b_0} does not contain points of either A or B. Therefore, the hyperplane H parallel to H_{a_0} and H_{b_0} that is halfway between them strongly separates A and B.

$\square$

The next theorem is the Separation Theorem, which was stated before Lemma 3.6.5.

Theorem 3.6.10. *(Separation Theorem)*

If A and B are convex subsets of $\mathbb{R}^n$ whose relative interiors are disjoint and if $\mathrm{aff}(A \cup B) = \mathbb{R}^n$, then A and B are separated by some hyperplane.

Proof. The proof uses the Strong Separation Theorem, and we are using the Euclidean norm for convenience, see the comments at the beginning of this section.

- Without loss of generality, we may assume that A and B are closed sets, since taking the closure does not add any relative interior points.
- If either A or B is bounded, we would be finished, since then one of A and B would be compact and the other closed, so that the Strong Separation Theorem would apply.
- If neither A nor B is bounded, the approach we will use is as follows:

   We first shrink A and B a little and then take a very large but nevertheless bounded subset of the shrunken A. The bounded set will be disjoint from the shrunken version of B, and so the Strong Separation Theorem will apply, yielding a hyperplane that separates the bounded set from the shrunken version of B. Of course, this hyperplane will likely not separate A and B, but it will "almost" separate them.

   We will repeat the process, taking larger and larger bounded subsets, at the same time, we shrink the sets less and less. Each time we do this, we will obtain a separating hyperplane. If we take vectors of length one that are orthogonal to these hyperplanes, then either one of these vectors p will occur infinitely often or else the set of vectors will have an accumulation point, p. The reason for the accumulation point is that the vectors are all subsets of the unit sphere, which is a compact set. It will turn out that some hyperplane orthogonal to p will separate A and B.

If A is not bounded, we may at least assume that $\overline{0}$ is relatively interior to A, otherwise, translate the coordinate system.

Now, let b_0 be a point in the relative interior of B. We will shrink A about the point $\overline{0}$ and shrink B about the point b_0.

For each positive integer m, let

$$A_m = \overline{B}(\overline{0}, m) \cap \left(1 - \frac{1}{m}\right) A$$

and

$$B_m = b_0 + \left(1 - \frac{1}{m}\right)(B - b_0).$$

For each m, the set A_m is convex, closed, and bounded; that is, A_m is a convex compact subset of $\mathbb{R}^n$, while the set B_m is a closed subset of $\mathbb{R}^n$.

Also, for each m, $A_m \cap B_m = \emptyset$ since

$$A_m \subset \operatorname{relint}(A) \qquad \text{and} \qquad B_m \subset \operatorname{relint}(B)$$

and $\operatorname{relint}(A) \cap \operatorname{relint}(B) = \emptyset$.

It is easily verified that

$$\overline{0} \in A_1 \subset A_2 \subset \cdots \qquad \text{and that} \qquad b_0 \in B_1 \subset B_2 \subset \cdots.$$

Moreover,

$$A = \bigcup_{m=1}^{\infty} A_m \qquad \text{and} \qquad B = \bigcup_{m=1}^{\infty} B_m.$$

For each $m \geq 1$, let H_m be a hyperplane strongly separating A_m and B_m, and let q_m be the point where the segment $[\overline{0}, b_0]$ intersects the hyperplane.

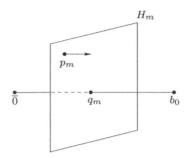

Let p_m be a vector of length one orthogonal to H_m, let p_0 be an accumulation point of the p_m's, and let q_0 be an accumulation point of the q_m's. The hyperplane passing through q_0 orthogonal to p_0 will separate A and B.

$\square$

Corollary 3.6.11. *If A and B are convex subsets of $\mathbb{R}^n$ such that $\text{int}(A) \neq \emptyset$ and $\text{int}(A) \cap B = \emptyset$, then there exists a hyperplane*

$$H = \{\, x \in \mathbb{R}^n \, : \, \langle p, x \rangle = \alpha \,\}$$

that separates A and B.

Proof. Since $\text{int}(A) \neq \emptyset$, then $\text{aff}(A \cup B) = \mathbb{R}^n$, by the Separation Theorem, since $\text{int}(A) \cap \text{int}(B) \subset \text{int}(A) \cap B = \emptyset$, there exists a hyperplane that separates A and B.

$\square$

Corollary 3.6.12. *If A and B are convex subsets of $\mathbb{R}^n$ such that $\text{int}(A) \neq \emptyset$ and $\text{int}(A) \cap B = \emptyset$, then there exists a hyperplane*

$$H_\alpha = \{\, x \in \mathbb{R}^n \, : \, \langle p, x \rangle = \alpha \,\}$$

that separates $\overline{A}$ and $\overline{B}$.

Proof. Since $\text{int}(A)$ is convex and has nonempty interior, and since $\text{int}(A) \cap B = \emptyset$, by the previous corollary, there exists a hyperplane

$$H_\alpha = \{\, x \in \mathbb{R}^n \, : \, \langle p, x \rangle = \alpha \,\}$$

that separates $\text{int}(A)$ and B. Thus,

$$\langle p, a \rangle \leq \alpha \leq \langle p, b \rangle$$

for all $a \in \text{int}(A)$ and all $b \in B$.

Let $a \in \overline{A}$, since $\overline{A} = \overline{\text{int}(A)}$, then $a \in \overline{\text{int}(A)}$. Hence, there exists a sequence of points $\{a_k\}$ in $\text{int}(A)$ such that $a_k \to a$ as $k \to \infty$, so that

$$\langle p, a_k \rangle \leq \alpha \leq \langle p, b \rangle$$

for all $b \in B$. The inner product is a continuous function, and letting $k \to \infty$, we obtain

$$\langle p, a \rangle \leq \alpha \leq \langle p, b \rangle$$

for all $a \in \overline{A}$ and $b \in B$.

Now, let $a \in \overline{A}$ and $b \in \overline{B}$, then there exists a sequence $\{b_k\}$ in B such that $b_k \to b$ as $k \to \infty$. Thus,

$$\langle p, a \rangle \leq \alpha \leq \langle p, b_k \rangle$$

for all $k \geq 1$, and letting $k \to \infty$, we have

$$\langle p, a \rangle \leq \alpha \leq \langle p, b \rangle$$

for all $a \in \overline{A}$ and $b \in \overline{B}$. Therefore, the hyperplane H_α separates $\overline{A}$ and $\overline{B}$.

Note that $\text{int}\left(\overline{A}\right) = \text{int}(A)$, so that if $a \in \text{int}\left(\overline{A}\right)$, then we must have $\langle p, a \rangle < \alpha$, since $\langle p, a \rangle = \alpha$ would imply that H_α contains an interior point of A, which is impossible.

□

Note. In the above corollary, if we assume that both A and B have nonempty interior and $\text{int}(A) \cap \text{int}(B) = \emptyset$, we can conclude that

$$\langle p, a \rangle < \alpha < \langle p, b \rangle,$$

for all $a \in \text{int}(A)$ and all $b \in \text{int}(B)$, that is, the hyperplane H_α *strictly separates* $\text{int}(A)$ and $\text{int}(B)$.

Example 3.6.13. *If we let*

$$A = \{ (x, y) \in \mathbb{R}^2 : x \leq 0, \ -1 < y \leq 1 \}$$

and

$$B = \{ (x, y) \in \mathbb{R}^2 : x = 0, \ -1 \leq y \leq 1 \},$$

then A and B are convex subsets of $\mathbb{R}^2$ with $\text{int}(A) \neq \emptyset$ and $\text{int}(A) \cap B = \emptyset$, so the hypotheses of the previous corollary hold, and the hyperplane

$$H_0 = \{ (x, y) \in \mathbb{R}^2 : x = 0, \ -\infty < y < \infty \}$$

separates $\overline{A}$ and $\overline{B}$ and hence A and B. Note that strict separation of A and B is not possible. Here, $\text{int}(B) = \emptyset$, while $\text{relint}(B) \neq \emptyset$.

The next example shows that it is possible to have two disjoint convex sets whose interiors are nonempty, which cannot be strictly separated by a hyperplane.

Example 3.6.14. *Let H be a hyperplane in $\mathbb{R}^n$ and let A and B be the open halfspaces determined by H. Let a and b be distinct points in H and let*

$$C_a = A \cup \{a\} \qquad and \qquad C_b = B \cup \{b\}.$$

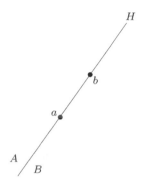

The sets C_a and C_b are disjoint convex sets with nonempty interiors, and H is the only hyperplane that separates them. However, they cannot be strictly separated.

3.6.1 Applications of the Separation Theorem

Theorem 3.6.15. *(Support Theorem)*

If $C \subset \mathbb{R}^n$ is a closed convex set with $\mathrm{int}(C) \neq \emptyset$ and $x_0 \in \mathrm{bdy}(C)$, then there exists a supporting hyperplane H for C at x_0.

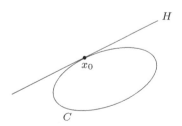

Proof. Let $A = C$ and $B = \{x_0\}$, then A and B are convex sets with $\mathrm{int}(A) \neq \emptyset$ and $\mathrm{int}(A) \cap B = \emptyset$, and by Corollary 3.6.12, there exists a hyperplane

$$H = \{\, x \in \mathbb{R}^n \,:\, \langle p, x \rangle = \alpha \,\},$$

where $p \neq \emptyset$, that separates C and x_0. Therefore, we may assume that $\langle p, x \rangle \leq \alpha$ for all $x \in C$ and $\langle p, x_0 \rangle \geq \alpha$.

Since $x_0 \in \text{bdy}(C)$, then every open ball centered at x_0 contains points in $\text{int}(C)$, so we can choose a sequence of interior points $\{x_k\}_{k \geq 1}$ of C such that $x_k \to x_0$, and so $\langle p, x_k \rangle \leq \alpha$ for all $k \geq 1$. Letting $k \to \infty$, we have $\langle p, x_0 \rangle \leq \alpha$.

Therefore, $\langle p, x \rangle \leq \alpha$ for all $x \in C$ and $\langle p, x_0 \rangle = \alpha$, and H is a hyperplane that supports C at x_0.

$\square$

Theorem 3.6.16. *If C is a subset of $\mathbb{R}^n$ which is contained in some halfspace, then $\overline{\text{conv}}(C)$ is the intersection of all closed halfspaces that contain C.*

Proof. Let A be the intersection of all closed halfspaces that contain C, so that $A \neq \emptyset$. We want to show that $\overline{\text{conv}}(C) = A$.

We already know that $\overline{\text{conv}}(C)$ is the intersection of all closed convex sets that contain C, so that $\overline{\text{conv}}(C) \subset A$.

Suppose that $x_0 \notin \overline{\text{conv}}(C)$, since $\{x_0\}$ is a compact convex set and $\overline{\text{conv}}(C)$ is a closed convex set, and $\{x_0\} \cap \overline{\text{conv}}(C) = \emptyset$, then the strong separation implies that $\{x_0\}$ and $\overline{\text{conv}}(C)$ can be strongly separated by some hyperplane H. Thus, H determines a closed halfspace that contains C but does not contain x_0, and therefore, $x_0 \notin A$. Thus, $A \subset \overline{\text{conv}}(C)$ and we have $A - \overline{\text{conv}}(C)$.

$\square$

Lemma 3.6.17. *A supporting hyperplane for the convex set $C \subset \mathbb{R}^n$ at a boundary point $x_0 \in C$ can contain no interior points of C.*

Proof. Let $x_0 \in C$ be a boundary point of the convex set C, and let

$$H = \{x \in \mathbb{R}^n : \langle p, x \rangle = \alpha\},$$

where $p \neq \overline{0}$, be a supporting hyperplane for C at x_0, such that $\langle p, x \rangle \leq \alpha$ for all $x \in C$ and $\langle p, x_0 \rangle = \alpha$.

If $x \in \text{int}(C)$, since $\text{int}(C) \subset C$, then $\langle p, x \rangle \leq \alpha$. Suppose that $\langle p, x \rangle = \alpha$, since $x \in \text{int}(C)$, then there exists an $\epsilon > 0$ such that $B(x, \epsilon) \subset C$.

Now, let

$$y = x + \eta \frac{p}{\|p\|^2} \quad \text{and} \quad z = x - \eta \frac{p}{\|p\|^2},$$

and choose η such that $0 < \eta < \epsilon \|p\|$, then $y, z \in B(x, \epsilon)$, so $y, z \in C$.

However,

$$\langle p, y \rangle = \langle p, x \rangle + \eta = \alpha + \eta > \alpha$$

and

$$\langle p, z \rangle = \langle p, x \rangle - \eta = \alpha - \eta < \alpha,$$

which contradicts the fact that $\langle p, u \rangle \leq \alpha$ for all $u \in C$. Therefore, $\langle p, x \rangle < \alpha$ for all $x \in \text{int}(C)$, and H does not contain any interior points of C.

$\square$

A halfspace S that contains a convex set C is said to be a ***supporting halfspace*** if the hyperplane defining S is a support hyperplane to C.

Theorem 3.6.18. *If C is a closed convex subset of $\mathbb{R}^n$, then C is the intersection of all the supporting halfspaces that contain C.*

Proof. Let B be the intersection of all supporting halfspaces of C, then clearly, $C \subset B$.

To complete the proof, we will show that no x in the complement of C can be in B. If $x \in \mathbb{R}^n \setminus C$, then there is a point $q \in C$ that is nearest to x, and letting $\delta = \|x - q\|$, we see that the interior of $\overline{B}(x, \delta)$ is disjoint from C. By the separation theorem, there is a hyperplane separating $\overline{B}(x, \delta)$ and C. Since the point q is common to both $\overline{B}(x, \delta)$ and C, it follows that the halfspace defined by the hyperplane that does not contain x is a supporting halfspace for C. If follows that x is not in the intersection of all supporting halfspaces for A, that is, $x \notin B$. Thus, we must have $C \subset B$.

$\square$

Note. If A is the open unit ball in $\mathbb{R}^n$, or if A is any open subset of $\mathbb{R}^n$, then A has no supporting hyperplanes, so the theorem above cannot be extended to arbitrary sets. That is, we cannot say that the closed convex hull of A is the intersection of all closed halfspaces that support A. In fact, we cannot even say this if the set A is closed (see the problems at the end of this section). It turns out, however, that the theorem can be extended to compact subsets of $\mathbb{R}^n$ (see Problems 3.6.2, #12).

We do have a partial converse to the previous theorem.

Theorem 3.6.19. *If the closed set $C \subset \mathbb{R}^n$ has nonempty interior, and if through each point of its boundary there passes a supporting hyperplane, then C is convex.*

Proof. Let B be the intersection of all closed halfspaces containing C that are determined by some supporting hyperplane to C. Clearly, B is a closed convex set containing C.

Suppose that $C \subsetneq B$, then there exists a point $x_0 \in B$ such that $x_0 \notin C$. Let y_0 be the closest point in C to x_0, so that $\|x_0 - y_0\|_2 = \text{dist}(x_0, C)$, then $y_0 \in \text{bdy}(C)$, and by hypothesis, there exists a supporting hyperplane to C, which passes through y_0.

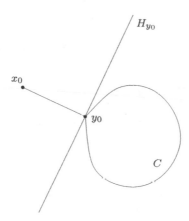

Now let H_{y_0} be the closed halfspace containing C, which is determined by this supporting hyperplane, then $x_0 \notin H_{y_0}$. However, this contradicts the fact that $x_0 \in B$, so that x_0 is in *every* closed halfspace containing C that is determined by some supporting hyperplane to C. Therefore, $C = B$ and C is convex.

$\square$

Note. We needed $\text{int}(C) \neq \emptyset$ to ensure that these supporting hyperplanes properly separated C and the support points in the boundary.

Example 3.6.20. *If K is a compact subset of $\mathbb{R}^n$, and H is any hyperplane in $\mathbb{R}^n$, then there exists a supporting hyperplane to K that is parallel to H.*

Solution. Suppose that $H = \{ x \in \mathbb{R}^n : \langle p, x \rangle = \alpha \}$, where $p \neq \overline{0}$, since K is compact, it is bounded, so there exists an $M > 0$ such that

$$\langle p, x \rangle \leq \|p\| \, \|x\| \leq \|p\| \, M$$

for all $x \in K$, and therefore,

$$\mu = \sup\{ \langle p, x \rangle : x \in K \} < \infty.$$

For each integer $k \geq 1$, the real number $\mu - \frac{1}{k}$ is no longer an upper bound for the set of real numbers $\{ \langle p, x \rangle : x \in K \}$, and so

$$F_k = \{ x \in \mathbb{R}^n : \langle p, x \rangle \geq \mu - \tfrac{1}{k} \} \cap K \neq \emptyset.$$

In fact, the sets $\{F_k\}_{k \geq 1}$ are compact and nested, and from the Nested Sets Theorem, we have

$$\bigcap_{k=1}^{\infty} F_k \neq \emptyset.$$

If x_0 is any point in this intersection, then $x_0 \in K$ so that $\langle p, x_0 \rangle \leq \mu$. Also, since

$$x_0 \in \{x \in \mathbb{R}^n : \langle p, x \rangle \geq \mu - \tfrac{1}{k}\} \cap K,$$

then $\langle p, x_0 \rangle \geq \mu - \tfrac{1}{k}$ for all $k \geq 1$, so that $\langle p, x_0 \rangle = \mu$.

Thus, $\langle p, x \rangle \leq \mu$ for all $x \in K$, and there exists a point $x_0 \in K$ such that $\langle p, x_0 \rangle = \mu$, so that

$$H_\mu = \{ x \in \mathbb{R}^n : \langle p, x \rangle = \mu \}$$

is a supporting hyperplane for the compact set K at x_0 and is parallel to H.

If we let

$$\nu = \inf\{ \langle p, x \rangle : x \in K \},$$

then a similar argument shows that the hyperplane H_ν supports K at a point x_1 where $\langle p, x_1 \rangle = \nu$ (see Example 2.6.15).

$\square$

3.6.2 Problems

1. Let A and B be bounded subsets of $\mathbb{R}^n$ and let $\lambda \geq 0$. If f is a linear functional on $\mathbb{R}^n$, show that

   (a) $\sup\{f(A+B)\} = \sup\{f(A)\} + \sup\{f(B)\}$.
   (b) $\sup\{f(\lambda A)\} = \lambda \sup\{f(A)\}$.

2. Let A and B be bounded subsets of $\mathbb{R}^n$ and let $\lambda \geq 0$. If f is a linear functional on $\mathbb{R}^n$, show that

   (a) $\inf\{f(A+B)\} = \inf\{f(A)\} + \inf\{f(B)\}$.
   (b) $\inf\{f(\lambda A)\} = \lambda \inf\{f(A)\}$.

3. If A is a bounded subset of $\mathbb{R}^n$ and f is a linear functional on $\mathbb{R}^n$, show that

   $$\sup\{f(A)\} = \sup\{f(\operatorname{conv}(A))\} \quad \text{and} \quad \inf\{f(A)\} = \inf\{f(\operatorname{conv}(A))\}.$$

4. Let f be a nonzero linear functional on $\mathbb{R}^n$, we define the **norm** of the linear functional f to be $\|f\| = \sup\{|f(x)| : x \in \overline{B}(\overline{0}, 1)\}$. Show that

   $$\inf\{f(\overline{B}(\overline{0}, 1))\} = -\|f\|.$$

5. Let f be a nonzero linear functional on $\mathbb{R}^n$ and let $x_0 \in \mathbb{R}^n$, if $\delta > 0$, show that

   (a) $\sup \left\{ f\left(\overline{B}(x_0, \delta)\right) \right\} = f(x_0) + \delta \|f\|$, and
   (b) $\inf \left\{ f\left(\overline{B}(x_0, \delta)\right) \right\} = f(x_0) - \delta \|f\|$.

6. Let f be a linear functional in $\mathbb{R}^n$ and let A be a bounded subset of $\mathbb{R}^n$. Show that if a_0 is an accumulation point of A, then

$$\inf\{f(A)\} \le f(a_0) \le \sup\{f(A)\}.$$

7. Show that the supremum and infimum of the linear functional f on $\mathbb{R}^n$ is the same for the open ball $B(\overline{0}, 1)$ as it is for the closed ball $\overline{B}(\overline{0}, 1)$, that is, show that

$$\sup \left\{ f\left(B(\overline{0}, 1)\right) \right\} = \|f\| \quad \text{and} \quad \inf \left\{ f\left(B(\overline{0}, 1)\right) \right\} = -\|f\|.$$

8. Show that if A is any subset of $\mathbb{R}^n$ such that

$$B(\overline{0}, 1) \subset A \subset \overline{B}(\overline{0}, 1),$$

   then

$$\sup\{f(A)\} = \|f\| \quad \text{and} \quad \inf\{f(A)\} = -\|f\|.$$

9. Let $A \subset \mathbb{R}^n$, and suppose that H is the hyperplane $f^{-1}(\beta)$. If $\sup\{f(A)\} = \alpha$ where $\alpha < \beta$, find $\text{dist}(A, H)$ in terms of α, β, and f.

10. Let $S = \{ (x, y) \in \mathbb{R}^2 : xy \ge 1, \ x > 0 \}$, and let $A = S \cup \{(0, 0)\}$. Show that $\text{conv}(A) \ne \overline{\text{conv}}(A)$.

11. A **convex body** in $\mathbb{R}^n$ is a compact convex set with nonempty interior. If A is a convex body, show that every boundary point of A is a support point for A.

12. If $A \subset \mathbb{R}^n$ is compact and f is a linear functional on $\mathbb{R}^n$, show that there is at least one point $a_0 \in A$ such that $f(a_0) = \sup\{f(A)\}$. Conclude that if A is compact, then the closed convex hull of A is the intersection of all closed halfspaces that support A.

13. If A and B are disjoint convex subsets of $\mathbb{R}^n$ such that $A \cup B = \mathbb{R}^n$ and A is closed, show that A is a halfspace.

14. Let A and B be compact subsets of $\mathbb{R}^n$, show that A and B can be strongly separated if and only if

$$\text{conv}(A) \cap \text{conv}(B) = \emptyset.$$

15. Let A be a bounded subset of $\mathbb{R}^n$.

   (a) Show that $\text{conv}\left(\overline{A}\right)$ is the intersection of all closed halfspaces containing A.
   (b) Show that part (a) is false if A is a proper subset of $\mathbb{R}^n$, which is not bounded.

16. Let A and B be convex subsets of $\mathbb{R}^n$. Show that A and B are strongly separated if and only if

$$\text{dist}(A, B) > 0.$$

17. Given a hyperplane $H = \{ x \in \mathbb{R}^n : \langle p, x \rangle = \alpha \}$, where $\|p\| = 1$, and two sets $A, B \subset \mathbb{R}^n$, the following are equivalent:

   (i) A and B are strongly separated by the hyperplane H,
   (ii) there exists an $\eta > 0$ such that

$$\langle p, a \rangle > \alpha + \eta \qquad \text{for all } a \in A$$

   and

$$\langle p, b \rangle < \alpha - \eta \qquad \text{for all } b \in B,$$

   (iii) there exists an $\epsilon > 0$ such that

$$\inf\{ \langle p, a \rangle : a \in A \} \geq \alpha + \epsilon$$

   and

$$\sup\{ \langle p, b \rangle : b \in B \} \leq \alpha - \epsilon.$$

18. If A and B are subsets of $\mathbb{R}^n$, then A and B are strongly separated if and only if there exists a $p \in \mathbb{R}^n$, with $p \neq \bar{0}$, such that

$$\inf\{ \langle p, a \rangle : a \in A \} > \sup\{ \langle p, b \rangle : b \in B \}.$$

19. Let C be a convex subset of $\mathbb{R}^n$, equipped with the norm $\| \cdot \|_2$, and let y_0 be a point such that $y_0 \notin C$.

   (a) Show that a point $x_0 \in C$ is a closest point in C to y_0 if and only if

$$\langle y_0 - x_0, x - x_0 \rangle \leq 0$$

   for all $x \in C$.
   (b) Show that if C is not convex, the above characterization of closest point need not hold.

20. Let A and B be disjoint open convex subsets of $\mathbb{R}^n$. Show that there exists a hyperplane H that strictly separates A and B.

21. Let A and B be nonempty subsets of $\mathbb{R}^n$, not necessarily convex.

   (a) Show that $A \cap B = \emptyset$ if and only if $\bar{0} \notin A - B$.
   (b) Show that A and B can be strongly separated if and only if $\bar{0}$ can be strongly separated from $A - B$.
   (c) Show that A and B can be separated if and only if $\bar{0}$ can be separated from $A - B$.

22. Suppose that A and B are convex subsets of $\mathbb{R}^n$ and that $\text{int}(A) = \text{int}(B) = \emptyset$. Show that if $\text{aff}(A \cup B) = \mathbb{R}^n$, while $\text{relint}(A) \cap \text{relint}(B) = \emptyset$, then either $A - B$ has nonempty interior or else A and B lie in distinct parallel hyperplanes.

3.7 EXTREME POINTS OF CONVEX SETS

3.7.1 Supporting Hyperplanes and Extreme Points

In the previous sections, since convex sets may not have smooth boundaries, we replaced the tangent planes by supporting hyperplanes. We recall some of the notions encountered earlier.

A point $x_0 \in \mathbb{R}^n$ is said to be a ***boundary point*** of a set A if and only if for every $\epsilon > 0$, the open ball $B(x_0, \epsilon)$ contains a point $x \in A$ and a point $y \notin A$, that is, if and only if given any $\epsilon > 0$, $B(x_0, \epsilon) \cap A \neq \emptyset$ and $B(x_0, \epsilon) \cap (\mathbb{R}^n \setminus A) \neq \emptyset$. Note that the definition allows us to have $x = x_0$ or $y = x_0$.

Also, if there exists an $\epsilon > 0$ such that the only point in $B(x_0, \epsilon)$ that is in A is x_0 itself, then we say that x_0 is an ***isolated point*** of A.

Example 3.7.1. *In $\mathbb{R}^2$, let*

$$A = \{\, x \in \mathbb{R}^2 : 0 < \|x\| < 1 \,\},$$

then $\overline{0}$ is a boundary point of A, and for each $0 < \epsilon < 1$, the only point in $B(\overline{0}, \epsilon)$ that is not in A is $\overline{0}$. The other boundary points of A are the points $x \in \mathbb{R}^2$ with $\|x\| = 1$.

A hyperplane $H = \{\, x \in \mathbb{R}^n : f(x) = \alpha \,\}$, where f is a nonzero linear functional on $\mathbb{R}^n$ and α is a scalar, is said to be a ***supporting hyperplane*** for a set A if and only if A is contained in a closed halfspace determined by H, say $f(x) \leq \alpha$ for all $x \in A$, and there is at least one point $x_0 \in A$ such that $f(x_0) = \alpha$. In this case, the hyperplane H is said to ***support A at x_0***.

Also, it is a ***nontrivial supporting hyperplane*** if and only if there exists a point $x_1 \in A$ such that $f(x_1) < \alpha$.

Example 3.7.2. *In $\mathbb{R}^3$, let $A = \{\, (x_1, x_2, x_3) \in \mathbb{R}^3 : 0 \leq x_1^2 + x_2^2 < 1, x_3 = 0 \,\}$. As a subset of $\mathbb{R}^3$, every point of A is a boundary point of A. The only supporting hyperplane at a point of A is the trivial one*

$$H = \{(x_1, x_2, x_3) \in \mathbb{R}^3 : x_3 = 0\}.$$

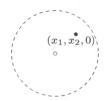

$(x_1, x_2, 0)$

Every point of the set $E = \{(x_1, x_2, x_3) : x_1^2 + x_2^2 = 1, x_3 = 0\}$ *is also a boundary point of A. At every point of E, there is a nontrivial supporting hyperplane for* $\overline{A}$ *with equation* $ax_1 + bx_2 = 1$ *for appropriate real numbers a and b.*

Theorem 3.7.3. *If K is a compact subset of* $\mathbb{R}^n$, *then for each nonzero linear functional* $f : \mathbb{R}^n \longrightarrow \mathbb{R}$, *there exists a scalar* α *such that*

$$H_\alpha = \{x \in \mathbb{R}^n : f(x) = \alpha\}$$

is a supporting hyperplane for K.

Proof. Since f is continuous on the compact set K, it attains its maximum at a point $x_0 \in K$, so that

$$f(x) \le f(x_0)$$

for all $x \in K$, and if we let $\alpha = f(x_0)$, then the hyperplane H_α supports K at x_0.

$\square$

Theorem 3.7.4. *If* $C \subset \mathbb{R}^n$ *is convex and* $x_0 \in C$ *is a boundary point of C, then there exists a supporting hyperplane H for C, which supports C at* x_0. *If* $\operatorname{int}(C) \ne \emptyset$, *the supporting hyperplane is nontrivial.*

Proof. From the support theorem, that is, Theorem 3.6.15, since C is convex, we have $\operatorname{int}(\overline{C}) = \operatorname{int}(C) \ne \emptyset$, so there exists a hyperplane H that supports $\overline{C}$ at x_0, and since $C \subset \overline{C}$ and $x_0 \in C \cap H$, then the hyperplane H supports C at x_0.

If $\operatorname{int}(C) \ne \emptyset$, then from Example 3.4.15, the hyperplane H can contain no interior points of C, and so the supporting hyperplane is nontrivial.

$\square$

Lemma 3.7.5. *If A is a subset of* $\mathbb{R}^n$, *then every point of A is either an interior point of A or a boundary point of A. Moreover, if A is a nonempty compact subset of* $\mathbb{R}^n$, *then* $\operatorname{bdy}(A) \ne \emptyset$.

Proof. Suppose that $x_0 \in A$ but $x_0 \notin \operatorname{int}(A)$, then given any $\epsilon > 0$, we have $B(x_0, \epsilon) \not\subset A$, so there exists a point $y \in B(x_0, \epsilon)$ such that $y \notin A$. Thus, the ball $B(x_0, \epsilon)$ contains points that are in A and points that are not in A. Therefore, every point of A is either an interior point of A or a boundary point of A.

Now suppose that A is a nonempty compact subset of $\mathbb{R}^n$. Since the norm function on $\mathbb{R}^n$ is continuous on the compact set A, it attains its maximum at some point $x_0 \in A$, so there exists an $M > 0$ such that $\|x\| \leq M$ for $x \in A$, and $\|x_0\| = M$.

The point x_0 is either an interior point or a boundary point of A. However, if $x_0 \in \text{int}(A)$, then there exists an $\epsilon > 0$ such that $B(x_0, \epsilon) \subset A$, and letting

$$z_0 = \frac{\epsilon}{2\|x_0\|} x_0,$$

we have

$$\|x_0 + z_0\| = \left(1 + \frac{\epsilon}{2\|x_0\|}\right) \|x_0\| = M + \frac{\epsilon}{2} > M$$

and $\|z_0\| = \frac{\epsilon}{2} < \epsilon$, so that $x_0 + z_0 \in B(x_0, \epsilon) \subset A$. However, this contradicts the fact that $\|x_0\| = M$ is the maximum over the set A.

Therefore, since $x_0 \notin \text{int}(A)$, from the first part of the lemma, we must have $x_0 \in \text{bdy}(A)$, so that $\text{bdy}(A) \neq \emptyset$.

$\square$

Finally, we have the following definition of an extreme point of a convex set.

If $C \subset \mathbb{R}^n$ is convex, a point $x_0 \in C$ is an **extreme point** of C if and only if x_0 is not an interior point of any closed interval $[u, v] \subset C$, that is, if and only if whenever

$$x_0 = (1 - \lambda)u + \lambda v$$

for $u, v \in C$ with $0 < \lambda < 1$, this implies that $u = v$. The set of all extreme points of a convex set C is denoted by $\text{ext}(C)$.

Theorem 3.7.6. *If C is a nonempty convex subset of $\mathbb{R}^n$, then x_0 is an extreme point of C if and only if $C \setminus \{x_0\}$ is convex.*

Proof. Let C be a nonempty convex subset of $\mathbb{R}^n$, and suppose that $x_0 \in \text{ext}(C)$. Let $x, y \in C \setminus \{x_0\}$ and $0 < \lambda < 1$, since C is convex, then $z = (1 - \lambda)x + \lambda y \in C$, and $(1 - \lambda)x + \lambda y \neq x_0$ since x_0 is an extreme point of C. Thus, $C \setminus \{x_0\}$ is convex.

Conversely, suppose that $C \setminus \{x_0\}$ is convex. If $x_0 = (1 - \lambda)u + \lambda v$ where $u, v \in C$ and $0 < \lambda < 1$, then both u and v cannot be in $C \setminus x_0$, since convexity of $C \setminus x_0$ would imply that $x_0 = (1 - \lambda)u + \lambda v \in C \setminus \{x_0\}$, which is a contradiction. Suppose $u \notin C \setminus \{x_0\}$, then this implies that $u = x_0$, so that

$$(1 - \lambda)x_0 + \lambda v = x_0,$$

that is, $\lambda(v - x_0) = \bar{0}$. Since $\lambda > 0$, this implies that $v = x_0$ also. Therefore, $x_0 \in \text{ext}(C)$.

$\square$

Example 3.7.7. *Let $C \subset \mathbb{R}^n$ be the closed ℓ_2 unit ball. Show that the set of extreme points of C is $S(\bar{0}, 1)$, the circumference of the ball.*

Solution. First, note that if $x \in C$ and $\|x\| < 1$, then x is an interior point of the line segment $\left[\bar{0}, x/\|x\|\right]$, so x is *not* an extreme point of C in this case. Thus, we have $\text{ext}(C) \subset S(\bar{0}, 1)$.

Next, suppose that $x \in S(\bar{0}, 1)$, so that $\|x\| = 1$, and that $x = (1 - \lambda)u + \lambda v$ where $u, v \in C$ and $0 < \lambda < 1$, then

$$1 = \|x\| \le (1 - \lambda)\|u\| + \lambda\|v\| \le (1 - \lambda) + \lambda = 1.$$

Thus, we must have $\|u\| = 1$ and $\|v\| = 1$, otherwise we would have $1 < 1$ in the inequality above. Now note that equality holds in the Cauchy–Schwarz inequality, since

$$\begin{aligned} 1 = \|x\|^2 &= (1 - \lambda)^2 \|u\|^2 + 2\lambda(1 - \lambda)\langle u, v \rangle + \lambda^2 \|v\|^2 \\ &\le (1 - \lambda)^2 \|u\|^2 + 2\lambda(1 - \lambda)\|u\|\,\|v\| + \lambda^2 \|v\|^2 \\ &= ((1 - \lambda)\|u\| + \lambda\|v\|)^2 \le 1. \end{aligned}$$

Therefore, $v = ku$ for some $k > 0$, and $x = (1 - \lambda + k\lambda)u$. Since $0 < \lambda < 1$, then

$$1 = \|x\| = (1 - \lambda + k\lambda)\,\|u\| = 1 - \lambda + k\lambda,$$

that is, $\lambda(k - 1) = 0$, and since $\lambda > 0$, then $k = 1$ and $u = v$. Therefore, x is an extreme point of C. Thus, we have $S(\bar{0}, 1) \subset \text{ext}(C)$, which completes the proof.

$\square$

The following theorem sometimes makes it easier to determine if a point is an extreme point of a convex set.

Theorem 3.7.8. *If $C \subset \mathbb{R}^n$ is convex, a point $x_0 \in C$ is an extreme point of C if and only if x_0 is not the midpoint of any line segment contained in C.*

Proof. If $x_0 \in \text{ext}(C)$, then x_0 cannot be the midpoint of any line segment contained in C.

Conversely, if $x_0 \notin \text{ext}(C)$, then $x_0 = (1 - \mu)u + \mu v$ for some $u, v \in C$ and $0 < \mu < 1$. Now choose a scalar k such that

$$0 < k < \frac{\min\{\|u - x_0\|, \|v - x_0\|\}}{\|v - u\|}$$

and let

$$u' = x_0 - \frac{k}{2}(v - u) \qquad \text{and} \qquad v' = x_0 + \frac{k}{2}(v - u),$$

then u' and v' are in the segment $[u, v] \subset C$, and $x_0 = \frac{1}{2}(u' + v')$ so that x_0 is the midpoint of a line segment contained in C.

$\square$

Also, it should be clear from Lemma 3.7.5 that the following is true.

Theorem 3.7.9. *If $C \subset \mathbb{R}^n$ is convex, with $\text{int}(C) \neq \emptyset$, and if $x_0 \in \text{int}(C)$, then x_0 is not an extreme point of C. Hence, $\text{ext}(C) \subset \text{bdy}(C)$.*

3.7.2 Existence of Extreme Points

Not every convex subset of $\mathbb{R}^n$ has an extreme point. For example, if $H \subset \mathbb{R}^n$ is a hyperplane, then $\text{ext}(H) = \emptyset$. The next theorem shows that this cannot happen for a compact convex subset of $\mathbb{R}^n$. First, a lemma.

Lemma 3.7.10. *If B and C are convex subsets of $\mathbb{R}^n$ and $C \subset B$, then any extreme point of B that is contained in C is also an extreme point of C.*

Proof. Let B and C be convex subsets of $\mathbb{R}^n$ with $C \subset B$ and let x_0 be an extreme point of B with $x_0 \in C$.

If x_0 is not an extreme point of C, then there exist points u and v in C with $u \neq x_0$ and $v \neq x_0$ such that

$$x_0 = \frac{1}{2}(u + v).$$

Since $u, v \in C$ and $C \subset B$, then $u, v \in B$, and x_0 is the midpoint of a line segment joining two points in B; however, this contradicts the fact that x_0 is an extreme point of B. Therefore, x_0 is an extreme point of C.

$\square$

Theorem 3.7.11. *If C is a nonempty compact convex subset of $\mathbb{R}^n$, then C has at least one extreme point.*

Proof. The notions of convex sets and extreme points of convex sets do not depend on the topology on $\mathbb{R}^n$; they depend only on the linear structure. Thus, in this proof, we assume that $\mathbb{R}^n$ is endowed with the Euclidean norm $\| \cdot \|$.

Let $f : \mathbb{R}^n \longrightarrow \mathbb{R}$ be defined by $f(x) = \|x\|$ for $x \in \mathbb{R}^n$, where $\|x\|$ is the Euclidean norm of x. If $x, y \in \mathbb{R}^n$, then from the triangle inequality, we have

$$|f(x) - f(y)| = \big|\|x\| - \|y\|\big| \leq \|x - y\|$$

for all $x, y \in \mathbb{R}^n$ and f is continuous on $\mathbb{R}^n$.

Since C is a nonempty compact subset of $\mathbb{R}^n$, then the continuous function f is bounded on C and attains its maximum on C at a point $x_0 \in C$.

Let $M = \max_{x \in C}\{f(x)\} = f(x_0) = \|x_0\|$, we will show that x_0 is an extreme point of the closed ball $B = \{x \in \mathbb{R}^n : \|x\| \leq M\}$.

Since $x_0 \in B$, if $x_0 = \dfrac{1}{2}(u + v)$ for some $u, v \in B$, then from the triangle inequality, we have

$$M = \|x_0\| \leq \frac{1}{2}\|u\| + \frac{1}{2}\|v\| \leq \frac{1}{2}M + \frac{1}{2}M = M.$$

If either $\|u\| < M$ or $\|v\| < M$, then the previous inequality implies that $M < M$, which is a contradiction, and therefore, $\|u\| = \|v\| = M$. Now, we have

$$\|u + v\| = 2\|x_0\| = \|u\| + \|v\|,$$

that is, equality holds in the triangle inequality. Therefore, $v = \lambda u$ for some $\lambda > 0$, and since $\|u\| = \|v\| = M$, we must have $\lambda = 1$, that is, $u = v$. Therefore, x_0 is an extreme point of B.

Since $x_0 \in C$ and $\|x\| \leq \|x_0\| = M$ for all $x \in C$, then $C \subset B$, and from the previous lemma, since x_0 is an extreme point of B, then x_0 is also an extreme point of C.

$$\square$$

Theorem 3.7.12. *If C is a nonempty compact convex subset of $\mathbb{R}^n$, then* $\text{ext}(C) \neq \emptyset$ *and every supporting hyperplane for C contains at least one extreme point of C.*

Proof. Let

$$H = \{ x \in \mathbb{R}^n : f(x) = \alpha \}$$

for some nonzero linear functional f and scalar α. Suppose that H is a supporting hyperplane for the set C. Since C is compact, then $C \cap H$ is a nonempty compact convex subset of $\mathbb{R}^n$, and by the previous theorem, $C \cap H$ has at least one extreme point x_0. It is easily seen that since H is a support functional, x_0 is an extreme point of $C \cap H$ if and only if x_0 is an extreme point of C.

$$\square$$

3.7.3 The Krein–Milman Theorem

Now we show that a compact convex subset of $\mathbb{R}^n$ is the closed convex hull of its set of extreme points.

Theorem 3.7.13. *(Krein–Milman Theorem)*
Let $C \subset \mathbb{R}^n$ *be a nonempty compact convex set, and let* $\text{ext}(C)$ *denote the set of extreme points of* C, *then*

$$C = \overline{\text{conv}}\,(\text{ext}(C)).$$

Proof. From the previous theorem, $\text{ext}(C) \neq \emptyset$, and since $\text{ext}(C) \subset C$ and C is convex, we have

$$\text{conv}\,(\text{ext}(C)) \subset \text{conv}(C) \subset C,$$

and since C is compact, we have

$$\overline{\text{conv}}\,(\text{ext}(C)) \subset \overline{C} = C.$$

Now we show that $C \subset \overline{\text{conv}}\,(\text{ext}(C))$. Suppose not, then there exists a point $x_0 \in C$ such that $x_0 \notin \overline{\text{conv}}\,(\text{ext}(C))$.

Since $\overline{\text{conv}}\,(\text{ext}(C))$ is a closed convex set, and $\{x_0\}$ is a compact convex set, then from the Strong Separation Theorem, there is a hyperplane

$$H = \{\, x \in \mathbb{R}^n \,:\, \langle p, x \rangle = \alpha \,\},$$

where $p \neq \overline{0}$, such that

$$\langle p, x \rangle < \alpha < \langle p, x_0 \rangle \tag{$*$}$$

for all $x \in \overline{\text{conv}}\,(\text{ext}(C))$, that is, x_0 and $\overline{\text{conv}}\,(\text{ext}(C))$ are strongly and, hence, strictly separated.

Now let $\beta = \max\{\langle p, x \rangle \,:\, x \in C\}$, since C is compact and the inner product is continuous, this maximum is attained at some point $z_1 \in C$, so that

$$H_1 = \{\, x \in \mathbb{R}^n \,:\, \langle p, x \rangle = \beta \,\}$$

is a supporting hyperplane to C at z_1. By Theorem 3.7.10, H_1 contains an extreme point x_1 of C, so that $x_1 \in \overline{\text{conv}}\,(\text{ext}(C)) \cap H_1$, and so

$$\beta = \langle p, x_1 \rangle < \alpha. \tag{$**$}$$

Since $x_0 \in C$, from $(*)$ and $(**)$, we have

$$\alpha < \langle p, x_0 \rangle \leq \beta,$$

which implies that $\beta < \alpha < \beta$, a contradiction. Therefore, $C \subset \overline{\text{conv}}\,(\text{ext}(C))$, and this completes the proof.

$\square$

Note. We have proved this theorem for $\mathbb{R}^n$ endowed with the Euclidean norm, and since all norms on $\mathbb{R}^n$ are equivalent, the theorem is true for any norm on $\mathbb{R}^n$.

Example 3.7.14. *Give an example of a nonempty compact convex subset of $\mathbb{R}^3$ whose set of extreme points is not closed.*

Solution. In $\mathbb{R}^3$, let

$$A = \{\, x \in \mathbb{R}^3 \,:\, (x_1 - 1)^2 + x_2^2 \leq 1,\ x_3 = 0 \,\}$$
$$B = \{\, x \in \mathbb{R}^3 \,:\, x_1 = 0,\ x_2 = 0,\ -1 \leq x_3 \leq 1 \,\}$$

as in the figure below.

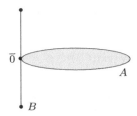

Let $C = \text{conv}(A \cup B)$.

Since A and B are compact subsets of $\mathbb{R}^3$, then $A \cup B$ is also compact, so that C is a compact convex set, and from the Krein–Milman theorem,

$$C = \overline{\text{conv}}\left(\text{ext}(C)\right).$$

Clearly, $\text{ext}(C)$ is a bounded set; however, it is not closed.

As is clear from the figure, we have

$$\text{ext}(C) = \{(0,0,1)\} \cup \{(0,0,-1)\}$$
$$\cup \{(x_1, x_2, x_3) \,:\, (1 - x_1)^2 + x_2^2 = 1,\ x_3 = 0,\ (x_1, x_2) \neq (0,0)\}.$$

The origin $\overline{0} = (0,0,0)$ is a limit point of $\text{ext}(C)$ but $\overline{0} \notin \text{ext}(C)$, so that $\text{ext}(C)$ is *not* closed.

$\square$

If $C \subset \mathbb{R}^n$ is a nonempty compact convex set, then C is bounded so that the set $\text{ext}(C)$ is also bounded. However, even if C is compact, the set of extreme points may not be closed, as in the above example. Thus, we can *not* conclude that $\text{conv}(\text{ext}(C))$ is compact.

However, in $\mathbb{R}^n$, the Krein–Milman theorem can be strengthened to the following.

Theorem 3.7.15. *(Krein–Milman Theorem)*

Let $C \subset \mathbb{R}^n$ be a nonempty compact convex set, and let $\text{ext}(C)$ denote the set of extreme points of C, then $C = \text{conv}\,(\text{ext}(C))$.

The above theorem can be proved using induction on $\dim(C)$, and the proof is omitted.

We also have a partial converse of the Krein–Milman theorem.

Theorem 3.7.16. *If C is a convex subset of $\mathbb{R}^n$ and A is a subset of $\mathbb{R}^n$ such that $\text{conv}(A) = C$, then $\text{ext}(C) \subset A$.*

Proof. Suppose that x_0 is an extreme point of C, which does not lie in A. From Theorem 3.7.6, the set $C \backslash \{x_0\}$ is a proper convex subset of C containing A, so it also contains $\text{conv}(A)$, that is, C. Thus, the assumption that C has an extreme point that does not lie in A leads to a contradiction, and hence, $\text{ext}(C) \subset A$.

$\square$

Note. If the set $A \subset \mathbb{R}^n$ in the previous theorem is compact, then $\text{conv}(A)$ is compact. Thus, if we write

$$C = \text{conv}(A),$$

then C is a compact convex set and $\text{ext}(C) \subset A$. In particular, this is true if the set A is finite, so that the set of extreme points of the compact convex set C is also finite.

3.7.4 Examples

In this section, we give some examples of well-known convex sets C and find the set of extreme points $\text{ext}(C)$.

Example 3.7.17. *Let $B \subset \mathbb{R}^2$ be the closed unit ℓ_1 ball. Show that*

$$\text{ext}(B) = \{\,(1,0),\,(0,1),\,(-1,0),\,(0,-1)\,\}.$$

Solution. Let $E = \{\,(1,0),\,(0,1),\,(-1,0),\,(0,-1)\,\}$. Suppose that $x = (x_1, x_2)$ is in $\text{bdy}(B)$ so that $|x_1| + |x_2| = 1$, but that $x \notin E$, then

$$0 < |x_1| < 1 \qquad \text{and} \qquad 0 < |x_2| < 1,$$

and x is in the interior of a line segment joining adjacent points in E, as in the following figure.

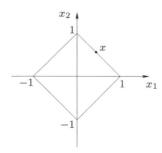

Thus, x is the midpoint of a line segment joining $y \in \mathrm{bdy}(B)$ and $z \in \mathrm{bdy}(B)$, and x is not an extreme point of B, so that $\mathrm{ext}(B) \subset E$.

Next, suppose that $x \in E$, for example, $x = (0,1)$, and suppose that

$$(0,1) = x = (1 - \lambda)y + \lambda z$$

for some $0 < \lambda < 1$, where $y = (y_1, y_2)$ and $z = (z_1, z_2)$ are in B, then

$$0 = (1 - \lambda)y_1 + \lambda z_1$$
$$1 = (1 - \lambda)y_2 + \lambda z_2.$$

From the second equation, the triangle inequality gives

$$\begin{aligned}
1 &= (1 - \lambda)y_2 + \lambda z_2 \\
&\leq (1 - \lambda)|y_2| + \lambda|z_2| \\
&\leq (1 - \lambda)(|y_1| + |y_2|) + \lambda(|z_1| + |z_2|) \\
&\leq (1 - \lambda) \cdot 1 + \lambda \cdot 1 \\
&= 1,
\end{aligned}$$

and if either $y_2 < 1$ or $z_2 < 1$, we get $1 < 1$, a contradiction. Therefore, $y_2 = z_2 = 1$.

Now we have

$$y = (y_1, 1) \qquad \text{and} \qquad z = (z_1, 1),$$

and since $|y_1| + 1 \leq 1$, this implies that $|y_1| \leq 0$, so that $y_1 = 0$. Similarly, $|z_1| + 1 \leq 1$ implies that $|z_1| \leq 0$, so that $z_1 = 0$.

Therefore, $x = y = z = (0,1)$, and $x = (0,1)$ is an extreme point of B. It is left as an exercise to show that the remaining points in E are extreme points of B, so that $E \subset \mathrm{ext}(B)$.

From the remark above, every point $x \in B$ can be written as a convex combination of the four points $(1,0)$, $(0,1)$, $(-1,0)$, and $(0,-1)$.

$\square$

Exercise 3.7.18. *Let $B \subset \mathbb{R}^2$ be the closed unit ℓ_∞ ball. Show that*

$$\text{ext}(B) = \{ (1,1),\ (1,-1),\ (-1,1),\ (-1,-1) \}.$$

Example 3.7.19. *In $\mathbb{R}^3$, let*

$$A = \{ (0,0,0),\ (1,0,0),\ (0,1,0),\ (0,0,1) \},$$

and let $x_0 = \left(\frac{1}{3}, \frac{1}{4}, \frac{1}{5}\right)$. Write x_0 as a convex combination of points from A. Explain why you need to use all of the points from A to accomplish this.

Solution.

(a) Note that $\text{conv}(A)$ is the intersection of the four closed halfspaces

$$H_1 = \{(x,y,z) \in \mathbb{R}^3 : x \geq 0\},$$
$$H_2 = \{(x,y,z) \in \mathbb{R}^3 : y \geq 0\},$$
$$H_3 = \{(x,y,z) \in \mathbb{R}^3 : z > 0\},$$
$$H_4 = \{(x,y,z) \in \mathbb{R}^3 : x+y+z \leq 1\},$$

and $x_0 = \left(\frac{1}{3}, \frac{1}{4}, \frac{1}{5}\right) \in \text{conv}(A)$, and therefore, x_0 is a convex combination of points from A.

(b) Note also that x_0 is not in the convex hull of any three of the points of A, since these sets consist of the sets given in the figure below.

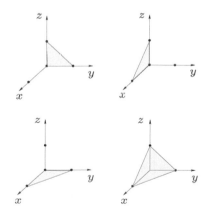

(c) We can write x_0 as a convex combination of all of the points of A as follows:

$$x_0 = \left(\tfrac{1}{3}, \tfrac{1}{4}, \tfrac{1}{5}\right) = \tfrac{13}{60}(0,0,0) + \tfrac{1}{3}(1,0,0) + \tfrac{1}{4}(0,1,0) + \tfrac{1}{5}(0,0,1).$$

□

3.7.5 Polyhedral Sets and Polytopes

A subset $P \subset \mathbb{R}^n$ is said to be a **polyhedral set** or a **polyhedron** if and only if it is the intersection of finitely many closed halfspaces.

Note. A polyhedron $P \subset \mathbb{R}^n$ is a closed convex subset of $\mathbb{R}^n$ and may be bounded or unbounded. If P is bounded, then P is a compact convex set, and by the Krein–Milman theorem, we have

$$P = \text{conv}(\text{ext}(P)).$$

Example 3.7.20. *Given the following linear inequalities:*

$$x + y \geq 10$$

$$4x - y \geq 0$$

$$y - x \geq 0,$$

sketch the polyhedral set in $\mathbb{R}^2$ determined by these inequalities and determine the extreme points of the polyhedron.

Solution. The bounding hyperplanes are

$$x + y = 10, \qquad 4x - y = 0, \qquad \text{and} \qquad x - y = 0,$$

and the region determined by the inequalities is the intersection of the three closed halfspaces (indicated by the arrows), as in the figure below.

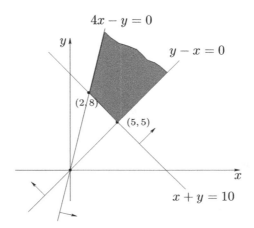

There are two extreme points:

(i) The intersection of the two bounding hyperplanes $x + y = 10$ and $4x - y = 0$, namely, $(2, 8)$.
(ii) The intersection of the two bounding hyperplanes $x + y = 10$ and $y - x = 0$, namely, $(5, 5)$.

In this example, the polyhedron P is unbounded.

□

Note. If $P = \mathrm{conv}\,(\{a_1, a_2, \ldots, a_k\})$ in $\mathbb{R}^n$, since the finite set $\{a_1, a_2, \ldots, a_k\}$ is compact, then P is compact, and $\mathrm{ext}(P) \subset \{a_1, a_2, \ldots, a_k\}$, that is, P has finitely many extreme points and each extreme point is one of the a_i's, where $1 \leq i \leq k$.

A **polytope** or **convex polytope** $P \subset \mathbb{R}^n$ is the convex hull of finitely many points.

Example 3.7.21. *Let P be the polyhedral set determined by the inequalities*

$$3x + 2y \geq 6$$
$$4x + \; y \leq 8$$
$$x \geq 0, \; y \geq 0.$$

Show that P is a polytope.

Solution. The bounding hyperplanes are $x = 0$, $4x + y = 8$, $3x + 2y = 6$, and the polyhedron P is shown in the figure below.

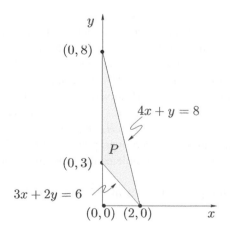

The extreme points of P are

$$u_1 = (2, 0), \qquad u_2 = (0, 3), \qquad u_3 = (0, 8).$$

In this example, the polyhedron P is bounded and hence compact; thus, from the Krein–Milman theorem,

$$P = \text{conv}(\{u_1, u_2, u_3\}).$$

Therefore, P is a polytope. $\qquad \square$

The result in the example above is true in general.

Theorem 3.7.22. *If $P \subset \mathbb{R}^n$ is a polyhedral set and P is bounded, then P is a polytope.*

Proof. Since P is bounded and closed, P is a convex compact set. Thus, to show that P is a polytope, from the Krein–Milman theorem, we only have to show that it has a finite number of extreme points.

We prove this by induction on the dimension n of the space.

For $n = 1$, P is either a single point or a closed line segment and so has only finitely many extreme points.

For the inductive step, assume that any bounded polyhedral set in an $(n-1)$-dimensional space has at most a finite number of extreme points. Now let $P \subset \mathbb{R}^n$ be a bounded polyhedral set, so that

$$P = \bigcap_{i=1}^{k} F_i,$$

where the F_i's are closed halfspaces, and let H_i be the hyperplane in $\mathbb{R}^n$ bounding the closed halfspace F_i for $1 \leq i \leq k$. If x_0 is an extreme point of P, then $x_0 \in \text{bdy}(P)$ so that $x_0 \in H_{i_0}$ for some $i_0 \in \{1, 2, \ldots, k\}$. Thus, x_0 is an extreme point of P which is in $P \cap H_{i_0}$, and by Lemma 3.7.10, x_0 is also an extreme point of $P \cap H_{i_0}$. However, $H_{i_0} \subset \mathbb{R}^n$ has codimension one, and by the inductive hypothesis, since $P \cap H_{i_0}$ is a bounded polyhedral set in an $(n-1)$-dimensional space, then $P \cap H_{i_0}$ has at most finitely many extreme points, and since the number of such hyperplanes is finite, P has only a finite number of extreme points. This completes the inductive proof.

Therefore, if P is any polyhedral set in $\mathbb{R}^n$ and P is bounded, then P is a polytope.

$\qquad \square$

The converse of the above theorem is also true; that is, every polytope P in $\mathbb{R}^n$ is a polyhedral set, which is bounded. However, before we prove this, we need to examine polytopes in a little more detail.

Recall that we defined a polytope in $\mathbb{R}^n$ to be the convex hull of a finite set of points, that is, $P \subset \mathbb{R}^n$ is a *polytope* if and only if

$$P = \text{conv}\left(\{x_1, x_2, \ldots, x_k\}\right)$$

for points $x_1, x_2, \ldots, x_k$ in $\mathbb{R}^n$.

The set $B = \{x_1, x_2, \ldots, x_k\}$ is a *minimal representation* of the polytope P if and only if

$$P = \text{conv}\left(\{x_1, x_2, \ldots, x_k\}\right)$$

and for each $i = 1, 2, \ldots, k$, no $x_i \in B$ is a convex combination of the other points from B.

Every polytope P has such a minimal representation, and this representation is unique because if P is written as the convex hull of a set of points B as above, and if some point x_i is a convex combination of the other points in B, then we can omit the point x_i from B without changing the convex hull. This process can be continued until the minimal representation is obtained.

Given a compact convex set $C \subset \mathbb{R}^n$, a set $F \subset C$ is called a *face* of C if there exists a supporting hyperplane H for C such that $F = C \cap H$.

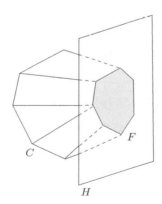

F is called a k-*face* of C if $\dim(F) = k$; while if P is a polytope, then P is called a k-*polytope* if $\dim(P) = k$.

Note. If C is a polytope in $\mathbb{R}^n$, then the faces of C are classified according to their dimension as follows:

- 0-faces of C are called **vertices** of C,
- 1-faces of C are called **edges** of C,
- $(n-1)$-faces are called **facets** of C.

Remark. From Theorem 3.7.6, it follows that the vertices of the polytope

$$P = \text{conv}(\{x_1, x_2, \ldots, x_k\})$$

are the extreme points of P if and only if $B = \{x_1, x_2, \ldots, x_k\}$ is the minimal representation of P. Otherwise, all we can say is that $\text{ext}(P) \subset B$.

A notion related to that of a vertex is an *exposed point*, and it may be considered as a generalization of a vertex for arbitrary convex sets.

A point p of a convex set $C \subset \mathbb{R}^n$ is said to be an **exposed point** of C if there is a hyperplane H that supports C at p and that misses $C \setminus \{p\}$.

For example, if C is the Euclidean closed unit ball in $\mathbb{R}^n$, then every boundary point of C is an exposed point of C. However, the same statement is not true for all of the boundary points of the ℓ_1 closed unit ball.

This is easily seen once we have the following lemma.

Lemma 3.7.23. *If C is a closed convex subset of $\mathbb{R}^n$ and p is an exposed point of C, then p is an extreme point of C.*

Proof. If H is a hyperplane that supports C at p, then there is a linear function f on $\mathbb{R}^n$ such that $H = f^{-1}(\alpha)$ with

$$C \subset \{x \in \mathbb{R}^n : f(x) \leq \alpha\}$$

and $f(p) = \alpha$. Since p is an exposed point of C, we have $f(x) < \alpha$ for all $x \in C \setminus \{p\}$.

Therefore, for any x and y in $C \setminus \{p\}$ and for any $\lambda \in [0, 1]$, we have

$$f((1 - \lambda)x + \lambda y) = (1 - \lambda)f(x) + \lambda f(y) < (1 - \lambda)\alpha + \lambda\alpha = \alpha,$$

so that p cannot equal $(1 - \lambda)x + \lambda y$. Thus, p is an extreme point of C.

$\square$

Note. If we denote the set of exposed points of a closed convex set C by $\exp(C)$, then the preceding lemma shows that $\exp(C) \subset \text{ext}(C)$ holds in general, not just for polytopes. However, as the next example shows, the reverse containment is not always true.

Example 3.7.24. *Give an example of a closed convex set $C \subset \mathbb{R}^2$ with an extreme point in C, which is not an exposed point of C.*

Solution. In $\mathbb{R}^2$, we let x_0 be a point external to a disk Γ and let the line segments $[x_0, a]$ and $[x_0, b]$ be the external tangents to the disk at a and b, respectively.

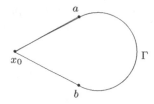

If we let

$$C = \text{conv}\left(\Gamma \cup [x_0, a] \cup [x_0, b]\right),$$

then a and b are extreme points of the convex set C but are *not* exposed points of C.

$\square$

For polytopes, we have the following.

Lemma 3.7.25. *If P is a polytope in $\mathbb{R}^n$, then a point x_0 is a vertex of P if and only if x_0 is an exposed point of P.*

Proof. By definition, a point x_0 is a vertex of the polytope P if and only if $\{x_0\}$ is a 0-face, so that x_0 is an exposed point of P.

Conversely, if x_0 is an exposed point of P, then there is a hyperplane H supporting P at precisely x_0, so that $\{x_0\}$ is a 0-face of P.

$\square$

Theorem 3.7.26. *Suppose that $P = \text{conv}(\{x_1, x_2, \ldots, x_k\})$ is a polytope in $\mathbb{R}^n$ and that the set $B = \{x_1, x_2, \ldots, x_k\}$ is the minimal representation of P, then the following statements are equivalent:*

(a) $x \in B$.
(b) x *is an extreme point of P.*
(c) x *is a vertex or exposed point of P.*

Proof. As we mentioned in the remark preceding the theorem, it follows from Theorem 3.7.6 that (a) and (b) are equivalent. We will show that (c) is equivalent to (a) and (b).

- (a) implies (c). Let $x \in B$ and let $Q = \mathrm{conv}(B \setminus \{x\})$, since B is a minimal representation of P, then $x \notin Q$, so that $\{x\}$ and Q are disjoint compact convex subsets of $\mathbb{R}^n$. From the Strong Separation Theorem, there is a hyperplane H' that strongly separates $\{x\}$ and Q. If we let H be the hyperplane parallel to H', which passes through x, then Q lies in one of the closed halfspaces H^+ bounded by H, so that $P \subset H^+$. Therefore, H supports P at x and because of the strong separation, $P \cap H = \{x\}$ and x is an exposed point or vertex of P.

- (c) implies (b). Suppose that $x_0 \in P$ is an exposed point of P, and let

$$H = \{x \in \mathbb{R}^n \ : \ f(x) = \alpha\}$$

be a supporting hyperplane for P at x_0, where f is a nonzero linear functional. Suppose that

$$f(x) \leq \alpha$$

for all $x \in P$ and $H \cap P = \{x_0\}$.

If there exist $x, y \in P$ and $0 < \lambda < 1$ such that

$$x_0 = (1 - \lambda)x + \lambda y,$$

then

$$\alpha = f(x_0) = (1 - \lambda)f(x) + \lambda f(y) \leq (1 - \lambda) \cdot \alpha + \lambda \cdot \alpha = \alpha.$$

If either $f(x) < \alpha$ or $f(y) < \alpha$, we get a contradiction in the above inequality. Therefore,

$$f(x) = \alpha \qquad \text{and} \qquad f(y) = \alpha,$$

that is, $x \in H \cap P$ and $y \in H \cap P$, so that $x = y = x_0$, and x_0 is an extreme point of P.

$$\square$$

Theorem 3.7.27. *If $P \subset \mathbb{R}^n$ is a polytope, then each face of P is also a polytope, and there are only a finite number of distinct faces.*

Proof. Let $P = \mathrm{conv}(\{x_1, x_2, \ldots, x_k\})$ be a polytope, where

$$B = \{x_1, x_2, \ldots, x_k\}$$

is the minimal representation of P, and let $F = P \cap H_\alpha$ be a face of P, where the supporting hyperplane is

$$H_\alpha = \{x \in \mathbb{R}^n \ : \ f(x) = \alpha\},$$

where f is a nonzero linear functional. We may assume without loss of generality that $\{x_1, x_2, \ldots, x_r\} \subset H_\alpha$ and that $f(x_i) > \alpha$ for $i = r+1, \ldots, k$.

Thus,
$$f(x_i) = \alpha$$
for $i = 1, 2, \ldots, r$ and
$$f(x_i) = \alpha + \epsilon_i$$
for some $\epsilon_i > 0$ when $i = r + 1, \ldots, k$.

Now let $x \in P$ be arbitrary, so that $x = \sum_{i=1}^{k} \lambda_i x_i$, where $\lambda_i \geq 0$ and $\sum_{i=1}^{k} \lambda_i = 1$.
From the above, we have

$$f(x) = \sum_{i=1}^{r} \lambda_i \alpha + \sum_{i=r+1}^{k} \lambda_i(\alpha + \epsilon_i)$$

$$= \alpha \sum_{i=1}^{k} \lambda_i + \sum_{i=r+1}^{k} \lambda_i \epsilon_i$$

$$= \alpha + \sum_{i=r+1}^{k} \lambda_i \epsilon_i,$$

and the point x is in H_α if and only if $f(x) = \alpha$, that is, if and only if $\sum_{i=r+1}^{k} \lambda_i \epsilon_i = 0$, and this is true if and only if $\lambda_i = 0$ for $i = r+1, \ldots, k$.
Thus, $x \in P \cap H_\alpha$ if and only if x is a convex combination of $\{x_1, \ldots, x_r\}$, that is,
$$P \cap H_\alpha = \text{conv}(\{x_1, \ldots, x_r\}),$$
and therefore, $F = P \cap H_\alpha$ is a polytope.

We see that P has only a finite number of distinct faces, since $B = \{x_1, \ldots, x_k\}$ has only a finite number of subsets, and each face of P corresponds to one of these subsets.

$\square$

And now the converse to Theorem 3.7.22.

Theorem 3.7.28. *If $P \subset \mathbb{R}^n$ is a polytope, then P is a bounded polyhedral set.*

Proof. Let $P = \text{conv}(\{x_1, x_2, \ldots, x_k\})$ where $B = \{x_1, x_2, \ldots, x_k\}$ is the minimal representation of P. We may assume without loss of generality that $\text{int}(P) \neq \emptyset$, that is, P is n-dimensional. Let $F_1, F_2, \ldots, F_m$ be the facets of P, and for each F_i, let H_i and H_i^+ denote the supporting hyperplane and closed supporting halfspace, respectively, such that $F_i = P \cap H_i$ and $P \subset H_i^+$. We will show that
$$P = H_1^+ \cap H_2^+ \cap \cdots \cap H_m^+.$$

Suppose that there exists a point $x_0 \in H_1^+ \cap H_2^+ \cap \cdots \cap H_m^+$ but $x_0 \notin P$. We let

$$D = \bigcup \{ \text{aff} \left(\{x_0\} \cup A \right) \, : \, A \subset B, \, |A| \leq n - 1 \},$$

where the union is over all sets A of $n - 1$ or fewer points of $B = \{x_1, x_2, \ldots, x_k\}$. Thus, D is a finite union of flats of dimension at most $n - 1$, and since $\dim(P) = n$, Problem 2.6.5 in Section 2.6.1 implies that $\text{int}(P) \not\subset D$.

Therefore, there exists a point $y_0 \in \text{int}(P) \setminus D$, as in the figure below.

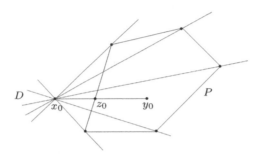

Since $y_0 \in \text{int}(P)$ and $x_0 \notin P$, there exists a point z_0 in the interior of the segment $[x_0, y_0]$ such that $z_0 \in \text{bdy}(P)$. We will show that z_0 belongs to a facet of P and does not belong to any face of P of lower dimension.

Suppose that z_0 belongs to a j-face of P, where $0 \leq j \leq n - 2$, then from Carathéodory's theorem, $z_0 \in \text{conv}(A_0)$, where A_0 is some subset of $n - 1$ or fewer points of $\{x_1, \ldots, x_m\}$. This implies that $z_0 \in D$, and since $x_0 \in D$, then the entire line through x_0 and z_0 lies in D, but this is a contradiction since $y_0 \notin D$.

Therefore, z_0 belongs to a facet F_j of P, which implies that $z_0 \in H_j$, and since $y_0 \in \text{int}(P) \subset H_j^+$, then $x_0 \notin H_j^+$; this contradicts our initial assumption that $x_0 \in H_1^+ \cap H_2^+ \cap \cdots \cap H_m^+$. Therefore,

$$H_1^+ \cap H_2^+ \cap \cdots \cap H_m^+ \subset P.$$

The reverse inclusion follows from the fact that $P \subset H_i^+$ for $i = 1, 2, \ldots, m$, so that P is a polyhedral set. Since P is a polytope, then it is compact and hence bounded. This completes the proof.

$\square$

Combining these results, we have the following characterization of polytopes in $\mathbb{R}^n$.

Theorem 3.7.29. *A set $P \subset \mathbb{R}^n$ is a polytope if and only if it is a bounded polyhedral set.*

Finally, we have a result that is important in linear programming, one that makes the simplex method such a powerful tool.

Theorem 3.7.30. *If P is a polyhedral subset of $\mathbb{R}^n$ that contains no lines and if f is a linear functional that is bounded above on P, then there exists an extreme point $x_0 \in P$ such that*

$$f(x_0) = \sup\{f(x) : x \in P\},$$

that is, f attains its maximum on P at an extreme point of P.

Proof. We may assume without loss of generality that $\dim(P) = n$, since otherwise we can work in a space of lower dimension, namely, one with dimension equal to $\dim(\mathrm{aff}(P))$. The proof is by induction on n.

If $n = 0$, then $\mathbb{R}^0 = \{\overline{0}\}$, and so P consists of the singleton $\{\overline{0}\}$, and clearly the theorem is true in this case.

For the inductive step, we assume that the theorem is true for all linear spaces of dimension less than n. Since P is closed, then $\mathrm{bdy}(P) \subset P$, and since f is bounded above on P, then

$$\alpha = \sup\{f(x) : x \in P\} < \infty.$$

Now, since f is linear, then

$$\alpha = \sup\{f(x) : x \in \mathrm{bdy}(P)\}.$$

The boundary of the polyhedral set P can be written as

$$\mathrm{bdy}(P) = (P \cap H_1) \cup (P \cap H_2) \cup \cdots \cup (P \cap H_m),$$

where $H_1, \ldots, H_m$ are the hyperplanes that bound the halfspaces whose intersection is P. Therefore,

$$\alpha = \sup\{f(x) : x \in P \cap H_i\}$$

for some i. The set $P \cap H_i$ is also polyhedral and since P contains no lines, then $P \cap H_i$ contains no lines. Since $\dim(P \cap H_i) < n$, the inductive hypothesis implies that $\alpha = f(x_0)$ for some extreme point $x_0 \in P \cap H_i$. However, any extreme point of $P \cap H_i$ is also an extreme point of P. This completes the proof by induction.

□

A similar result holds for the minimum of a linear function that is bounded below on a polyhedral set that contains no lines.

3.7.6 Birkhoff's Theorem

An $n \times n$ matrix is called a ***permutation*** matrix if there is exactly one 1 in each row and column, and all other entries are 0.

An $n \times n$ matrix $A = (a_{ij})$ is called ***doubly stochastic*** if $a_{ij} \geq 0$ for all $1 \leq i, j \leq n$, and the sum of the entries in any row or column is 1.

The next theorem, due to Dénes König and Garrett Birkhoff, shows that the set of doubly stochastic matrices is the convex hull of its extreme points.

Theorem 3.7.31. *(Birkhoff's Theorem)*
If Γ_n denotes the set of all doubly stochastic $n \times n$ matrices, then Γ_n is convex and $\text{ext}(\Gamma_n)$ is the set of all $n \times n$ permutation matrices.

Proof. If $A = (a_{ij})$ and $B = (b_{ij})$ are doubly stochastic $n \times n$ matrices and $0 \leq \lambda \leq 1$, then

$$\lambda A + (1 - \lambda)B = \big(\lambda a_{ij} + (1 - \lambda)b_{ij}\big)$$

and $\lambda a_{ij} + (1 - \lambda)b_{ij} \geq 0$ since $a_{ij} \geq 0$ and $b_{ij} \geq 0$.

Also,

$$\sum_{i=1}^{n} \big(\lambda a_{ij} + (1 - \lambda)b_{ij}\big) = \lambda \sum_{i=1}^{n} a_{ij} + (1 - \lambda) \sum_{i=1}^{n} b_{ij} = \lambda + (1 - \lambda) = 1$$

for $j = 1, \ldots, n$.

Similarly,

$$\sum_{j=1}^{n} \big(\lambda a_{ij} + (1 - \lambda)b_{ij}\big) = \lambda \sum_{j=1}^{n} a_{ij} + (1 - \lambda) \sum_{j=1}^{n} b_{ij} = \lambda + (1 - \lambda) = 1$$

for $i = 1, \ldots, n$.

Therefore, $\lambda A + (1 - \lambda)B$ is also doubly stochastic, and the set of all doubly stochastic $n \times n$ matrices is convex.

Let P be an $n \times n$ permutation matrix and suppose that $P = \frac{1}{2}(A + B)$ where A and B are in Γ_n.

For any entry $p_{ij} = 0$ in P, we have

$$\tfrac{1}{2}(a_{ij} + b_{ij}) = 0$$

with $a_{ij} \geq 0$ and $b_{ij} \geq 0$, so that $a_{ij} = 0$ and $b_{ij} = 0$.

For any entry $p_{ij} = 1$ in P, we have

$$\tfrac{1}{2}(a_{ij} + b_{ij}) = 1,$$

so if either $a_{ij} < 1$ or $b_{ij} < 1$, we would have $2 = a_{ij} + b_{ij} < 1 + 1 = 2$, which is a contradiction. Thus, $a_{ij} = 1$ and $b_{ij} = 1$, so that $P = A = B$.

Therefore, P is not the midpoint of a line segment whose endpoints are in Γ_n, and so P is an extreme point of Γ_n.

Now, we have to show that if $X \in \Gamma_n$ is an extreme point of Γ_n, then X is a permutation matrix. We will use the proof in the book *Linear Algebra and Its Applications*, by Peter D. Lax.

Let $X \in \Gamma_n$ and suppose there exists an entry in $X = (x_{ij})$ such that

$$0 < x_{i_0 j_0} < 1, \tag{00}$$

we will show that this implies that X is not an extreme point of Γ_n.

We construct a sequence of entries, all of which lie between 0 and 1 and which lie alternately on the same row or on the same column.

- Choose a column j_1 such that

$$0 < x_{i_0 j_1} < 1, \tag{01}$$

this is possible since the sum of the elements in the i_0th row must be 1 and since (00) holds.

- Choose a row i_1 such that

$$0 < x_{i_1 j_1} < 1, \tag{11}$$

this is possible since the sum of the elements in the j_1st column must be 1 and since (01) holds.

- Choose a column j_2 such that

$$0 < x_{i_1 j_2} < 1, \tag{12}$$

this is possible since the sum of the elements in the i_1st row must be 1 and since (11) holds.

- Choose a row i_2 such that

$$0 < x_{i_2 j_2} < 1, \tag{22}$$

this is possible since the sum of the elements in the j_2st column must be 1 and since (12) holds.

$$\vdots$$

Continuing in this manner, since the number of rows and columns is finite, we eventually come back to a position (i_m, j_m) in the matrix we have visited earlier.

Therefore, we have constructed a closed chain of entries

$$x_{i_k j_k} \longrightarrow x_{i_k j_{k+1}} \longrightarrow \quad \cdots \quad \longrightarrow x_{i_m j_m} = x_{i_k j_k}.$$

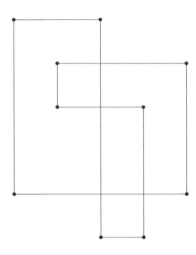

Now define a matrix N as follows:

(a) The entries of N are zero except for those points that lie on the chain.
(b) The entries of N at points of the chain are $+1$ and -1, in succession.

The matrix N has the following property:

(c) The row sums and the column sums of N are zero.

Now define two matrices A and B by

$$A = X + \varepsilon N \qquad \text{and} \qquad B = X - \varepsilon N,$$

where $\varepsilon > 0$ is to be determined.

From (c), it follows that the row sums and the column sums of both A and B are 1, while from the construction of the chain and from (a), it follows that the elements of X are positive at all points where N has a nonzero entry.

Therefore, we may choose $\varepsilon > 0$ so small that both A and B have nonnegative entries, so that both A and B are doubly stochastic, that is, $A, B \in \Gamma_n$.

However,

$$X = \frac{A+B}{2},$$

so that X is not an extreme point of Γ_n.

Thus, if X is an extreme point of Γ_n, then the entries of X are either 0 or 1, and since X is doubly stochastic, then each row and each column of X has exactly one 1, that is, X is a permutation matrix.

$\square$

Note: If we identify the linear space of all real $n \times n$ matrices with $\mathbb{R}^{n^2}$ endowed with the Euclidean norm, then Γ_n is a closed and bounded subset of $\mathbb{R}^{n^2}$, and from the Krein–Milman theorem, each $X \in \Gamma_n$ can be written as a convex combination of permutation matrices. However, this representation is not unique unless $n = 2$. The number of $n \times n$ permutation matrices is $n!$, which is a huge number for large n. We will show that every doubly stochastic matrix can be written as a convex combination of $n^2 - 2n + 2$ or fewer permutation matrices.

Recall that a set H in $\mathbb{R}^n$ is a hyperplane if and only if it is a translate of an $(n-1)$-dimensional subspace, that is, if and only if it is an $(n-1)$-dimensional flat. Thus, H is a hyperplane if and only if there exist scalars $\alpha_0, \alpha_1, \ldots, \alpha_n$, where not all of $\alpha_1, \ldots, \alpha_n$ are zero, such that

$$H = \{ (x_1, \ldots, x_n) \in \mathbb{R}^n : \alpha_0 + \alpha_1 x_1 + \cdots + \alpha_n x_n = 0 \}.$$

Lemma 3.7.32. *In $\mathbb{R}^n$, if F is a flat with $\dim(F) = k$, where $1 \leq k \leq n-1$, then F is the intersection of $n - k$ hyperplanes; thus, F is the solution set of some system of $n - k$ linear equations.*

Proof. Let $B_k = \{a_0, a_1, \ldots, a_k\}$ be an affine basis for F and extend this to an affine basis $B_n = \{a_0, a_1, \ldots, a_k, a_{k+1}, \ldots, a_n\}$ for $\mathbb{R}^n$. If $x \in \mathbb{R}^n$, then x can be written uniquely as

$$x = \lambda_0 a_0 + \lambda_1 a_1 + \cdots + \lambda_n a_n,$$

where $a_0 + a_1 + \cdots + a_n = 1$. Therefore, F is the intersection of the $n - k$ hyperplanes with equations $\lambda_{k+1} = 0, \ldots, \lambda_n = 0$.

The flat F is the subset of $\mathbb{R}^n$ consisting of all those x's whose barycentric coordinates $\lambda_{k+1}, \ldots, \lambda_n$ are all zero. Since each of the sets

$$\{x \in \mathbb{R}^n : \lambda_i = 0\}$$

is the hyperplane $H_i = \text{aff}\{ a_0, \ldots, a_{i-1}, a_{i+1}, \ldots, a_n \}$, then $F = \bigcap_{i=1}^{k} H_i$.

$\square$

Corollary 3.7.33. *Every doubly stochastic $n \times n$ matrix can be written as a convex combination of $n^2 - 2n + 2$ or fewer permutation matrices.*

Proof. The set of all doubly stochastic matrices Γ_n consists of all those points $x = (x_{ij}) \in \mathbb{R}^{n^2}$ satisfying the relations

$$x_{ij} \geq 0 \qquad \text{for} \quad i, j = 1, 2, \ldots, n; \qquad (*)$$

$$\sum_{j=1}^{n} x_{ij} = 1 \qquad \text{for} \quad i = 1, 2, \ldots, n; \qquad (**)$$

$$\sum_{i=1}^{n} x_{ij} = 1 \qquad \text{for} \quad j = 1, 2, \ldots, n - 1. \qquad (***)$$

Note that the linear equation

$$x_{1n} + x_{2n} + \cdots + x_{nn} = 1$$

follows directly from the $2n - 1$ equations $(**)$ and $(***)$; hence, each $x \in \Gamma_n$ is in the intersection of only $2n - 1$ hyperplanes. From the lemma, $\mathrm{aff}(\Gamma_n)$ has dimension $n^2 - (2n - 1) = n^2 - 2n + 1$. From Birkhoff's theorem and Carathéodory's theorem, every doubly stochastic $n \times n$ matrix can be written as a convex combination of at most $n^2 - 2n + 2$ permutation matrices.

$\square$

3.7.7 Problems

1. Let $a, b \in \mathbb{R}^n$, with $a \neq b$. Show that the line segment $[a, b]$ contains no interior points, but it does contain relative interior points. Show that a and b are the only extreme points of $[a, b]$.

2. Let S be the unit sphere in the ℓ_1 norm for $\mathbb{R}^n$.

    (a) Using the Euclidean norm, find the closest point x_0 in S to $\overline{0}$. Is x_0 an extreme point of S?
    (b) What is the radius of the largest closed Euclidean ball that is contained within S? Does it contain any extreme points of S?

3. Let $C = A \cup B$, where $A = [(-1, 0), (1, 0)]$ and $B = [(0, -1), (0, 1)]$ are closed intervals in $\mathbb{R}^2$.

    (a) Find the extreme points of A and B.
    (b) Are these also extreme points of C?
    (c) Find all the extreme points of C.

4. Give an example in $\mathbb{R}^2$ of an exposed point a of a compact convex set A such that there is *no* point b for which a is a farthest point of A from b.

5. If $C \subset \mathbb{R}^2$ is a closed convex set, show that $\text{ext}(C)$ is closed.

6. If $A \subset \mathbb{R}^n$, show that $\text{ext}(\text{conv}(A)) \subset A$.

7. If $A \subset \mathbb{R}^n$, show that the set of extreme points of $\text{conv}(A)$ are precisely those points $a \in A$ such that $a \notin \text{conv}(A \setminus \{a\})$.

8. Let H be an open halfspace in $\mathbb{R}^n$, and let a and b be distinct points in $\text{bdy}(H)$.

   (a) Show that $\text{ext}(H) = \emptyset$.

   (b) Show that $H \cup \{a\}$ is convex and find $\text{ext}((H \cup \{a\})$.

   (c) Show that $H \cup [a, b]$ is convex and find $\text{ext}(H \cup [a, b])$.

9. Let x_0 be a fixed point in $\mathbb{R}^n$ and let $\delta > 0$. Let $C(x_0, \delta)$ denote the closed cube with center at x_0 and length of a side equal to 2δ, that is,

$$C(x_0, \delta) = \{ x \in \mathbb{R}^n : \|x - x_0\|_\infty \le \delta \}.$$

The vertices of the cube are the 2^n points of the form

$$x_0 + \delta(\epsilon_1, \epsilon_2, \ldots, \epsilon_n)$$

where the ϵ_i take on only the values ± 1.

   (a) Show that the set of 2^n vertices is the set of extreme points of $C(x_0, \delta)$.

   (b) Show that each extreme point is actually a "vertex," that is, an exposed point of $C(x_0, \delta)$.

   (c) Show that $C(x_0, \delta) = \text{conv}(\exp(C(x_0, \delta)))$.

10. Let C be a compact convex subset of $\mathbb{R}^n$, and let x_0 be a farthest point in C from some point y_0. Show that x_0 is an exposed point of C.

11. Show that every compact convex subset of $\mathbb{R}^n$ has an exposed point in every open halfspace which meets it.

12. Prove the following theorem of Straszewicz: Every compact convex subset of $\mathbb{R}^n$ is the closed convex hull of its exposed points, that is, if $C \subset \mathbb{R}^n$ is a compact convex set, then $C = \overline{\text{conv}}(\exp(C))$.

13. Prove that each extreme point of a closed convex set in $\mathbb{R}^n$ is the limit of some sequence of exposed points of the set.

14. Show that the compactness condition is needed in the Krein–Milman theorem by finding

   (a) a closed convex set in $\mathbb{R}^2$ with no extreme points,

   (b) a bounded convex set in $\mathbb{R}^2$ with no extreme points,

   (c) a closed and bounded convex set in

$$C_0(\mathbb{R}) = \{ f : \mathbb{R} \longrightarrow \mathbb{R} : f \text{ is continuous}, f(x) \to 0 \text{ as } x \to \pm\infty \}$$

   with $\|f\| = \max |f(x)|$, which has no extreme points.

15. Show that if $C \subset \mathbb{R}^n$ is a compact convex set with $\text{int}(C) \ne \emptyset$, then each $x_0 \in \text{int}(C)$ can be expressed as a convex combination of $n + 1$ or fewer exposed points of C.

16. Let A and B be polytopes in $\mathbb{R}^n$, and let $\alpha \in \mathbb{R}$. Show that $A + B$ and αA are polytopes.

17. Let $A_1, A_2, \ldots, A_k$ be polytopes in $\mathbb{R}^n$, and let $\lambda_1, \lambda_2, \ldots, \lambda_k \in \mathbb{R}$. Show that $\lambda_1 A_1 + \lambda_2 A_2 + \cdots + \lambda_k A_k$ is a polytope, called a **zonotope**.

18. If A and B are subsets of $\mathbb{R}^n$ and $C = A + B$, and if $x_0 \in \text{ext}(C)$, show that there is just one point $a_0 \in A$ and one point $b_0 \in B$ such that $x_0 = a_0 + b_0$ with $a_0 \in \text{ext}(A)$ and $b_0 \in \text{ext}(B)$.

19. Show that the n-cube C in $\mathbb{R}^n$ with side length 1, that is,

$$C = \{(x_1, x_2, \ldots, x_n) \in \mathbb{R}^n : 0 \le x_i \le 1, \ i = 1, 2, \ldots, n\}$$

is the polytope

$$P = \text{conv}(\{\overline{0}, e_1\}) + \text{conv}(\{\overline{0}, e_2\}) + \cdots + \text{conv}(\{\overline{0}, e_n\}),$$

where $\{e_1, e_2, \ldots, e_n\}$ is the standard orthonormal basis for $\mathbb{R}^n$.

20. If the hyperplane H supports the convex set C at x_0, then x_0 is an extreme point of C if and only if x_0 is an extreme point of $H \cap C$.

21. Show that if P_1 and P_2 are polytopes in $\mathbb{R}^n$, then

$$P_1 + P_2 = \text{conv}(\text{ext}(P_1)) + \text{conv}(\text{ext}(P_2)).$$

What are the extreme points of $P_1 + P_2$?

4 HELLY'S THEOREM

Note: **Throughout this chapter, we assume that the linear space $\mathbb{R}^n$ is equipped with the Euclidean norm.**

4.1 FINITE INTERSECTION PROPERTY

4.1.1 The Finite Intersection Property

We say that a family $\mathcal{F}$ of subsets of $\mathbb{R}^n$ has the *finite intersection property* if every finite subfamily of $\mathcal{F}$ has a nonempty intersection.

Example 4.1.1. *Let $\mathcal{F} = \{\, F_1,\, F_2,\, \ldots \,\}$ be a nested family, that is, $F_{i+1} \subset F_i$ for $i = 1, 2, \ldots$, then $\mathcal{F}$ has the finite intersection property.*

Example 4.1.2. *Let $\mathcal{F}$ be a family of closed balls $\overline{B}(x, 1)$ in $\mathbb{R}^n$, for an infinite set of points $x \in \mathbb{R}^n$ with $\|x\| = 1$. The family $\mathcal{F}$ is not a nested family, but it does have the finite intersection property.*

Solution. Note that the origin is common to each of the closed balls $\overline{B}(x, 1)$ where $\|x\| = 1$. □

Compactness can be reformulated using the finite intersection property. We have the following theorem.

Theorem 4.1.3. *Let K be a subset of $\mathbb{R}^n$. The following statements are equivalent.*

(i) *The set K is compact.*

(ii) *Whenever $\mathcal{F}$ is a family of closed subsets of K with the finite intersection property, $\mathcal{F}$ has a nonempty intersection.*

Geometry of Convex Sets, First Edition. I. E. Leonard and J. E. Lewis.
© 2016 John Wiley & Sons, Inc. Published 2016 by John Wiley & Sons, Inc.

Proof. (i) implies (ii). Let K be a compact subset of $\mathbb{R}^n$ and let $\{F_\alpha : \alpha \in I\}$ be any collection of closed subsets of K. For each α in the index set I, let U_α be the complement of F_α, that is, $U_\alpha = \mathbb{R}^n \setminus F_\alpha$. Since F_α is a closed set, then U_α is an open set.

Suppose that the intersection of the F_α's is empty, that is, $\bigcap\limits_{\alpha \in I} F_\alpha = \emptyset$, then

$$\mathbb{R}^n = \mathbb{R}^n \setminus \emptyset = \mathbb{R}^n \setminus \bigcap_{\alpha \in I} F_\alpha = \bigcup_{\alpha \in I} (\mathbb{R}^n \setminus F_\alpha) = \bigcup_{\alpha \in I} U_\alpha.$$

Therefore, $K \subset \bigcup\limits_{\alpha \in I} U_\alpha$, and since K is compact and $\{U_\alpha : \alpha \in I\}$ is an open cover for K, from Theorem 2.5.9, there are finitely many indices $\alpha_1, \alpha_2, \ldots, \alpha_r$ such that

$$K \subset \bigcup_{k=1}^{r} U_{\alpha_k},$$

so that $\bigcap\limits_{k=1}^{r} F_{\alpha_k} \subset \mathbb{R}^n \setminus K$. However, if $x \in \bigcap\limits_{k=1}^{r} F_{\alpha_k}$, this implies that $x \notin K$, which contradicts the fact that each $F_{\alpha_i} \subset K$. Therefore, we must have $\bigcap\limits_{k=1}^{r} F_{\alpha_k} = \emptyset$.

We have shown that if $K \subset \mathbb{R}^n$ is compact and $\{F_\alpha : \alpha \in I\}$ is a family of closed subsets of K and $\bigcap\limits_{\alpha \in I} F_\alpha = \emptyset$, then there exists a finite subfamily $\{F_{\alpha_1}, F_{\alpha_2}, \ldots, F_{\alpha_r}\}$ such that $\bigcap\limits_{k=1}^{r} F_{\alpha_k} = \emptyset$.

Therefore, if $K \subset \mathbb{R}^n$ is compact and $\{F_\alpha : \alpha \in I\}$ is a family of closed subsets of K with the finite intersection property, then $\bigcap\limits_{\alpha \in I} F_\alpha \neq \emptyset$.

(ii) implies (i). Suppose that whenever $\mathcal{F}$ is a family of closed subsets of K with the finite intersection property, then $\mathcal{F}$ has a nonempty intersection.

Suppose also that K is not compact, so there exists an open cover $\mathcal{U} = \{U_\alpha : \alpha \in I\}$ for K which has no finite subcover. Let $\{U_{\alpha_1}, U_{\alpha_2}, \ldots, U_{\alpha_r}\}$ be any finite subfamily of the open cover $\mathcal{U}$, so that this subfamily does not cover K. Hence,

$$K \setminus \bigcup_{k=1}^{r} U_{\alpha_k} \neq \emptyset.$$

For each $\alpha \in I$, let $K_\alpha = K \setminus U_\alpha$ and let $\mathcal{F} = \{K_\alpha : \alpha \in I\}$, then $\mathcal{F}$ is a family of closed subsets of K which has the finite intersection property. By hypothesis, $\mathcal{F}$ has nonempty intersection, that is,

$$\bigcap_{\alpha \in I} K_\alpha = \bigcap_{\alpha \in I} (K \setminus U_\alpha) = K \setminus \bigcup_{\alpha \in I} U_\alpha \neq \emptyset,$$

and so

$$K \not\subset \bigcup_{\alpha \in I} U_\alpha,$$

which contradicts the assumption that $\mathcal{U}$ is an open cover for K. Therefore, K must be compact.

$\square$

There is a generalization of the nested sets theorem (Theorem 2.5.8) that is sometimes useful.

Corollary 4.1.4. *If $\mathcal{F}$ is a family of compact subsets of $\mathbb{R}^n$ that has the finite intersection property, then $\mathcal{F}$ has a nonvoid intersection.*

Proof. Choose any set T from $\mathcal{F}$. By hypothesis, every member F of $\mathcal{F}$ intersects T. Let $\mathcal{T}$ be the family $\{T \cap F : F \in \mathcal{F}\}$. Since each member of $\mathcal{F}$ is closed, it follows that each subset $T \cap F$ of T is closed. Since $\mathcal{F}$ has the finite intersection property, it follows that $\mathcal{T}$ also has the finite intersection property. The corollary now follows since $\bigcap \{F : F \in \mathcal{F}\} \equiv \bigcap \{T \cap F : F \in \mathcal{F}\}$.

$\square$

The theorems involving the finite intersection property actually use a new idea, the idea that we can sometimes determine a property for an infinite situation by examining what is happening for all corresponding finite situations.

If $\mathcal{F}$ is a family of compact subsets of $\mathbb{R}^n$ with the finite intersection property, then there is a point in common to all of the sets. In other words, to verify that the family $\mathcal{F}$ has a nonempty intersection, we need only to check that every finite subfamily of $\mathcal{F}$ has a nonempty intersection.

4.1.2 Problems

1. Show that if $\{F_k\}_{k \geq 1}$ is a sequence of nonempty closed subsets of $\mathbb{R}^n$ such that $F_{k+1} \subset F_k$ for $k \geq 1$ and $\operatorname{diam}(F_k) \to 0$, then there is exactly one point in $\bigcap F_k$.
2. Let $\{x_k\}_{k \geq 1}$ be a sequence in $\mathbb{R}^n$, which converges to x_0, and for each $k \geq 1$, let $A_k = \{x_k, x_{k+1}, \ldots\}$. Show that $\{x_0\} = \bigcap_{k=1}^{\infty} \overline{A_k}$.

3. Let C be a compact subset of $\mathbb{R}^n$ and let $\{x_k\}_{k\geq 1}$ be a Cauchy sequence, with $x_k \in C$ for all $k \geq 1$. Show that $x_0 = \lim\limits_{k\to\infty} x_k$ exists and that $x_0 \in C$.

4. If $C \subset \mathbb{R}^n$, then a point $x_0 \in C$ is said to be *isolated* in the set C if there is an open ball B containing x_0 such that $B \cap C = \{x_0\}$. Show that x_0 is isolated in C if and only if there exists an $\epsilon > 0$ such that for all $x \in C$ with $x \neq x_0$, we have $\|x - x_0\| > \epsilon$.

5. A subset of $\mathbb{R}^n$ is said to be *discrete* if all of its points are isolated. Show that a discrete set is compact if and only if it is finite.

6. Let $\{U_k\}_{k\geq 1}$ be a sequence of nonempty open sets in $\mathbb{R}^n$ such that $\overline{U_k}$ is compact and $\overline{U_{k+1}} \subset U_k$ for all $k \geq 1$. Show that $\bigcap\limits_{k=1}^{\infty} U_k \neq \emptyset$.

7. Show that there exists a sequence of distinct positive integers $\{n_k\}_{k\geq 1}$, with $\lim\limits_{k\to\infty} n_k = \infty$, such that both $\lim\limits_{k\to\infty} \sin n_k$ and $\lim\limits_{k\to\infty} \cos n_k$ exist.

8. Show that in $\mathbb{R}^n$, a set is compact if and only if every *countable* open cover has a finite subcover.

9. Suppose that $A \subset \mathbb{R}^n$ is *not* compact. Show that there exists a sequence $\{F_k\}_{k\geq 1}$ of closed sets with $F_{k+1} \subset F_k$ and $F_k \cap A \neq \emptyset$ for all $k \geq 1$ such that $A \cap \bigcap\limits_{k=1}^{\infty} F_k = \emptyset$.

10. Show that any closed subset $F \subset \mathbb{R}^n$ is a countable intersection of open sets.

    Hint. Let $U_k = \{x \in \mathbb{R}^n : \|x - x_0\| < 1/k \text{ for some } x_0 \in F\}$.

11. A set $N \subset \mathbb{R}^n$ is said to be *nowhere dense* if for any nonempty open set U, we have $U \cap \overline{N} \neq U$. Show that N is nowhere dense if and only if $\text{int}(\overline{N}) = \emptyset$.

12. (Baire Category theorem for $\mathbb{R}^n$). Show that $\mathbb{R}^n$ cannot be written as a countable union of nowhere dense sets.

    Hint. If $\mathbb{R}^n = \bigcup\limits_{k=1}^{\infty} N_k$ where each N_k is nowhere dense, find a noncon-vergent Cauchy sequence $\{x_k\}$ by carefully choosing nested closed balls $\overline{B}(x_k, r_k)$ with $\overline{B}(x_k, r_k) \subset \mathbb{R}^n \setminus (N_i \cup \cdots \cup N_k)$.

4.2 HELLY'S THEOREM

If $\mathcal{F}$ is a family of compact *convex* subsets of $\mathbb{R}^n$, we can make a stronger statement than that in the previous section. In order to conclude that $\mathcal{F}$ has a nonempty intersection, we need not check *all* finite subfamilies, but only those subfamilies of size $n + 1$ or less. This result is part of Helly's theorem, and it is one of the most useful theorems in finite-dimensional geometry.

Theorem 4.2.1. *(Helly's Theorem)*

Let $\mathcal{F}$ be a family of convex subsets of $\mathbb{R}^n$, and suppose either that $\mathcal{F}$ is finite or that all of the members of $\mathcal{F}$ are compact. If every $n+1$ members of $\mathcal{F}$ have a point in common, then there is a point in common to all members of $\mathcal{F}$.

Proof. We will prove the theorem for the case $n = 1$ and then use mathematical induction to complete the proof.

- The case $n = 1$.

The convex subsets of $\mathbb{R}$ are intervals. We will prove the theorem first for the case where the members of $\mathcal{F}$ are compact. We let A and B denote, respectively, the set of all left-hand endpoints of the intervals in $\mathcal{F}$ and the set of right-hand endpoints of the intervals in $\mathcal{F}$. Let $a = \sup A$ and $b = \inf B$.

We must have $a \leq b$, since if $b < a$, then by definition of b, there would be an interval $[a_1, b_1] \in \mathcal{F}$ with

$$b \leq b_1 < \tfrac{1}{2}(a + b).$$

Similarly, by definition of a, there would be an interval $[a_2, b_2] \in \mathcal{F}$ with

$$\tfrac{1}{2}(a + b) < a_2 \leq a.$$

However, this implies that there are two intervals, $[a_1, b_1]$ and $[a_2, b_2]$, in $\mathcal{F}$ with empty intersection, which is a contradiction. Thus, $a \leq b$, and it then follows that every member of $\mathcal{F}$ contains $[a, b]$, which is either a point or a proper interval.

For the case where $\mathcal{F}$ is finite, the argument is similar. The only problem will occur if $a = b$ and if one of the members of $\mathcal{F}$ has left-hand endpoint a, respectively, right-hand endpoint b, and is open at that endpoint. Of course, the finiteness of $\mathcal{F}$ and the definition of b, respectively, a, implies that there must also be a member of $\mathcal{F}$, which has right-hand endpoint b, respectively, left-hand endpoint a. But then there would again be two members of $\mathcal{F}$ with empty intersection.

- The general case.

Because of the finite intersection property, the compact case reduces to a finite case. We need only show that every finite subfamily of $\mathcal{F}$ has nonempty intersection. That is, the compact case reduces to the case where we have a finite family $\mathcal{F}$ of compact convex subsets of $\mathbb{R}^n$, where every $n+1$ members of the family have nonempty intersection.

For the general finite case, the sets need not be compact. However, in this case, the theorem can also be reduced to the compact finite case. To see why, suppose that $\mathcal{F}$ is a finite family of convex subsets of $\mathbb{R}^n$. Consider all possible ways of choosing $n + 1$ members of $\mathcal{F}$. The hypotheses guarantee that such a family has a point in common. Select such a point for every subfamily of size $n + 1$, and let B be the set of chosen points. Replace each C in $\mathcal{F}$ by the convex hull C' of $B \cap C$. Since B is finite, each C' is a compact convex polytope, and it follows from the way that B was obtained that every $n + 1$ of the sets C' has a nonempty intersection. Furthermore, since $C' \subset C$, any point that belongs to all of the sets C' must also belong to all of the sets C. This shows that the general finite case reduces to the *finite compact* case.

The theorem is proved by induction. We may, therefore, assume that the theorem is true for dimension $n - 1$. Thus, we assume that in any $(n - 1)$-dimensional space, if every n members of a finite family of compact convex sets have a nonempty intersection, then there is a point in common to the entire family.

Suppose now that the theorem fails in $\mathbb{R}^n$, that is, the family $\mathcal{F}$ has empty intersection, then there must be a minimal subfamily

$$\mathcal{G} = \{ C_1, C_2, \ldots, C_k \}$$

(which may be $\mathcal{F}$ itself, but with $k > n + 1$ at least) with the following properties.

(i) Every $n + 1$ members of $\mathcal{G}$ have a point in common.
(ii) $C_1 \cap C_2 \cap \cdots \cap C_{k-1} \neq \emptyset$.
(iii) $C_1 \cap C_2 \cap \cdots \cap C_{k-1} \cap C_k = \emptyset$.

Letting $A = C_1 \cap C_2 \cap \cdots \cap C_{k-1}$, properties (ii) and (iii) imply that A and C_k are disjoint compact convex sets in $\mathbb{R}^n$ and so can be strongly separated by a hyperplane H as in the figure below.

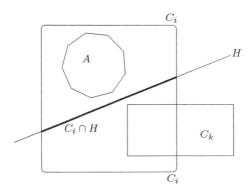

Now, given C_i with $i < k$, we have $A \subset C_i$ and $C_i \cap C_k \neq \emptyset$, since every $n + 1$ members of $\mathcal{G}$ have nonempty intersection. In other words, each C_i with

$i < k$ contains points in both of the closed halfspaces determined by H, and it follows that $C_i \cap H$ is a nonempty convex subset of H. Let

$$\mathcal{E} = \{ C_i \cap H : 1 \leq i \leq k - 1 \}.$$

We claim further that every n members of $\mathcal{E}$ have a nonempty intersection. This is clear, since if

$$C_{i_1} \cap H, C_{i_2} \cap H, \ldots, C_{i_n} \cap H$$

are any n members of $\mathcal{E}$, then

$$A \subset C_{i_1} \cap C_{i_2} \cap \cdots \cap C_{i_n}$$

and

$$(C_{i_1} \cap C_{i_2} \cap \cdots \cap C_{i_n}) \cap C_k \neq \emptyset,$$

again by virtue of the fact that every $n + 1$ members of $\mathcal{G}$ have nonempty intersection. This means that $C_{i_1} \cap C_{i_2} \cap \cdots \cap C_{i_n}$ contains points on both sides of H, which shows that

$$\{ C_{i_1} \cap H, C_{i_2} \cap H, \ldots, C_{i_n} \cap H \}$$

has nonempty intersection.

By the induction hypothesis, the theorem is true in the $(n - 1)$-dimensional affine subspace H, so it follows that $\mathcal{E}$ has nonempty intersection. But this means that A intersects the hyperplane H, since

$$A \cap H = (C_1 \cap C_2 \cap \cdots \cap C_{k-1}) \cap H = (C_1 \cap H) \cap (C_2 \cap H) \cap \cdots \cap (C_{k-1} \cap H) \neq \emptyset.$$

This contradicts the fact that H strongly separates A and C_k and completes the proof.

$\square$

Example 4.2.2. *If the closed sets in Helly's theorem are not convex, then there is a family of four subsets of $\mathbb{R}^2$ such that each three have a point in common, but with no point in common to all four.*

Solution. Consider the four subsets of $\mathbb{R}^2$ show in the figure below, where three of the sets A, B, and C are circular disks, while the fourth set D is an annulus, that is, the region on and between two concentric circles.

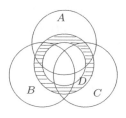

Clearly, every three of these four subsets have a point in common, but there is no point common to all four sets.

Exercise 4.2.3. *If the closed convex sets in Helly's theorem are not bounded, then there is an infinite family of closed convex subsets of $\mathbb{R}^2$ such that each three have a point in common, but with no point in common to all members of the family.*

Exercise 4.2.4. *If the convex sets in Helly's theorem are not closed, then there is an infinite family of bounded convex subsets of $\mathbb{R}^2$ such that each three have a point in common, but with no point in common to all members of the family.*

Note. In Helly's theorem, two different but related hypotheses lead to the same conclusion. In other words, Helly's theorem is really two separate theorems, and it is useful to state the two theorems separately.

Theorem 4.2.5. *(Helly's Theorem for a Finite Family)*

Let $\mathcal{F} = \{C_1, C_2, \ldots, C_k\}$ be a finite family of convex subsets of $\mathbb{R}^n$. If every $n + 1$ members of $\mathcal{F}$ have a point in common, then there is a point in common to all members of $\mathcal{F}$.

Theorem 4.2.6. *(Helly's Theorem for Compact Convex Sets)*

Let $\mathcal{F}$ be a family of compact convex subsets of $\mathbb{R}^n$. If every $n + 1$ members of $\mathcal{F}$ have a point in common, then there is a point in common to all members of $\mathcal{F}$.

Both theorems are about families of convex sets. The first one places no other restrictions on the sets—they may be open, closed, neither open nor closed; they may be bounded or unbounded—but requires that the family contain only a finite number of sets. The second one removes the restriction on the size of the family but restricts the sets to being both closed and bounded.

- When $n = 1$, $\mathbb{R}^n$ is just the real line, and compact convex subsets of the real line are simply closed intervals. Thus, for the real line, the second version of Helly's theorem becomes the following.

  Theorem 4.2.7. *(Helly's Theorem for Intervals in $\mathbb{R}$)*

  Let $\mathcal{F}$ be a family of closed intervals of the real line. If every two of the intervals have a point in common, then there is a point in common to all of the intervals.

- When $n = 2$, as it will be for most of the examples in the next section, Helly's theorem becomes the following.

Theorem 4.2.8. *(Helly's Theorem for the Plane)*

Let $\mathcal{F}$ be a family of convex subsets of the plane, and suppose that either $\mathcal{F}$ has only finitely many members or that every member of $\mathcal{F}$ is compact. If every three members of $\mathcal{F}$ have a point in common, then there is a point in common to all members of $\mathcal{F}$.

4.3 APPLICATIONS OF HELLY'S THEOREM

4.3.1 The Art Gallery Theorem

The first application of Helly's theorem is concerned with *star-shaped sets*. A subset S of $\mathbb{R}^n$ is said to be *star-shaped from* p or simply *star shaped* if there is a point $p \in S$ such that every ray emanating from p intersects S in a convex set. In other words, the intersection of S and every line through p is a line segment.

The figure below illustrates some sets that are star shaped and some sets that are not star shaped.

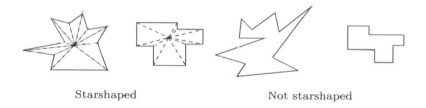

Starshaped   Not starshaped

Clearly, a convex set is star shaped, but a star-shaped set may fail to be convex as the figure shows.

Another way to describe the notion of a star-shaped set is in terms of *visibility*.

Let p be a point of the subset S of $\mathbb{R}^n$. A point $x \in S$ is said to be *visible from* p if the segment $[p, x]$ is contained in S. With this terminology, a set $S \subset \mathbb{R}^n$ is star shaped from p if and only if every point of S is visible from p.

We leave the proof of the following theorem as an exercise. By a polygon in the plane, we mean as usual the polygon together with its interior.

Theorem 4.3.1. *A polygon in the plane is star shaped if and only if for every three edges of the polygon, there is a point in the set from which all three edges are visible.*

The first application is a result due to Krasnosselsky in 1946.

Theorem 4.3.2. *(The Art Gallery Theorem)*

Let P be a simple polygon, together with its interior, in $\mathbb{R}^2$. If whenever three points are chosen from P, it is possible to find a point q of P from which the three points are visible, then P is star shaped.

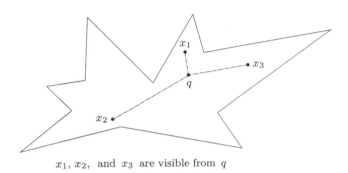

$x_1,\ x_2,$ and x_3 are visible from q

The reason this is called the art gallery theorem* is because it can be formulated as follows:

> *If for each three paintings in an art gallery, there is a place from which all three can be viewed, then there must be a place in the gallery from which all of its paintings can be viewed.*

Proof. First note that the conclusion of the theorem is true if and only if all of the boundary points of P are visible from some common point P.

We begin by directing the edges of P so that as we walk in the given direction along an edge, the interior of the polygon is to the left of the edge, as in the figure below.

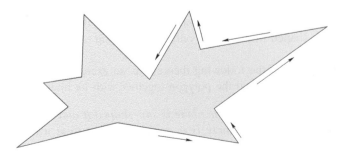

---

* It is a safe bet that from now on, you will never be able to enter an art gallery without thinking of this theorem.

By hypothesis, none of the edges of P cross each other, so this orientation is well defined.

Now, each edge is contained in a line, and the direction of the edge assigns a direction to this line. For each edge e_r of P, let H_r be the closed halfspace to the left of the directed line. Note that every point that is outside the polygon is outside at least one of these halfspaces.

Each H_r is convex set, and there are only a finite number of such sets since a polygon has only a finite number of edges.

We claim that every three of the convex sets H_r have a point in common. To see this, consider three typical halfspaces H_i, H_j, and H_k. Choose points x_i, x_j, x_k from the respective three corresponding edges e_i, e_j, e_k. By hypothesis, there is a point p of P from which x_i, x_j, and x_k are visible. But this means that

$$[p, x_i] \subset H_i, \quad [p, x_j] \subset H_j, \quad \text{and} \quad [p, x_k] \subset H_k.$$

Consequently, p must be in $H_i \cap H_j \cap H_k$, which proves out claim.

Thus, the sets H_r satisfy the following: they are convex subsets of $\mathbb{R}^2$, there are only a finite number of them, and every three of them have a point in common. From Helly's theorem, there is a point q common to all of them. We only have to show that P is star shaped from q.

First, q is a point of P, since if not, it would have to be to the right of one of the directed lines determined by one of the edges. But this would mean that q would fail to be in one of the H_r, which is contrary to the construction of q.

Second, every point in the boundary of P is visible from q. If this were not the case, a segment joining q to one of the boundary points would have to pass exterior to the polygon, and it follows from this that q would have to be outside one of the halfspaces H_r, again contradicting the construction of q.

$\square$

As another application of Helly's theorem in the plane, we also have the following result.

Theorem 4.3.3. *If F is a finite subset of $\mathbb{R}^2$, then there is a point p in $\mathbb{R}^2$ with the property that every closed halfspace with p in its boundary contains at least $n/3$ points of F.*

Proof. Since F is finite, it is bounded, and so F is contained in some closed ball D.

Now consider all closed halfspaces H with the property that H contains more than $2n/3$ of the points of F. We know that there must be some such H, because any halfspace that contains D contains all of F.

For each of these closed halfspaces H, let $H' = H \cap D$, and let

$$\mathcal{F} = \{H' : H' = H \cap D \text{ and } H \text{ contains more than } 2n/3 \text{ points of } F\}.$$

Note that each $H' \in \mathcal{F}$ is convex since it is the intersection of two convex sets, each H' is closed since it is the intersection of two closed sets, and each H' is bounded since it is a subset of D. Thus, $\mathcal{F}$ is a nonempty family of compact convex subsets of $\mathbb{R}^2$.

We claim that every three members of $\mathcal{F}$ have a point in common. To see this, suppose that H', G', and K' are three members of $\mathcal{F}$, corresponding, respectively, to three closed halfspaces H, G, and K. We have to show that $H \cap G \cap K$ contains a point $p \in F$. Since $p \in F \subset D$, this would show that p was also in $H \cap D$, $G \cap D$, and $K \cap D$, that is, $p \in H' \cap G' \cap K'$, and would prove our claim.

In order to show that $H \cap G \cap K$ contains at least one point of F, it is sufficient to show that the complement $\mathbb{R}^2 \setminus (H \cap G \cap K)$ does not contain all of F.

Now, we have

$$\mathbb{R}^2 \setminus (H \cap G \cap K) = (\mathbb{R}^2 \setminus H) \cup (\mathbb{R}^2 \setminus G) \cup (\mathbb{R}^2 \setminus K),$$

and by definition of H, G, and K, each contains more than $2n/3$ points of F. Thus, each of

$$\mathbb{R}^2 \setminus H, \quad \mathbb{R}^2 \setminus G, \quad \text{and} \quad \mathbb{R}^2 \setminus K$$

contains less than $n/3$ points of F, and it follows that $\mathbb{R}^2 \setminus (H \cap G \cap K)$ cannot contain all of F, which completes the proof of our claim that every three members of $\mathcal{F}$ have a point in common.

To summarize, $\mathcal{F}$ is a family of compact convex subsets of $\mathbb{R}^2$, and every three of them have a point in common. By Helly's theorem, there is a point p in common to all members of $\mathcal{F}$.

Finally, we will show that if H is any closed halfspace in the plane with p in its boundary, then H contains at least $n/3$ points of F.

Suppose to the contrary that some H with p in its boundary contains fewer than $n/3$ points of F, then the open halfspace $U = \mathbb{R}^2 \setminus H$ contains more than $2n/3$ points of F.

Now, since F is finite, there is a point q of F in U that is closest to ℓ, the line forming the boundary of H. Let m be a line parallel to ℓ passing through q

and let M be the closed halfspace determined by m that lies in U, as in the figure below.

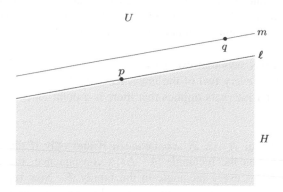

The halfspace M contains all of the points of F that lie in U, that is, M is a closed halfspace that contains more than $2n/3$ points, yet M does not contain p. This contradicts the construction of p and completes the proof.

$\square$

Helly's theorem is frequently applied by "mapping" the problem at hand into a different space. We will illustrate this with one example. Later, when we discuss transversals we will see more examples using this technique. The following example shows how Helly's theorem can be "improved" for families of rectangles in the plane whose corresponding edges are parallel by using this idea.

Example 4.3.4. *Suppose that $\mathcal{F}$ is a family of rectangles in the plane whose edges are parallel to the coordinate axes. If every two of the rectangles have a point in common, then there is a point in common to all of the rectangles in $\mathcal{F}$.*

Solution. For each rectangle $A \in \mathcal{F}$, let A_x and A_y denote the projections of A onto the x and y coordinate axes, respectively, as shown in part (a) of the figure below.

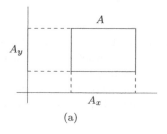

(a)

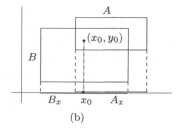

(b)

We will focus on the projections onto the x-axis. Each of the sets A_x is a closed interval in the x-axis. Thus, we have transformed the family $\mathcal{F}$ of rectangles in $\mathbb{R}^2$ into a family $\mathcal{F}_x$ of closed intervals in $\mathbb{R}^1$. Note that if the rectangles A and B in $\mathcal{F}$ have the point (x_0, y_0) in common, then the intervals A_x and B_x have the point x_0 in common, as in part (b) of the figure.

Now, the hypothesis says that every two of the rectangles in $\mathcal{F}$ have a point in common, and so every two of the intervals in $\mathcal{F}_x$ have a point in common. Helly's theorem for intervals implies that there is a point x_0 common to all of the intervals in $\mathcal{F}_x$.

Note that an interval A_x of $\mathcal{F}_x$ contains x_0 if and only if the vertical line through x_0 intersects the rectangle A. As a consequence, the vertical line through x_0 intersects every rectangle in the family $\mathcal{F}$. Now repeat the same argument for the projections onto the y-axis. In this case, we obtain a point y_0 on the y-axis with the property that the horizontal line through y_0 intersects all of the rectangles in $\mathcal{F}$.

Finally, note that if A is a rectangle in $\mathcal{F}$ and if the vertical line through x_0 meets A and the horizontal line through y_0 meets A, then the point (x_0, y_0) is in A. In other words, every member of $\mathcal{F}$ contains the point (x_0, y_0).

$\square$

Example 4.3.5. *Let C be a convex subset of $\mathbb{R}^n$. Let $\mathcal{G}$ be a family of convex subsets of $\mathbb{R}^n$ such that either $\mathcal{G}$ has only finitely many members or that C and all members of $\mathcal{G}$ are compact. If some translate of C intersects every $(n + 1)$-membered subfamily of $\mathcal{G}$, then some translate of C intersects all members of $\mathcal{G}$.*

Solution. For each $A \in \mathcal{G}$, let

$$A' = \{x \in \mathbb{R}^n : (x + C) \cap A \neq \emptyset\},$$

we will show that

(i) each A' is convex, and
(ii) every $n + 1$ of the sets A' have a nonempty intersection.

(i) To show that A' is convex, let x and y be points in A' and λ be a fixed but arbitrary number with $0 \leq \lambda \leq 1$, we want to show that $z = (1 - \lambda)x + \lambda y \in A'$, that is, we want to show that $(z + C) \cap A \neq \emptyset$. From the definition of A', if $x \in A'$, then there exists a point $u \in (x + C) \cap A$. Similarly, if $y \in A'$, then there exists a point $v \in (y + C) \cap A$.

Since $u \in x + C$, then $u - x \in C$, and similarly, since $v \in y + C$, then $v - y \in C$, and since C is convex, this implies that

$$(1 - \lambda)(u - x) + \lambda(v - y) = (1 - \lambda)u + \lambda v - [(1 - \lambda)x + \lambda y]$$
$$= (1 - \lambda)u + \lambda v - z$$

is in C. Thus, $(1 - \lambda)u + \lambda v \in z + C$.

Also, since A is convex and u and v are in A, then $(1 - \lambda)u + \lambda v \in A$, so that $(1 - \lambda)u + \lambda v \in (z + C) \cap A$, and $(z + C) \cap A$ is nonempty, so that $z \in A'$, and A' is convex.

(ii) Let $A'_{i_1}, A'_{i_2}, \ldots, A'_{i_{n+1}}$ be $n + 1$ of the sets A'. By hypothesis, there is some translate of C, say $p + C$, that meets the members $A_{i_1}, A_{i_2}, \ldots, A_{i_{n+1}}$ of $\mathcal{G}$. But by definition of the A', this means that p is in every one of the sets $A'_{i_1}, A'_{i_2}, \ldots, A'_{i_{n+1}}$. This completes the proof that every $n + 1$ of the sets A' have a nonempty intersection.

Helly's theorem for a Finite Family (Theorem 4.2.5) now implies that there is a point q common to all of the sets A', where $A \in \mathcal{G}$. This means that $q + C$ intersects each member A of $\mathcal{G}$ and proves the result for the case where $\mathcal{G}$ is finite.

For the case where C and all of the members of $\mathcal{G}$ are compact convex subsets of $\mathbb{R}^n$, we have already shown that each A' is convex and that each $n + 1$ of the A' has a nonempty intersection. Thus, we need only show that for each $A \in \mathcal{G}$, the set A' is compact.

Let $\{x_i\}_{i \geq 1}$ be a sequence of points in A'. By definition of A', for each x_i, there must be a point u_i in $(x_i + C) \cap A$. By compactness of A, the sequence $\{u_i\}_{i \geq 1}$ must have a convergent subsequence $\{u_{i_k}\}_{k \geq 1}$, and its limit u_0 must belong to A.

Since $u_{i_k} \in x_{i_k} + C$, then $u_{i_k} - x_{i_k} \in C$, and since C is compact, the sequence $\{u_{i_k} - x_{i_k}\}_{k \geq 1}$ has a subsequence, which we again denote by $\{u_{i_k} - x_{i_k}\}_{k \geq 1}$, converging to a point z_0 of C.

Thus,

$$\lim_{k \to \infty}(u_{i_k} - x_{i_k}) = \lim_{k \to \infty} u_{i_k} - \lim_{k \to \infty} x_{i_k} = z_0$$

so that $u_0 - \lim_{k \to \infty} x_{i_k} = z_0$, that is, $\lim_{k \to \infty} x_{i_k} = u_0 - z_0$. Letting $x_0 = u_0 - z_0$, this shows that the subsequence $\{x_{i_k}\}_{k \geq 1}$ of $\{x_i\}_{i \geq 1}$ converges to x_0.

To show that A' is compact, we need only show that $x_0 \in A'$. However, since $z_0 \in C$, then $x_0 + z_0 \in x_0 + C$, so that $u_0 = x_0 + z_0 \in x_0 + C$. Since u_0 is in both $x_0 + C$ and A, this implies that $(x_0 + C) \cap A \neq \emptyset$, so that $x_0 \in A'$.

Helly's theorem for Compact Convex Sets (Theorem 4.2.6) now implies that there is a point q common to all of the sets A' where $A \in \mathcal{G}$. Again, this means that $q + C$ intersects each member A of $\mathcal{G}$ and proves the result for the case where C and every member of $\mathcal{G}$ is compact.

$\square$

4.3.2 Vincensini's Problem

Another application of Helly's theorem, which we will examine in considerable detail, arises from a problem of Vincensini that asks whether Helly's theorem can be extended to sets having a common line rather than a common point:

Question 4.3.6. *Suppose that $\mathcal{G}$ is a family of convex subsets of $\mathbb{R}^n$. Is there a "Helly number" k such that if every k-members of $\mathcal{G}$ are simultaneously intersected by some straight line, then the entire family $\mathcal{G}$ is simultaneously intersected by some straight line?*

The answer, without rather severe restrictions on $\mathcal{G}$, is known to be negative. Thus, the question becomes: Under what restrictions on $\mathcal{G}$, or even on the intersecting lines, will the answer be affirmative?

It is convenient to introduce the jargon that has grown up around this problem.

A straight line L that intersects each member of a family $\mathcal{G}$ is called a ***common transversal*** for $\mathcal{G}$, or simply, a ***transversal*** for $\mathcal{G}$.

If a family $\mathcal{G}$ has a common transversal, we say that $\mathcal{G}$ has ***property*** **T**.

If every m-membered subfamily of $\mathcal{G}$ has property **T**, we say that $\mathcal{G}$ has ***property*** **T**(m).

Using this terminology, the problem becomes

Find conditions such that there is some natural number m so that property **T**(m) *implies property* **T**.

It should be mentioned that even in $\mathbb{R}^2$, the full story is still unfolding. We will begin with a simple fact that has proven useful on several occasions.

Lemma 4.3.7. *In $\mathbb{R}^n$, let $\mathcal{G}$ be a family of convex subsets, which either has a finite number of members or for which every member is compact. If there is a direction v such that every n members of $\mathcal{G}$ admit a transversal in the direction of v, then $\mathcal{G}$ has a transversal in the direction v.*

Proof. By a line in the direction v, we simply mean a line that is parallel to the nonzero vector v, that is, a line whose equation is of the form

$$x = a + \lambda v,$$

where $-\infty < \lambda < \infty$.

If we let H be a hyperplane perpendicular to v, then H is an affine subspace of dimension $n - 1$.

Now, for each A in $\mathcal{G}$, let A' be the projection of A along v onto H, as in the figure below.

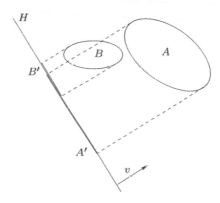

Since a projection along v onto H is a linear mapping, it preserves convexity; thus, A' is convex for each $A \in \mathcal{G}$.

Since projections are continuous linear mappings, it follows that if A is compact, then A' is also compact. Thus, if A is closed and bounded, then A' is also closed and bounded.

In other words, the family

$$\mathcal{G}' = \{A' : A \in \mathcal{G}\}$$

inherits from $\mathcal{G}$ the property that it either has only finitely many members, each of which is convex, or each member of $\mathcal{G}'$ is a compact convex set.

Now, the fact that every n members of $\mathcal{G}$ have a transversal parallel to v implies that every n members of $\mathcal{G}'$ have a point in common. Applying Helly's theorem to the $(n-1)$-dimensional affine subspace H allows us to conclude that there is a point p in common to all members of $\mathcal{G}'$. But then the line through p parallel to v must be a common transversal for $\mathcal{G}$.

$\square$

A family of compact convex sets in $\mathbb{R}^2$ is said to be ***totally separated*** if there is a direction v such that each line in that direction intersects *at most* one member of the family, as in the figure below.

The convexity of the sets imply that the sets lie in mutually disjoint parallel strips in the plane.

In the same paper in which Vincensini raised the question about extending Helly's theorem, he proved the following theorem.

Theorem 4.3.8. *(Vincensini)*

For a totally separated family $\mathcal{G}$ of compact convex sets in $\mathbb{R}^2$, property $\mathbf{T}(4)$ *implies property* $\mathbf{T}$.

Proof. Without loss of generality, we may assume that every line parallel to the x-axis intersects at most one member of G. This means that if A and B are members of G, then A and B do not have a common horizontal transversal.

If we let $[A, B]$ denote the set of all positive angles that the common transversals for A and B make with the x-axis, then it follows that $[A, B]$ is an interval whose left-hand endpoint is no smaller than 0 and whose right-hand endpoint is less than π. In other words, the set $[A, B]$ is a compact convex subset of $\mathbb{R}$.

Now let $\mathcal{F}$ be the family of all such intervals generated by all possible pairs of $\mathcal{G}$. The fact that every four members A, B, C, D of $\mathcal{G}$ have a common transversal means that every two members $[A, B]$, $[C, D]$ of $\mathcal{F}$ have nonempty intersection. Applying Helly's theorem to the line $\mathbb{R}$, it follows that all members of $\mathcal{F}$ have a point α in common. However, this means that every two members of $\mathcal{G}$ have a transversal that makes an angle α with the x-axis.

By the lemma, it follows that all members of $\mathcal{G}$ have a transversal that makes an angle α with the x-axis, and this finishes the proof.

□

Victor Klee later showed that the Helly number 4 could be reduced to 3.* In the next theorem, we show how Klee's reduction works.

Theorem 4.3.9. *(Klee)*

For a totally separated family $\mathcal{G}$ of compact convex sets in $\mathbb{R}^2$, property $\mathbf{T}(3)$ implies property $\mathbf{T}$.

Proof. Again, without loss of generality, we may assume that every line parallel to the x-axis intersects at most one member of G. As in the previous theorem, we let $[A, B]$ denote the set of all positive angles that the common transversals for A and B make with the x-axis. As before, it follows that $[A, B]$ is a compact convex subset of $\mathbb{R}$, and we let $\mathcal{F}$ be the family of all such intervals generated by all possible pairs of $\mathcal{G}$.

By Theorem 4.3.8, it suffices to show that $\mathbf{T}(3)$ implies $\mathbf{T}(4)$. In other words, if A, B, C, and D are four members of $\mathcal{G}$, it suffices to show that $\mathbf{T}(3)$ implies that A, B, C, and D have a common transversal. Thus, we wish to show that

$$[A, B] \cap [A, C] \cap [A, D] \cap [B, C] \cap [B, D] \cap [C, D] \neq \emptyset.$$

To show this, by Helly's theorem for $\mathbb{R}$, it suffices to show that every two have a point in common. Now by $\mathbf{T}(3)$, it is clear that each two of the type $[U, V]$ and $[V, W]$ have a point in common, we need only to worry about pairs like $[A, B]$ and $[C, D]$. We will show that even in this case, the two sets $[A, B]$ and $[C, D]$ must have a point in common.

Suppose that this were not the case, then there must be a point α that lies strictly between $[A, B]$ and $[C, D]$.

Now note that each $[A, C]$, $[A, D]$, $[B, C]$, and $[B, D]$ have points in common with $[A, B]$ and $[C, D]$. This means that α is in each one of

$$[A, C], \quad [A, D], \quad [B, C], \quad [B, D].$$

Consequently, each of the pairs

$$\{A, C\}, \quad \{A, D\}, \quad \{B, C\}, \quad \{B, D\}$$

has a transversal that makes an angle α with the x-axis.

On the other hand, the fact that α is in neither $[A, B]$ nor $[C, D]$ means that both A and B and C and D are separated by a line that makes an angle α with the x-axis, as shown in the figure below (except that the positions of A, B, C, and D may have to be permuted).

---

* Note that 3 is the smallest Helly number for transversal problems because every family has property $\mathbf{T}(2)$.

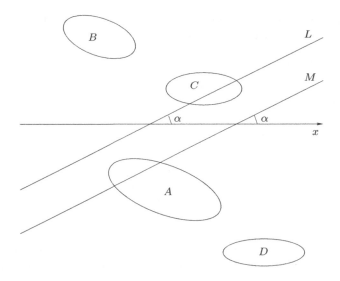

However, this means that at least one of the pairs

$$\{A,C\}, \quad \{A,D\}, \quad \{B,C\}, \quad \{B,D\}$$

must also be separated by a line making an angle α with the x-axis. But this is a direct contradiction to the previous paragraph, and so we conclude that $[A, B]$ and $[C, D]$ must have a point in common. This completes the proof.

$\square$

The tactic used in the proof of Theorem 4.3.8 is a common one: transversals or lines are "transformed" into points in some suitable finite dimensional space F, and sets of transversals are turned into convex subsets of F. Helly's theorem is then used to obtain a point that is common to all of the convex sets, and this point is then converted by the "inverse transformation" into a common transversal.

This description is deliberately vague. It is a tactic—not a guaranteed solution, and the "transformation" that is used will change from one situation to another. The next theorem illustrates another way in which this tactic is used.

Theorem 4.3.10. *Let $\mathcal{G}$ be a finite family of parallel line segments in $\mathbb{R}^2$. If $\mathcal{G}$ has property $\mathbf{T}(3)$, then $\mathcal{G}$ has property $\mathbf{T}$.*

Proof. Draw lines L_1 and L_2 that are parallel to the segments and that contain the segments between them. Now coordinatize the lines. There is no loss in generality in assuming that L_1 and L_2 are vertical; that is, in Cartesian coordinates, L_1 has the equation $x = x_1$ while L_2 has the equation $x = x_2$.

We may also assume that no vertical line intersects two of the segments, since if this were the case, it would follow from $\mathbf{T}(3)$ that the line contains all of the segments, and the theorem would be proved.

For each segment S, we are, therefore, interested only in those lines that intersect S but are not parallel to S. Such a line M must intersect the coordinate lines at points $(x_1, a_1) \in L_1$ and $(x_2, a_2) \in L_2$. Thus, we can establish a direct and unique correspondence between the line M and at the point (a_1, a_2) of $\mathbb{R}^2$. For the segment S, let S' denote the collection of all points (a_1, a_2) such that the line through (x_1, a_1) on L_1 and (x_2, a_2) on L_2 intersects S, as shown in the figure below.

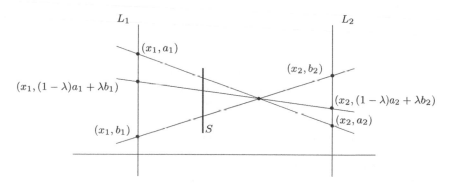

To show that the set S' is convex, let (a_1, a_2) and (b_1, b_2) be points in S' and let λ be any number between 0 and 1. We will show that the point $(1 - \lambda)(a_1, a_2) + \lambda(b_1, b_2)$ is in S', that is, we will show that the point $((1 - \lambda)a_1 + \lambda b_1, (1 - \lambda)a_2 + \lambda b_2)$ yields a line that intersects the segment S.

For definiteness, suppose that S has endpoints (x_0, s) and (x_0, t) with $x_1 < x_0 < x_2$ and with $s < t$. Let μ be the number such that

$$x_0 = (1 - \mu)x_1 + \mu x_2.$$

Now, the line through (x_1, a_1) and (x_2, a_2) intersects S at (x_0, y_0) where $y_0 = (1 - \mu)a_1 + \mu a_2$, so that

$$s \le (1 - \mu)a_1 + \mu a_2 \le t$$

and similarly,

$$s \le (1 - \mu)b_1 + \mu b_2 \le t.$$

We want to show that

$$s \le (1 - \mu)\left[(1 - \lambda)a_1 + \lambda b_1\right] + \mu\left[(1 - \lambda)a_2 + \lambda b_2\right] \le t.$$

This can be obtained by multiplying the first inequality by $(1 - \lambda)$, multiplying the second inequality by λ, and adding

$$(1-\lambda)s + \lambda s \le (1-\lambda)\left[(1 - \mu)a_1 + \mu a_2\right] + \lambda\left[(1 - \mu)b_1 + \mu b_2\right] \le (1-\lambda)t + \lambda t,$$

that is,

$$s \le (1 - \lambda)\left[(1 - \mu)a_1 + \mu a_2\right] + \lambda\left[(1 - \mu)b_1 + \mu b_2\right] \le t,$$

which shows that the point $(1 - \lambda)(a_1, a_2) + \lambda(b_1, b_2)$ does indeed yield a line that passes through S.

Now the fact that the family of segments S has property $\mathbf{T}(3)$ means that the family of compact convex subsets S' has the property that every three members have a point in common. Applying Helly's theorem in $\mathbb{R}^2$, it follows that there is a point in common to all members S'. This point yields a unique line, and since the point belongs to every set S', the corresponding line passes through every segment $S \in \mathcal{G}$.

$\square$

Remark. The preceding theorem can be obtained as a corollary to Theorem 4.3.8. However, there is often an added benefit to constructing a different proof. Once the proof of the above theorem is understood, it becomes clear that it can be used almost verbatim to prove the next theorem, and the next theorem certainly does not follow from Theorem 4.3.8.

Theorem 4.3.11. *Let $\mathcal{G}$ be a finite family of line segments in $\mathbb{R}^n$, each of which has a negative slope (not necessarily the same slope). Suppose that every three members of $\mathcal{G}$ have an ascending transversal, then $\mathcal{G}$ has an ascending transversal.*

Proof. By an ascending line, we mean one that is horizontal, vertical, or one that has a positive slope.

Let L_1 and L_2 be as in the proof of the previous theorem. For each segment S of $\mathcal{G}$, let S' be the set of points (y_1, y_2) such that $y_1 \le y_2$ and such that the line M passing through (x_1, y_1) and (x_2, y_2) is an ascending transversal for the set S. Since M is ascending and S is descending, it can be shown that S' is convex, and the rest of the proof follows exactly as in the previous theorem.

$\square$

Corollary 4.3.12. *Let $\mathcal{G}$ be a finite family of rectangles in $\mathbb{R}^2$, each of which has its edges parallel to the coordinate axes. If every three members of $\mathcal{G}$ have an ascending transversal, then $\mathcal{G}$ has an ascending transversal.*

Proof. It is evident that an ascending line intersects one of the rectangles if and only if it intersects the diagonal of the rectangle with negative slope.

$\square$

Corollary 4.3.13. *(Santalo)*

Let G be a finite family of rectangles, each of which has its edges parallel to the coordinate axes. If G has property $\mathbf{T}(6)$, then G has property $\mathbf{T}$.

Proof. If not, then G has neither an ascending nor a descending transversal. By the previous corollary, there must be some three members that fail to have an ascending transversal, and by an obvious analogue of the previous corollary, there are some three that fail to have a descending transversal. Together, then, there are some six (or fewer) sets that fail to admit either an ascending or a descending transversal.

$\square$

The preceding result is due to Santalo, who proved it in a different manner. It should be noted that the Helly number 6 cannot be reduced unless further restrictions are imposed on the family.[*]

Although these are only a few of the applications of Helly's theorem to the problem of existence of common transversals, it should be enough to illustrate the spirit and power of the theorem. We have restricted ourselves to two dimensions, but Helly's theorem has also been applied to transversal problems in higher dimensions.

For three or more dimensions, there is great scope for interpreting what is meant by the word *transversal*. The word could be interpreted as meaning a straight line that intersects all of the sets, or it could mean a hyperplane that intersects all of the sets. Results have been obtained for both interpretations. Of course, a *transversal* could also be defined as being a flat (other than a straight line or a hyperplane) that intersects all members of the family. Very little is known about this case.

4.3.3   Hadwiger's Theorem

There are many more results about common transversals in two dimensions. And, there are still questions that have not been answered. Many of the questions deal with families of convex bodies that are pairwise disjoint. In this case, the interest is focused on finite families. The reason for this is the following.

Lemma 4.3.14. *Let G be a family of pairwise disjoint compact convex subsets of $\mathbb{R}^2$. If every finite subfamily of G has a common transversal, then G has a common transversal.*

---

[*]   One such restriction is to have the family consist of mutually disjoint translates of a rectangle. In this case, Branko Grünbaum has shown that $\mathbf{T}(5)$ implies $\mathbf{T}$.

Proof. For each pair of members A and B of $\mathcal{G}$, let $T(A, B)$ be the set of points x of the unit circle such that there is a transversal for A and B parallel to the line through $\overline{0}$ and x.

Note that if $x \in T(A, B)$, then $-x \in T(A, B)$ also, that is, $T(A, B)$ is not in any way "convex." The compactness of A and B implies that $T(A, B)$ is compact, and the fact that every finite subfamily of $\mathcal{G}$ has a common transversal means that the collection of compact sets $\{T(A, B) : A, B \in \mathcal{G}\}$ has the finite intersection property.

Consequently, there is a point in common to all of the sets $T(A, B)$. However, this means that there is a fixed direction such that every pair of members of $\mathcal{G}$ have a transversal in that direction. The lemma now follows from Lemma 4.3.7.

$\square$

Another reason that the interest is in finite families is due to a result of Hadwiger. However, before stating Hadwiger's theorem, we give a lemma concerning the central projection of points in the plane onto the unit circle $C \subset \mathbb{R}^2$.

Lemma 4.3.15. *For nonzero points a and b in the plane, let $\tilde{a} = \dfrac{a}{\|a\|}$ and $\tilde{b} = \dfrac{b}{\|b\|}$ be the central projections of a and b onto the unit circle C.*

For $\epsilon > 0$ and $\delta > 0$, if $\|a - b\| < \delta$ and $\|a\| > \dfrac{2\delta}{\epsilon}$, then $\|\tilde{a} - \tilde{b}\| < \epsilon$.

Proof. We have

$$
\begin{aligned}
\|\tilde{a} - \tilde{b}\| &= \left\| \frac{a}{\|a\|} - \frac{b}{\|b\|} \right\| \\
&\le \left\| \frac{a}{\|a\|} - \frac{b}{\|a\|} \right\| + \left\| \frac{b}{\|a\|} - \frac{b}{\|b\|} \right\| \\
&= \frac{\|a - b\|}{\|a\|} + \frac{\|b\| - \|a\|}{\|a\|} \\
&\le \frac{\|a - b\|}{\|a\|} + \frac{\|b - a\|}{\|a\|} \\
&\le \frac{2\delta}{\|a\|} \\
&< \epsilon.
\end{aligned}
$$

Note that the lemma says that if two points a and b in the plane are less than some fixed distance apart and if either of them is far from the origin, then their projections onto the unit circle C are very close together.

$\square$

Theorem 4.3.16. *(Hadwiger I)*

Let $\mathcal{G}$ be an infinite family of pairwise disjoint congruent compact convex bodies in $\mathbb{R}^2$. If $\mathcal{G}$ has property $\mathbf{T}(3)$, then $\mathcal{G}$ has property $\mathbf{T}$.

Proof. The proof is accomplished by showing that there is a direction such that every two members of $\mathcal{G}$ admit a transversal in that direction.

• Explanation of the Proof.

First, we note that because the family consists of congruent convex bodies, each member of the family has a definite area. This means that no matter how large a ball we take, that ball can only contain a finite number of members of $\mathcal{G}$. Also, again since the members of $\mathcal{G}$ are congruent, they all have the same maximum diameter, say δ. This means that if a member of $\mathcal{G}$ intersects the ball $\overline{B}(\overline{0}, r)$, then that member must lie inside the ball $\overline{B}(\overline{0}, r + 2\delta)$. Together, these facts imply that the number of members of $\mathcal{G}$ that intersect any ball must be finite. An additional consequence is that $\mathcal{G}$ is countably infinite, that is, the members of $\mathcal{G}$ can be put into a one-to-one correspondence with set of positive integers, so we can write $\mathcal{G} = \{A_1, A_2, \ldots, A_n, \ldots\}$.

The preceding discussion implies that, given any number m, no matter how large, there must be members of $\mathcal{G}$ separated by a distance of at least m. It follows from this that if there is a transversal for $\mathcal{G}$, then any two transversals for $\mathcal{G}$ must be parallel (since two nonparallel transversals diverge, if we go out far enough along one line, say to a point x, it is clear that any member of $\mathcal{G}$ that intersects the line beyond x must miss the other line, that is, the other line would not be a transversal).

The object of the proof is to find a direction of the (hypothetical) parallel transversals. We will be guided by the intuitive fact that if two members of $\mathcal{G}$ are very far apart, then any transversal for those two members must have a direction that is very close to the direction of any common transversal. In fact, suppose that we fix any point p in the plane. It seems intuitively clear that if we take any member A of $\mathcal{G}$ that is extremely far from p, then any line through p and A must also have a direction that is very close to the direction of the hypothetical common transversal. Thus, we will fix a point p in the plane and will use the sets that are very far from p to obtain lines that "converge" to the desired direction. Since any point p seems as good as any other, we will choose p to be the origin $\overline{0}$.

• The Proof Itself.

Let C denote the unit circle in $\mathbb{R}^2$. For each A_n in $\mathcal{G}$, let $\tilde{A}_n$ be the central projection of A_n onto the circle, as in the following figure.

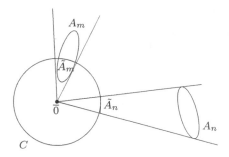

Since the diameter of each A_n is fixed, if A_n is far enough away from C, then from Lemma 4.3.15, the diameter of $\tilde{A}_n$ must be very small, that is, $\lim\limits_{n \to \infty} \operatorname{diam}(\tilde{A}_n) = 0$.

For each $n \geq 1$, pick $x_n \in \tilde{A}_n$, then the set $X = \{x_n : n = 1, 2, \dots\}$ is an infinite subset of the compact set C, and by the Bolzano–Weierstrass theorem, X has an accumulation point v. It follows that given any $\epsilon > 0$, there must be infinitely many indices such that the projections $\tilde{A}_n$ are inside $B(v, \epsilon)$. The reason that this is true is that infinitely many x_n must be inside $B(v, \epsilon/2)$, and if n is large enough, $\operatorname{diam}(\tilde{A}_n) < \epsilon/2$. The triangle inequality now shows that no point of $\tilde{A}_n$ can be further from v than $\epsilon/2 + \epsilon/2$. Let $\mathcal{G}'$ be the subfamily consisting of all those members of $\mathcal{G}$ that are inside $B(v, \epsilon/2)$.

We will complete the proof by showing that every two members of $\mathcal{G}$ have a transversal in the direction v. Let B and D be any two members of $\mathcal{G}$, we will use points in C to identify the directions of transversals for B and D.

Let $\epsilon > 0$ be given. We will first show that B and D have a transversal whose direction w is within ϵ of v: For n large enough, there must be some member A_n of $\mathcal{G}'$ that is different from B and D, since B and D are contained in some large ball, and only finitely many A_n can intersect that ball. Also, for n large enough, every point of $\tilde{A}_n$ must be within $\epsilon/2$ of v.

Now, by hypothesis, B, D, and A_n have a common transversal, say L. Let $x \in L \cap A_n$ and $y \in L \cap B$. Since B is fixed, every point y of B is within a fixed distance δ of the origin. So from Lemma 4.3.15, letting $a = x$ and $b = x - y$, it follows that if x is far enough from the origin, then the distance between $x/\|x\|$ and $(x - y)/\|x - y\|$ will be smaller than $\epsilon/2$. We can guarantee that x is far enough from the origin by taking n sufficiently large. Now, $x/\|x\| \in \tilde{A}_n$ and $(x - y)/\|x - y\|$ is the direction of L. Thus, it follows that

$$\left\| v - \frac{x - y}{\|x - y\|} \right\| \leq \left\| v - \frac{x}{\|x\|} \right\| + \left\| \frac{x}{\|x\|} - \frac{x - y}{\|x - y\|} \right\| < \epsilon.$$

This shows that B and D have a transversal whose direction is within ϵ of v. Since B and D are compact, we can use the following limiting argument to show that B and D have a common transversal in the direction v.

For each positive integer n, let L_n be a transversal for B and D that is within $1/n$ of v. Pick points $x_n \in L_n \cap B$ and $y_n \in L_n \cap D$. By compactness of B and D, we can pick corresponding subsequences $\{x_{n_k}\}_{k \geq 1}$ and $\{y_{n_k}\}_{k \geq 1}$ such that

$$\lim_{k \to \infty} x_{n_k} = x_0 \quad \text{and} \quad \lim_{k \to \infty} y_{n_k} = y_0,$$

where $x_0 \in B$ and $y_0 \in D$. Now, the transversal through x_0 and y_0 must be within $1/n$ of v for every n, that is, B and D have a common transversal in the direction v.

The theorem now follows from Lemma 4.3.7.

$\square$

It should be mentioned that the hypotheses cannot be weakened, as the following four examples show.

Example 4.3.17. *The conclusion of Hadwiger's theorem may fail if the hypothesis that the family be infinite is omitted. The four rectangles in the figure below have property* $\mathbf{T}(3)$*, but do not have property* $\mathbf{T}$*.*

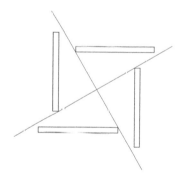

Example 4.3.18. *The conclusion of Hadwiger's theorem may fail if the hypothesis that the sets have nonempty interior is omitted. In the above figure, fill one of the rectangles with infinitely many line segments, each segment being a translate of the long edge of the rectangle. The family still has property* $\mathbf{T}(3)$ *but does not have property* $\mathbf{T}$*.*

Example 4.3.19. *The conclusion of Hadwiger's theorem may fail if the hypothesis of congruency is replaced by similarity. Inscribe a square in a circle centered at the origin, forming four circular segments. Choose a pair of opposite segments. Form a rosette of circular segments by rotating the pair about*

$\overline{0}$ *through angles of $\pi/4$, $2\pi/4$, $3\pi/4$, and so on, and at the same time, expand the pair out from $\overline{0}$ by factors of 2, 4, 8, and so on, as in the figure below. This forms an infinite family of pairwise disjoint similar compact convex bodies with property* **T**(3), *but without property* **T**.

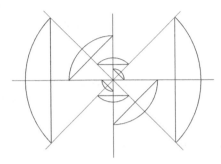

Example 4.3.20. *The conclusion of Hadwiger's theorem may fail if the sets are not mutually disjoint. Let a, b, c, and d be the vertices of a square. Using disks of radius slightly less than half the length of one edge of the square, center a disk at each of the vertices a, b, and c. Now center infinitely many disks at vertex d, as shown in the figure below. This forms an infinite family of congruent disks with property* **T**(3), *but without property* **T**.

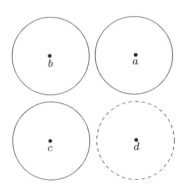

The next theorem is also due to Hadwiger. If you hear someone referring to Hadwiger's theorem, the person is probably referring to this one rather than Theorem 4.3.16. Although this is a Helly-type theorem, the proof does not use Helly's theorem.

Theorem 4.3.21. *(Hadwiger II)*
Let $\mathcal{G}$ be a finite or countably infinite family of pairwise disjoint compact convex sets of the plane. If $\mathcal{G}$ can be ordered in such a way that every three members of $\mathcal{G}$ are intersected by a line in the given order, then $\mathcal{G}$ has common transversal.

Proof. According to Lemma 4.3.14, we need only to prove the theorem for the case where the family is finite. The proof uses a "shrinking" process, which can be described as follows. The family $\mathcal{G}$ will be replaced by a family $\mathcal{G}'$, which is obtained by shrinking each member of $\mathcal{G}$ about one of its points. The shrinking factor will be the same for each member of $\mathcal{G}$. The family $\mathcal{G}'$ inherits the ordering of $\mathcal{G}$, and the shrinking is done in such a manner that $\mathcal{G}'$ retains the property that every three members of $\mathcal{G}'$ can still be intersected by a line in the given order.

Here is the process in detail: Assume that $\mathcal{G}$ consists of m sets A_i and that the ordering is $\{A_1, A_2, \ldots, A_m\}$, that is, A_i "precedes" A_j if $i < j$. The hypothesis of the theorem says that if we have three indices i, j, and k, with $i < j < k$, then there is some transversal for A_i, A_j, and A_k that meets the sets in the order (i, j, k): the transversal meets A_i first, then A_j, and then A_k.

For each i, pick a point $x_i \in A_i$. For this theorem only, we will refer to the point x_i as the **center** of A_i. By shrinking A_i, we mean replacing A_i by a set

$$A_i' = x_i + \lambda(A_i - x_i),$$

where $0 \leq \lambda \leq 1$. Note that by the convexity of A_i, we have $A_i' \subset A_i$, so that any transversal for the shrunken family will also be a transversal for the original family.

For each triple (A_i, A_j, A_k) with $i < j < k$, let $\lambda_{(i,j,k)}$ be the smallest scalar such that the entire family $\mathcal{G}$ can be shrunk by the factor $\lambda_{(i,j,k)}$ while still preserving the property that the three shrunken sets A_i', A_j', and A_k' are intersected by a transversal in the order (i, j, k) (compactness of each A_m together with a convergence argument will prove the existence of a smallest $\lambda_{(i,j,k)}$).

Since there are only finitely many triples, among all the factors $\lambda_{(i,j,k)}$, there is a largest one, which we will denote by λ_0. If it so happens that $\lambda_0 = 0$, then each A_m' will be a point, and the fact that every three are intersected by some line implies that they are all collinear—and the theorem would be proved. If it happens that $\lambda_0 > 0$, then the reason must be that there is some triple (A_i', A_j', A_k') from the shrunken family that cannot be shrunk further.

This means that there is a line L meeting A_i', A_j', and A_k' in the order (i, j, k), but that if the sets are shrunk further, no line intersects them in the order (i, j, k). Choose points p_i, p_j, and p_k from the intersection of L with A_i', A_j', and A_k', respectively. Let U and V denote the two closed halfspaces determined by L. Now, only one of A_i' and A_j' can contain interior points of U.

If both A_i' and A_j' contained interior points of U, it would be possible to find points $u_i \in A_i'$ and $u_j \in A_j'$, both interior to U, so it would be possible to find a line M through p_k that intersects the segments (u_i, p_i) and (u_j, p_j) as in the figure below.

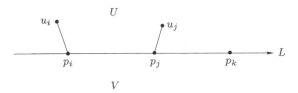

Thus, no matter where x_k is, there would be a line parallel to M that intersects the three segments (u_i, p_i), (u_j, p_j), and (x_k, p_k) in the order (i, j, k). However, this means that it would be possible to shrink A'_i, A'_j, and A'_k even further, which is a contradiction.

A similar argument applied to the closed halfspace V shows that only one of A'_i and A'_j can contain interior points of V. In other words, A'_i and A'_j lie in opposite halfspaces. Similarly, A'_j and A'_k must also lie in opposite halfspaces, so we can conclude that A'_i and A'_k lie in the same closed halfspace, say U, while A'_j lies in the opposite closed halfspace, V. The situation is as depicted in the figure below.

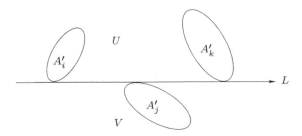

It is evident that if H is any transversal for the pair A'_i and A'_j and H is directed so that it meets the pair in the order (i, j), then any point on H before A'_i is in U, and any point after A'_j is in V. Similarly, if M is any transversal for the pair A'_j and A'_k, then on M, any point before A'_j is in V and any point after A'_k is in U. Finally, given any transversal for A'_i and A'_k, the points on the transversal between A'_i and A'_k must lie in U.

Now let A'_n be any other member of $\mathcal{G}'$. If $n < i$, then there is a transversal for A'_n, A'_i, and A'_j intersecting them in the order (n, i, j), so A'_n contains points of U. Also, there is a transversal for A'_n, A'_j and A'_k intersecting them in the order (n, j, k), so A'_n contains points of V. Thus, A'_n contains points in both of the closed halfspaces U and V, which means that A'_n intersects L. Similarly, if $i < n < j$, then consideration of the triples (A'_i, A'_n, A'_k) and (A'_j, A'_n, A'_k) will show that A'_n intersects L. Similarly, when $j < n < k$ and when $k < n$, it follows that A'_n intersects L. This shows that all members of $\mathcal{G}'$ intersect L and completes the proof.

$\square$

The preceding theorems have dealt in depth with the problem of extending Helly's theorem to common transversals. We have seen that the extension is only possible when restrictions are imposed on the family of sets. There are not many results concerning line transversals in $\mathbb{R}^n$ for $n \geq 3$, For the most part, the question is wide open in spaces of dimension 3 or more.

There are extensions of Helly's theorem to other settings. The following theorem is truly a full generalization. The proof depends on Helly's original theorem, that is, Theorem 4.2.1.

Theorem 4.3.22. *Let C be a convex subset of $\mathbb{R}^n$. Let $\mathcal{G}$ be a family of at least $n + 1$ convex subsets of $\mathbb{R}^n$ such that either $\mathcal{G}$ has only finitely many members or that C and all the members of $\mathcal{G}$ are compact. If for each $(n+1)$-membered subfamily of $\mathcal{G}$, there is a translate of C that intersects [contains; is contained in] the $n + 1$ members of $\mathcal{G}$, then there is a translate of C that intersects [contains; is contained in] all members of $\mathcal{G}$.*

There are three separate theorems here, all of which may be proven using the same approach. The idea is as follows, where we use the symbol "o" to denote the words "intersects" or "contains" or "is contained in."

For each $A \in \mathcal{G}$, let

$$A' = \{x \in \mathbb{R}^n : (x + C) o A\}.$$

Then it can be shown that A' is convex (for example, see Problems 1 and 2 in Section 3.2.1).

It follows from the hypotheses of the theorem that every $n + 1$ of the sets A' has a point in common, and then Helly's theorem applies and the theorem follows.

This is explored in more detail in Example 4.3.5, Theorem 4.3.33, and in Problems 4 and 5 in Problem set 4.3.9.

4.3.4 Theorems of Radon and Carathéodory

The next two theorems are closely related to Helly's theorem. In fact, it can be proved that all three theorems are logically equivalent, but we leave that as an exercise.

Theorem 4.3.23. *(Radon)*

If S is any subset of $\mathbb{R}^n$ containing at least $n + 2$ elements, then S can be decomposed into two disjoint subsets whose convex hulls intersect.

Proof. We know from Linear Algebra that any m vectors $v_1, v_2, \ldots, v_m$ in $\mathbb{R}^n$ are linearly dependent if $m > n$, that is, there are scalars $\alpha_1, \alpha_2, \ldots, \alpha_m$, not all zero, such that

$$\alpha_1 v_1 + \alpha_2 v_2 + \cdots + \alpha_m v_m = \overline{0}.$$

Now let $X = \{x_1, x_2, \ldots, x_{n+1}, x_{n+2}\}$ be any subset of S of size $n + 2$. It suffices to show that we can partition X into disjoint sets Y and Z, that is, $X = Y \cup Z$ with $Y \cap Z = \emptyset$, such that

$$\operatorname{conv}(Y) \cap \operatorname{conv}(Z) \neq \emptyset,$$

since the remaining elements, namely, $S \setminus X$, can be adjoined to either Y or Z.

Since $n + 2 > n + 1$, the system of $n + 1$ linear equations in $n + 2$ unknowns (the α_i's) below

$$\alpha_1 x_1 + \alpha_2 x_2 + \cdots + \alpha_{n+2} x_{n+2} = \overline{0}$$
$$\alpha_1 + \alpha_2 + \cdots + \alpha_{n+2} = 0$$

has a nontrivial solution. Hence, there exist scalars $\alpha_1, \alpha_2, \ldots, \alpha_{n+2}$, not all zero, which satisfy the $n + 2$ equations above.

Since $\sum_{i=1}^{n+2} \alpha_i = 0$, and not all of the α_i's are zero, if we let

$$I = \{i : 1 \leq i \leq n + 2, \ \alpha_i > 0\} \quad \text{and} \quad J = \{i : 1 \leq i \leq n + 2, \ \alpha_i \leq 0\},$$

then $I \neq \emptyset$, $J \neq \emptyset$, $I \cap J = \emptyset$, and there exists an index $j \in J$ such that $\alpha_j < 0$.

Now let

$$\sum_{i \in I} \alpha_i = \alpha = -\sum_{i \in J} \alpha_i.$$

Clearly, $\alpha > 0$ and

$$\sum_{i=1}^{n+2} \alpha_i x_i = \overline{0},$$

so that

$$\sum_{i=1}^{n+2} \left(\frac{\alpha_i}{\alpha} \right) x_i = \overline{0}$$

also.

Now let

$$x_0 = \sum_{i \in I} \left(\frac{\alpha_i}{\alpha} \right) x_i = - \sum_{i \in J} \left(\frac{\alpha_i}{\alpha} \right) x_i.$$

If

$$Y = \{x_i : i \in I\} \quad \text{and} \quad Z = \{x_i : i \in J\},$$

then $x_0 \in \text{conv}(Y) \cap \text{conv}(Z)$.

Thus, $X = Y \cup Z$, where $Y \cap Z = \emptyset$ and $\text{conv}(Y) \cap \text{conv}(Z) \neq \emptyset$.

$\square$

We know that if $x \in \mathbb{R}^n$ is in the convex hull of a set $S \subset \mathbb{R}^n$, then we can find a finite subset X of S (depending on x) such that x is a convex combination of members of X. In fact, Carathéodory's theorem (Theorem 3.3.10) says that in $\mathbb{R}^n$, not only can we find a finite subset X but also we can find one with no more than $n + 1$ members. The standard proof of this result is derived from Radon's theorem.

Theorem 4.3.24. *(Carathéodory)*
If $S \subset \mathbb{R}^n$, then each point of $\text{conv}(S)$ is a convex combination of $n + 1$ or fewer points of S.

Proof. If $y \in \text{conv}(S)$, then y is a convex combination of some finite subset of S, that is,

$$y = \sum_{i=1}^m \alpha_i x_i,$$

where $x_i \in S$ and $\alpha_i \geq 0$ for $1 \leq i \leq m$, and $\sum_{i=1}^m \alpha_i = 1$. We will show that if $m \geq n+2$, then it is possible to write y as a convex combination of at most $m - 1$ elements of S.

From Radon's theorem, if we have $m \geq n + 2$, then we can partition the set $\{x_1, x_2, \ldots, x_m\}$ into two disjoint subsets whose convex hulls intersect. By relabeling the indices if necessary, we may assume that the convex hulls of $\{x_1, \ldots, x_k\}$ and $\{x_{k+1}, \ldots, x_m\}$ have a point in common.

Thus, there are nonnegative scalars β_i, for $1 \leq i \leq k$, and γ_i, for $k+1 \leq i \leq m$, such that

$$\sum_{i=1}^k \beta_i = 1 \quad \text{and} \quad \sum_{i=k+1}^m \gamma_i = 1,$$

and

$$\sum_{i=1}^k \beta_i x_i = \sum_{i=k+1}^m \gamma_i x_i,$$

so that

$$\sum_{i=1}^{k} \beta_i x_i - \sum_{i=k+1}^{m} \gamma_i x_i = \overline{0}.$$

Now, for $k+1 \leq i \leq m$, let $\beta_i = -\gamma_i$. The preceding equation then shows that there are scalars β_i, not all zero, such that

$$\sum_{i=1}^{m} \beta_i x_i = \overline{0}$$

$$\sum_{i=1}^{m} \beta_i = 0.$$

Now let k_0 be such that $|\alpha_{k_0}/\beta_{k_0}|$ is the smallest among all those $|\alpha_i/\beta_i|$ with $\beta_i \neq 0$. By replacing each β_i by $-\beta_i$, if necessary, we may assume that β_{k_0} is negative. Thus, we may write y as

$$y = \sum_{i=1}^{m} \left(\alpha_i + \frac{\alpha_{k_0}}{\beta_{k_0}} \beta_i \right) x_i.$$

The coefficients satisfy

$$\alpha_i + \frac{\alpha_{k_0}}{\beta_{k_0}} \beta_i \geq 0 \quad \text{and} \quad \sum_{i=1}^{m} \left(\alpha_i + \frac{\alpha_{k_0}}{\beta_{k_0}} \beta_i \right) = 1,$$

while the coefficient of x_{k_0} is 0. Thus, y is a convex combination of $m-1$ points, and we can repeat this argument until y is a convex combination of at most $n+1$ elements of S. This completes the proof.

$\square$

4.3.5 Kirchberger's Theorem

The next result was discovered by P. Kirchberger in 1902, and his proof did not make use of Helly's theorem; in fact, his proof was almost 24 pages long. The following proof was discovered by H. Rademacher and I. J. Schoenberg in 1950. It is extremely brief and illustrates the power of Helly's theorem.

Theorem 4.3.25. *(Kirchberger)*
Let A and B be finite subsets of $\mathbb{R}^n$ such that $A \cup B$ contains at least $n+2$ points. If for each set $C \subset A \cup B$ containing $n+2$ points, the sets $A \cap C$ and $B \cap C$ can be strictly separated, then A and B can be strictly separated.

Proof. For each $a \in A$ and $b \in B$, let $\tilde{a} = (a_1, a_2, \ldots, a_n, 1) \in \mathbb{R}^{n+1}$ and $\tilde{b} = (b_1, b_2, \ldots, b_n, 1) \in \mathbb{R}^{n+1}$. Let H_a and H_b be the open halfspaces in $\mathbb{R}^{n+1}$ defined by

$$H_a = \{x \in \mathbb{R}^{n+1} : \langle \tilde{a}, x \rangle < 0\} \quad \text{and} \quad H_b = \{x \in \mathbb{R}^{n+1} : \langle \tilde{b}, x \rangle > 0\}.$$

Let $\mathcal{F}$ be the family of these open halfspaces H_a and H_b for $a \in A$ and $b \in B$.

Now let C be a set of $n+2$ points of $A \cup B$. Since $A \cap C$ and $B \cap C$ can be strictly separated, there exist $\tilde{c} = (c_1, c_2, \ldots, c_n, c_{n+1}) \in \mathbb{R}^{n+1}$ such that

$$\langle \tilde{c}, \tilde{a} \rangle < 0 \quad \text{and} \quad \langle \tilde{c}, \tilde{b} \rangle > 0$$

for $a \in A \cap C$ and $b \in B \cap C$. Thus, $\tilde{c} \in H_a$ when $a \in A \cap C$ and $\tilde{c} \in H_b$ when $b \in B \cap C$.

Summarizing, $\mathcal{F}$ is a finite family of convex subsets of $\mathbb{R}^{n+1}$ that contains at least $n+2$ members and every $n+2$ members of $\mathcal{F}$ have nonempty intersection. By Helly's theorem in $\mathbb{R}^{n+1}$, there is a point $\tilde{d} = (d_1, d_2, \ldots, d_n, d_{n+1})$ belonging to all members of $\mathcal{F}$. Since the intersection of the finitely many members of $\mathcal{F}$ is open, we may assume that not all of $d_1, d_2, \ldots, d_n$ are zero. Thus,

$$\langle \tilde{d}, \tilde{a} \rangle < 0$$

for all $a \in A$, while

$$\langle \tilde{d}, \tilde{b} \rangle > 0$$

for all $b \in B$.

Therefore, the hyperplane in $\mathbb{R}^n$ given by

$$H_d = \{x \in \mathbb{R}^n : d_1 x_1 + d_2 x_2 + \cdots + d_n x_n + d_{n+1} = 0\}$$

strictly separates A and B.

$\square$

In the book *Convex Sets* by F. A. Valentine, Kirchberger's theorem is introduced with the following example.

Example 4.3.26. *If a field contains a number of stationary black and white sheep, what simple condition will guarantee that a straight line exists which separates the black sheep from the white sheep?*

Solution. Now that we have Kirchberger's theorem, the answer is obvious.

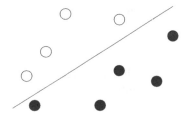

With $n = 2$, if A is the set of black sheep and B is the set of white sheep, the theorem states that if for every set $C \subset A \cup B$ containing $n + 2 = 4$ sheep, the sets $A \cap C$ and $B \cap C$ can be strictly separated, then A and B can be strictly separated.

□

4.3.6   Helly-type Theorems for Circles

In this section, we mean a circle, not a disk. The theorems involve *arcs of circles* instead of convex sets. Of course, all arcs are assumed to lie on the same circle. Basically, the idea is as follows: a circle is something like a line, and arcs on the circle are like segments on a line. Thus, one might expect a theorem that says something similar to the following:

> *If every two arcs of a family of arcs intersect, then there is a point in common to all of the arcs in the family.*

Example 4.3.27. *Using diagrams, give a counterexample to the above statement.*

Solution. Consider the family of three arcs in the figure below.

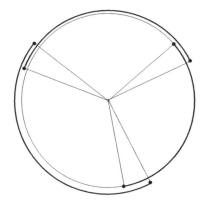

Clearly, every two arcs intersect, but the intersection of all three arcs is empty, that is, there is no point in common to all of the three arcs.

□

However, we will show in the following pages that there are Helly-type theorems for arcs. Some of the results are achieved by imposing restrictions on the size of the arcs to prevent them from "wrapping" around the circle.

Note. In the following, a semicircle S in the plane contains its endpoints.

Theorem 4.3.28. *Suppose that $\mathcal{F}$ is a family of arcs, all smaller than a semi-circle. Suppose also that either $\mathcal{F}$ has only finitely many members or that all of the arcs are closed. If each three of the arcs have a point in common, then all of the arcs in the family $\mathcal{F}$ have a point in common.*

Proof. We use the classic *transform-solve-invert* strategy, that is, first transform the problem into a different setting, then solve the problem in that setting, and finally, invert the solution back to the original setting.

Given an arc $S \in \mathcal{F}$, consider the set $S' = \text{conv}(S)$, and let $\mathcal{F}'$ be the family of all the sets S' such that $S \in \mathcal{F}$. The sets S' never contain the center O of the circle, so if p is any point in S', then $p \neq \overline{0}$.

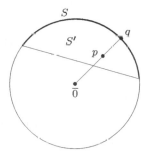

Thus, the ray from $\overline{0}$ through p pierces the circle at some point q, and q must be a point of S. So if all the sets S' contain a point p, then all of the arcs S contain the corresponding point q.

With S' defined as above, note that if S is closed, then S' is compact; or, if $\mathcal{F}$ is finite, then $\mathcal{F}'$ is also finite. Therefore, $\mathcal{F}'$ is either a family of compact convex sets, or a finite family of convex sets, such that every three of the sets in $\mathcal{F}'$ have a point in common.

By applying Helly's theorem, all of the sets S' in the family $\mathcal{F}'$ have a point p_0 in common, and therefore, all of the arcs S in the family $\mathcal{F}$ have a point q_0 in common.

$\square$

Example 4.3.29. *Draw a diagram showing that four semicircles can have the property that every three intersect, while there is no point in common to all four.*

Solution. Note that the intersection of two semicircles is not necessarily convex. The figure below has the property that every three semicircles have a point in common, but not all four.

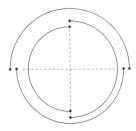

$\square$

Example 4.3.30. *Draw a diagram of five arcs with the property that every four of them intersect, while there is no point in common to all five.*

Solution. Let C be the circle, in this case, for each arc S_i, $1 \le i \le 5$, we consider the ***complement*** S_i' of the arc that we want, that is, $S_i' = C \setminus S_i$.

Note that if $\displaystyle\bigcup_{i=1}^{5} S_i' = C$, then $\displaystyle\bigcap_{i=1}^{5} S_i = \emptyset$, and if the union of any four of the S_i' does not contain the entire circle C, then the intersection of the four corresponding S_i cannot be empty.

The figure below has the property that the union of all five of the complements is the entire circle C, but the union of any four of the complements does not contain C.

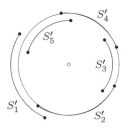

$\square$

Theorem 4.3.31. *Suppose that $\mathcal{F}$ is a family of arcs of a circle, all smaller than a third of the circle. Suppose also that $\mathcal{F}$ is finite or that all of the arcs are closed. If each two of the arcs have a point in common, then all arcs of the family have a point in common.*

Proof. There are two strategies that we can use to prove the theorem. One of them is to show that if every two of the arcs intersect, then every three of them intersect, and then apply Theorem 4.3.28. We leave this as an exercise.

The second strategy is to show that the union of all the arcs cannot be the entire circle; this means that they leave some point of the circle uncovered. We then cut the circle at this point, unroll the circle into a straight line segment, and apply Helly's theorem for intervals in $\mathbb{R}$ (Theorem 4.2.7). It is this one that we will use.

Let A be one of the arcs, and suppose that the angular measure of the arc is θ, so that $\theta < 2\pi/3$. Without loss of generality, we may assume that the arc extends from 0 to θ. If B is any other arc in the family, by hypothesis, it must intersect A.

Since B is smaller than a third of the circumference of the circle, the clockwise endpoint of B must be between $-2\pi/3$ and θ and the counterclockwise endpoint must be between 0 and $\theta + 2\pi/3$. Thus, B must fall within the counterclockwise arc X from $-2\pi/3$ to $\theta + 2\pi/3$. Since $\theta < 2\pi/3$, it follows that the angular measure of X is strictly less than 2π. Thus, there is a point p on the circle that is not contained in X, as shown in the figure below.

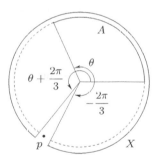

Thus, none of the arcs of the family contain p, so we cut the circle at this point, unroll it and the arcs into a straight line segment, and apply Helly's theorem for intervals in $\mathbb{R}$.

$\square$

Theorem 4.3.32. *Suppose that $\mathcal{F}$ is a family of arcs of a circle and that $\mathcal{F}$ is finite or that all of the arcs are closed. If each two of the arcs have a point in common, then there is an antipodal pair of points such that each of the arcs includes at least one point of the pair. Thus, there is a diameter of the circle that intersects all of the arcs.*

Proof. We may assume that the circle is the unit circle in the plane. Consider the perpendicular projection of arc onto the x-axis. It is an interval, and it is closed if the arc is closed.

Now consider what happens to the interval if the circle and arc are together rotated about $\bar{0}$ through some angle α. The figure below indicates what happens for selected values of α.

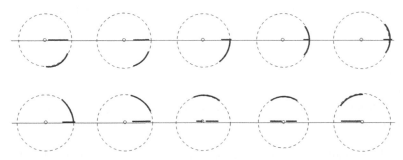

It should be clear that as α changes continuously, the projected interval varies continuously and the intervals for the angle α and $\alpha + \pi$ are precisely the negatives of each other.

Consider what happens when we project all of the arcs from $\mathcal{F}$ onto the x-axis. In this case, we get a family of intervals, and every pair of these intersect since every pair of the arcs intersect. Thus, Helly's theorem for $\mathbb{R}^1$ applies, and so we know that the intersection of all of the intervals is either a point or a nonempty interval.

For each rotation angle α, we denote the interval as I_α, and note that the sets I_0 and I_π are the negatives of each other. It follows from the previous considerations that I_α varies continuously with α. Consequently, there must be at least one angle α_0 such that I_{α_0} contains $\bar{0}$. But this means that after rotating the circles and arcs through the angle α_0, the orthogonal projection of every arc onto the x-axis contains $\bar{0}$. In other words, after the rotation, every arc contains either the point $(0, 1)$ or the point $(0, -1)$. Thus, before the rotation, every arc contains either the point $(\sin \alpha_0, \cos \alpha_0)$ or its antipode $(-\sin \alpha_0, -\cos \alpha_0)$.

$\square$

4.3.7 Covering Problems

Helly's theorem also applies to a variety of other problems in the plane. We will discuss two types that have drawn a lot of attention. The first type is commonly called a *covering problem*. Broadly stated a covering problem asks

Under what circumstances can one set or a family of sets be covered by a congruent copy of a given set?

We do have one rather general result.

Theorem 4.3.33. *Suppose that $\mathcal{G}$ is a family of at least $n+1$ convex subsets of $\mathbb{R}^n$ and that C is a convex subset of $\mathbb{R}^n$. Suppose also that either $\mathcal{G}$ has only finitely many members or that C and all of the members of $\mathcal{G}$ are compact. If for each $n+1$ members of $\mathcal{G}$, there is a translate of C that contains the $n+1$ members, then there is a translate of C that contains all members of $\mathcal{G}$.*

Proof. For each $A \in \mathcal{G}$, let $A' = \{x \in \mathbb{R}^n : A \subset (x + C)\}$. From Section 3.2.1, Problem 1(b), each set A' is convex. It follows from the hypotheses of the theorem that every $n+1$ of the sets A' have a point in common. Therefore, Helly's theorem implies that there is a point x_0 common to all of the sets A', and this means that $A \subset x_0 + C$ for each $A \in \mathcal{G}$.

$\square$

Problem. *Given a subset of the plane with diameter d, what is the diameter of the smallest circular disk that contains the set?*

Most people, upon first reading the problem, immediately conclude that a disk of the same diameter will do the job. Alas, this is not the case: an equilateral triangle whose sides are of unit length cannot be contained in a disk of diameter 1. The correct answer is given by the following theorem, which is the planar version of what is known as **Jung's theorem**.

Theorem 4.3.34. *If a subset S of the plane has diameter 1, then it can be contained in a closed disk of radius $\dfrac{1}{\sqrt{3}}$.*

Proof. Note that if D is a closed disk in the plane and if every three points of S are contained in some translate of D, then by the previous theorem, the entire set is contained in some translate of D. As a consequence, we need only to prove that every three points that form a set of diameter 1 can be contained in a disk of radius $1/\sqrt{3}$.

In other words, we have to prove that a triangle whose longest side is of length 1 has a circumradius no greater than $1/\sqrt{3}$.

A glance at part (a) of the following figure makes this obvious, if the side $[a, b]$ is of length 1, then the third vertex of the triangle must be within the shaded area, and as shown in part (b) of the following figure, the shaded area is contained in a disk of radius $1/\sqrt{3}$.

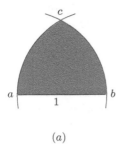

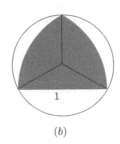

(a) (b)

$\square$

The result for $\mathbb{R}^n$ was proved by H. W. E. Jung in 1901 and is one of the oldest and most well-known results in combinatorial geometry. It states that every subset of $\mathbb{R}^n$ with diameter d is contained in some closed ball $\overline{B}(x_0, R)$ of radius

$$R = \sqrt{\frac{n}{2(n+1)}}\, d.$$

Here, we are assuming that $\mathbb{R}^n$ is endowed with the Euclidean norm.

Before proving Jung's theorem, we first prove a lemma concerning a compact convex subset $C \subset \mathbb{R}^n$ with nonempty interior and diameter d. We will show that for such a set, there exists a unique point x_0 and a unique positive number R such that $C \subset \overline{B}(x_0, R)$. The point x_0 is called the **circumcenter** of C and the number R is called the **circumradius** of C and is defined to be

$$R = \inf\{\rho > 0 : C \subset B, \text{ where } B \text{ is a closed ball of radius } \rho \text{ in } \mathbb{R}^n\},$$

while the ball $\overline{B}(x_0, R)$ itself is called the **circumball** of C.

We note first that since $\text{int}(C) \neq \emptyset$ and C is bounded, that $0 < R < \infty$.

Lemma 4.3.35. *If C is a compact convex subset of $\mathbb{R}^n$ with nonempty interior, that is, a convex body, then C is contained in a unique closed ball of radius R, where R is the greatest lower bound of the radii of all closed balls that contain C.*

Proof. Since R is a well-defined positive real number, from the properties of the infimum, for each integer $j \geq 1$, there exists a point $x_j \in \mathbb{R}^n$ and a real number $r_j > 0$ such that $C \subset \overline{B}(x_j, r_j)$ and $R \leq r_j < R + \dfrac{1}{j}$.

Clearly, the sequence $\{r_j\}_{j \geq 1}$ converges to R. Also, each of the closed balls $\overline{B}(x_j, r_j)$ contains C and their radii converge to R; from this, it follows that the sequence $\{x_j\}$ must be bounded. Therefore, since bounded sequences in $\mathbb{R}^n$ have convergent subsequences, there exists a subsequence $\{x_{j_k}\}_{k \geq 1}$ that converges to a point $x_0 \in \mathbb{R}^n$; and the corresponding subsequence $\{r_{j_k}\}_{k \geq 1}$ converges to R since $\{r_j\}_{j \geq 1}$ converges to R.

Now, suppose that $x \in C$. For each $k \geq 1$, we have $C \subset \overline{B}(x_{j_k}, r_{j_k})$, so that

$$\|x - x_{j_k}\| \leq r_{j_k} < R + \frac{1}{j_k}.$$

Letting $k \to \infty$, we have $\|x - x_0\| \leq R$, so that $x \in \overline{B}(x_0, R)$, and since $x \in C$ was arbitrary, then $C \subset \overline{B}(x_0, R)$.

To show that this *circumball* is unique, suppose that C is contained in two different closed balls $\overline{B}(x_0, R)$ and $\overline{B}(y_0, R)$, then for each $x \in C$, we have

$$\|x - \tfrac{1}{2}(x_0 + y_0)\|^2 = \tfrac{1}{2}\|x - x_0\|^2 + \tfrac{1}{2}\|x - y_0\|^2 - \tfrac{1}{4}\|x_0 - y_0\|^2$$
$$\leq R^2 - \tfrac{1}{4}\|x_0 - y_0\|^2.$$

This follows immediately from the *parallelogram law*:

$$\|a + b\|^2 + \|a - b\|^2 = 2\|a\|^2 + 2\|b\|^2$$

for $a, b \in \mathbb{R}^n$ by taking $a = \tfrac{1}{2}(x - x_0)$ and $b = \tfrac{1}{2}(x - y_0)$.

Therefore, C is also contained in the closed ball centered at $z_0 = \tfrac{1}{2}(x_0 + y_0)$ with radius $r = \sqrt{R^2 - \tfrac{1}{4}\|x_0 - y_0\|^2}$, which is less than R. Thus, we must have $x_0 = y_0$, and there is exactly one closed ball of radius R with $C \subset \overline{B}(x_0, R)$.

□

Similarly, if C is a compact convex subset of $\mathbb{R}^n$ with nonempty interior and diameter d, we define

$$r = \sup\{\rho > 0 : B \subset C, \text{ where } B \text{ is a closed ball of radius } \rho \text{ in } \mathbb{R}^n\},$$

then $0 < r \leq d/2$, and r is called the *inradius* of the convex body C. Any closed ball of radius r contained in C is called an *inball* of C and its center is called an *incenter* of C.

Although the circumball and circumcenter of a convex body are unique, the incenter of a convex body is not unique, as a nonsquare rectangle in the plane shows.

Note also that the definitions of *circumradius* and *circumcenter* above do *not* coincide with those used in elementary Euclidean geometry. For example, if we consider an isosceles triangle with base angles equal to $\pi/6$ and side lengths $a, a, a\sqrt{3}$. According to the definitions above, the circumradius is $\sqrt{3}$ and its circumcenter is the midpoint of its longest side.

Theorem 4.3.36. *(Jung's Theorem)*

Let C be a subset of $\mathbb{R}^n$ endowed with the Euclidean norm. If $\mathrm{diam}(C) = d$, then C lies in a closed ball of radius

$$R = \sqrt{\frac{n}{2(n+1)}}\, d.$$

Proof. We suppose first that C contains only finitely many points, let us assume that C has at most $n+1$ points and that the circumcenter of C is $\overline{0}$. Let $B = \{x_0, x_1, \ldots, x_m\}$ be the points of C such that

$$\|x_i - \overline{0}\| = \|x_i\| = R$$

for $i = 0, 1, \ldots, m$, where R is the circumradius of C, so that $0 < m \leq n$.

Note that the minimality of R implies that

$$\overline{0} \in \mathrm{conv}(\{x_0, x_1, \ldots, x_m\}) = \mathrm{conv}(B),$$

since otherwise, if we take v to be the nearest point of $\mathrm{conv}(B)$ to $\overline{0}$, then for $\epsilon > 0$ and sufficiently small, the point ϵv would be the center of a closed ball containing C with radius less than R.

Since $\overline{0} \in \mathrm{conv}(B)$, there exist nonnegative scalars $\lambda_0, \lambda_1, \ldots, \lambda_m$ with

$$\lambda_0 + \lambda_1 + \cdots + \lambda_m = 1$$

such that

$$\lambda_0 x_0 + \lambda_1 x_1 + \cdots + \lambda_m x_m = \overline{0}.$$

For each $j = 0, 1, \ldots, m$, we have

$$d^2(1 - \lambda_j) = d^2 \sum_{i=0}^{m} \lambda_i - d^2 \lambda_j = d^2 \sum_{i \neq j} \lambda_i$$

$$\geq \sum_{i \neq j} \lambda_i \|x_i - x_j\|^2 = \sum_{i=0}^{m} \lambda_i \|x_i - x_j\|^2$$

$$= \sum_{i=0}^{m} \lambda_i \left(\|x_i\|^2 - 2\langle x_i, x_j \rangle + \|x_j\|^2 \right)$$

$$= \sum_{i=0}^{m} \lambda_i \left(2R^2 - 2\langle x_i, x_j \rangle \right),$$

so that

$$d^2(1 - \lambda_j) \geq 2R^2 - 2\langle (\sum_{i=0}^{m} \lambda_i x_i), x_j \rangle$$
$$= 2R^2,$$

since $\sum_{i=0}^{m} \lambda_i x_i = \overline{0}$.

Therefore, for each $j = 0, 1, \ldots, m$, we have

$$2R^2 \leq d^2(1 - \lambda_j),$$

and adding these inequalities, we obtain

$$2(m + 1)R^2 \leq md^2,$$

so that

$$R \leq \sqrt{\frac{m}{2(m + 1)}} \, d \leq \sqrt{\frac{n}{2(n + 1)}} \, d.$$

This completes the proof for the case where C contains at most $n + 1$ points.

In the case where C has more than $n + 1$ points, we let $\mathcal{F}$ be the family of all closed balls of radius $\sqrt{\frac{n}{2(n + 1)}} \, d$ having their centers in C. From the first part of the proof, every $n + 1$ members of $\mathcal{F}$ have a common point, the circumcenter of the set of their centers. Thus, Helly's theorem shows that there is some point x_0 belonging to every member of $\mathcal{F}$.

Therefore, C lies in the closed ball with center x_0 and radius $\sqrt{\frac{n}{2(n + 1)}} \, d$, which completes the proof.

$\square$

A compact convex subset K of $\mathbb{R}^n$ is said to be a **universal cover** if any set $S \subset \mathbb{R}^n$ of diameter 1 can be covered by a congruent copy of K.

The previous theorems assert that the closed disk of radius $1/\sqrt{3}$ is a universal cover in $\mathbb{R}^2$ and that the closed ball of radius $\sqrt{\frac{n}{2(n + 1)}}$ is a universal cover in $\mathbb{R}^n$ for $n > 2$.

There is, however, a "better" universal cover in the plane.

Theorem 4.3.37. *(Pál's Theorem)*
The regular hexagon whose incircle has diameter 1 is a universal cover in $\mathbb{R}^2$.

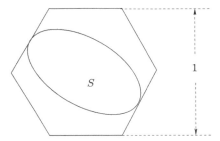

Proof. We will use the proof in the book *Geometric Études in Combinatorial Mathematics* by V. Boltyanski and A. Soifer. It does not use Helly's theorem.

Given a subset S of the plane with diameter 1, draw an initial ray m_0 from which to measure angles and construct two parallel support lines L_1 and L_2 to the set S, which make an angle ϕ_0 with the initial ray m_0 (see Example 2.6.15).

If x_1 and x_2 are the two support points in S, then $\|x_1 - x_2\| \leq 1$, and therefore, the perpendicular distance between L_1 and L_2 is less than or equal to 1. If this distance is less than 1, then we move these lines apart until the distance between them is equal to 1 (the dotted lines in the figure below).

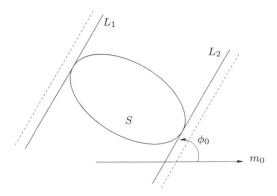

Now we construct two strips $S(\phi_0)$ and $S(\phi_0 + 60°)$. Their intersection is the rhombus containing S that has an angle of $60°$ at one vertex and a distance 1 between its opposite sides, as in the figure below.

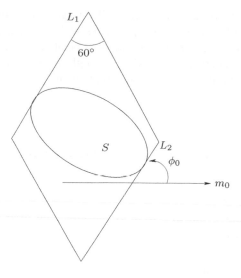

Finally, we construct the strip $S(\psi_0 + 120°)$, the dotted lines in the figure below. The intersection of the three strips is the hexagon $H(\phi)$ with angles of $120°$ containing S, as shown in the figure.

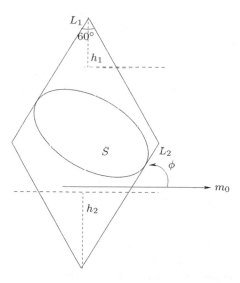

This hexagon is regular if the lengths of h_1 and h_2 are equal, and irregular otherwise. Let us assume that $h_1 \neq h_2$, say $h_1 < h_2$. Now we increase the angle ϕ from ϕ_0 to $\phi_0 + 180°$. Then, h_1 and h_2 will change continuously with

ϕ, and therefore, the distance $h_1 - h_2$ will change continuously. But when ϕ reaches $\phi_0 + 180°$, the lengths h_1 and h_2 will switch their roles. Thus, the difference, being negative at first, becomes positive in the end, and by the intermediate value theorem, there exists an angle ϕ_1 such that $h_1 - h_2 = 0$. At this angle ϕ_1, the hexagon $H(\phi_1)$ is regular, its incircle has diameter 1, and it contains S.

$\square$

Not much is known about universal covers in higher dimensions. This has some interesting consequences, as we will see later. Even for the two-dimensional case, there is a famous unsolved problem:*

Lebesgue Covering Problem. *What is the minimum area that a universal cover in $\mathbb{R}^2$ can have?*

4.3.8 Piercing Problems

The "Helly Number" for Helly's theorem is $n + 1$, and the conclusion of the theorem will not hold if this number is reduced. However, this does not mean that nothing can be concluded if the number $n + 1$ is replaced by a smaller one.

In the plane, one asks what can be concluded if every *two* members of the family have a point in common. The next application can be stated vividly in the following manner:

> *If every two members of a family of congruent disks in the plane can be pierced by a needle, then with three needles we can pierce all members of the family.*

The proof does not use Helly's theorem directly, but instead uses one of the previous applications of Helly's theorem.

Theorem 4.3.38. *Let $\mathcal{G}$ be a family of congruent closed disks in the plane. If every two members of $\mathcal{G}$ have a point in common, then there are three points such that every member of $\mathcal{G}$ contains at least one of those points.*

Proof. We may assume that the disks all have a diameter of 1 unit. Since every pair of disks have a point in common, an application of the triangle inequality shows that no two points in the union of $\mathcal{G}$ can be farther apart than 2 units.

In other words, $\bigcup \mathcal{G}$ forms a set of diameter at most two units. By Pál's theorem (Theorem 4.3.37), this means that $\bigcup \mathcal{G}$ can be covered by a hexagon H circumscribing a disk of diameter 2.

---

* See [4] for the latest upper and lower bounds as of the date of publication.

Now let a, b, and c be the vertices of an equilateral triangle as in the figure below.

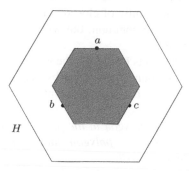

The center of each disk in $\mathcal{G}$ must be within the shaded region, otherwise the disk would not be contained in H.

The theorem now follows, since every point in the shaded region is at a distance $\delta \leq 1/2$ from at least one of the vertices a, b, or c.

$\square$

Remark. The theorem above is a special case of a conjecture of T. Gallai, and results of this nature are sometimes called ***Gallai-type theorems***. The "piercing number" 3 cannot be reduced, as the nine disks in the figure show.

4.3.9 Problems

1. Show that if the closed convex sets in Helly's theorem are not bounded, then there is an infinite family of closed convex subsets of $\mathbb{R}^2$ such that each three have a point in common, but with no point in common to all members of the family.

2. Show that if the convex sets in Helly's theorem are not closed, then there is an infinite family of bounded convex subsets of $\mathbb{R}^2$ such that each three have a point in common, but with no point in common to all members of the family.

3. Prove Theorem 4.3.1: *A polygon in the plane is star shaped if and only if for every three edges of the polygon, there is a point in the set from which all three edges are visible.*

4. Let C be a convex subset of $\mathbb{R}^n$. Let $\mathcal{G}$ be a family of convex subsets of $\mathbb{R}^n$ such that either $\mathcal{G}$ has only finitely many members or that C and all members of $\mathcal{G}$ are compact. If for each $n+1$ members of $\mathcal{G}$, there is a translate of C that is contained in all of the $n+1$ members, then there is a translate of C that is contained in all members of $\mathcal{G}$.

5. Let C be a convex subset of $\mathbb{R}^n$. Let $\mathcal{G}$ be a family of convex subsets of $\mathbb{R}^n$ such that either $\mathcal{G}$ has only finitely many members or that C and all members of $\mathcal{G}$ are compact. If for each $n+1$ members of $\mathcal{G}$, there is a translate of C that contains all of the $n+1$ members, then there is a translate of C that contains all members of $\mathcal{G}$.

6. Let $\mathcal{G}$ be a family of parallelograms with corresponding sides being parallel. If every two members of $\mathcal{G}$ have a point in common, then there is a point in common to all members of $\mathcal{G}$.

7. Use Theorem 4.3.32 to show that if a family $\mathcal{G}$ of compact convex sets in the plane is such that every two of its members have a point in common, then through each point of the plane there is a line that intersects all of the sets in the family.

8. Use Theorem 4.3.32 to show that if a family $\mathcal{G}$ of compact convex sets in the plane is such that every two of its members have a point in common, then for each line in the plane, there is a parallel line that intersects all of the sets in the family.

9. Show that any quadrilateral Q in the plane that is circumscribed about a parallelogram with altitudes a and b can be covered by a strip of width $a + b$.

10. Show that if a compact convex subset C of the plane can be covered by two strips of width a and b, then it can be covered by a single strip of width $a + b$.

11. Show that if a compact convex subset of the plane cannot be covered by any strip of width 1, then it contains a circle of radius $r = 1/3$.

12. Show that a compact convex subset C of the plane with nonempty interior can be covered by three of its translates C_1, C_2, and C_3 in such a way that $C \subset \operatorname{int}(C_1 \cup C_2 \cup C_3)$ if and only if C is *not* a parallelogram.

13. Show that the set of all points P such that a compact subset C of the plane is star shaped from a point in P is either empty or a compact convex subset of the plane.

4.4 SETS OF CONSTANT WIDTH

It is a fact that for a car to move parallel to the ground, it must have circular wheels. This is so obvious that no one wants to prove it. Of course, sometimes you have to move things that do not have wheels. Or sometimes the thing you are moving is so heavy that you have to use rollers instead of wheels. If we want to move a heavy concrete block smoothly along some rollers, then again it is obvious that the rollers should be circular in shape—like very smooth round logs—otherwise, the block will rise and fall as we push it over the rollers and make our work much more difficult.

Actually, the obvious "fact" that the rollers must be circular is false. There are noncircular rollers that will still let us do the job. The cross sections of the rollers in the figure below are sets with the property that any two distinct parallel supporting lines are the same constant distance apart.

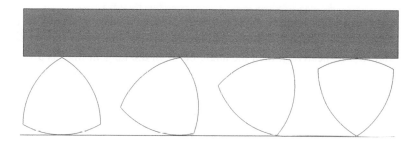

In general, any compact set C in $\mathbb{R}^n$ is said to be of **constant breadth** δ or **constant width** δ if every pair of parallel supporting hyperplanes are separated by a distance of δ units.

4.4.1 Reuleaux Triangles

In the plane, noncircular sets of constant width were first studied by Euler in the 1770s. The simplest example was given by Franz Reuleaux (1829–1905), who introduced them in machine design.

Example 4.4.1. *(Reuleaux Triangle)*
*Let $\triangle ABC$ be an equilateral triangle in the plane with the length of each side equal to a. With each vertex as center, draw the smaller arc of radius a joining the other two vertices. The union of these three arcs is the boundary of a **Reuleaux triangle**.*

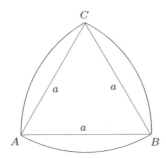

(a) *Find the angle between the tangent lines at each of the corner points A, B, and C.*

(b) *Show that if ℓ and m are two distinct parallel support lines to the Reuleaux triangle, then one of ℓ and m must pass through a vertex.*

(c) *Show that the Reuleaux triangle has constant width, that is, the distance between any two distinct parallel support lines is the same.*

(d) *Find the length of the perimeter of the Reuleaux triangle.*

Solution.

(a) Let ℓ and m be tangent lines to the Reuleaux triangle as shown in the figure below. Since ℓ and m are perpendicular to the radius vectors BC and AC, respectively, then the angle between the tangent lines is $\theta = \dfrac{\pi}{3}$.

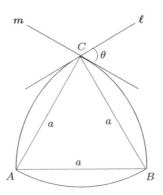

(b) Let ℓ and m be lines that support the Reuleaux triangle at the points x and y on different arcs, respectively, then these support lines are tangent to the circular arcs, which make up the Reuleaux triangle, and so the angle between the tangent lines is the angle between the rays from a vertex to a support point.

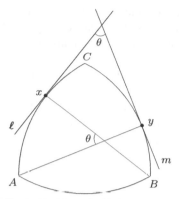

Therefore, if neither x nor y is a vertex, then the angle θ between the support lines through these points satisfies $0 < \theta < \dfrac{2\pi}{3}$, and ℓ and m are not parallel.

(c) We have shown above that if ℓ and m are parallel support lines, then one of them must pass through a vertex and the other is tangent to the circular arc opposite the vertex, so the distance between the support lines is a, the radius of the circular arc. Therefore, the Reuleaux triangle has constant width.

(d) The perimeter of the Reuleaux triangle consists of three circular arcs, each subtending an angle of $\dfrac{\pi}{3}$, so the perimeter is

$$p = 3 \cdot \frac{\pi a}{3} - \pi a.$$

There is a more general result: *Barbier's theorem* states all convex bodies in $\mathbb{R}^2$ of constant width a have perimeter πa.

$\square$

4.4.2 Properties of Sets of Constant Width

Let A be a nonempty compact subset of $\mathbb{R}^n$, and let ℓ be a line in $\mathbb{R}^n$. The *width of A in the direction of* ℓ is the distance between two parallel supporting hyperplanes of A that are perpendicular to ℓ and that contain A between them.

Lemma 4.4.2. *If $A \subset \mathbb{R}^n$ is a nonempty compact set, then the maximum width w of the set A is equal to the diameter of A, that is, $w = \mathrm{diam}(A)$.*

Proof. Let w be the maximum width of A and let H_1 and H_2 be parallel supporting hyperplanes of A at a distance w apart. If x and y are points of A lying on H_1 and H_2, respectively, we claim that $\|x - y\| = w$.

Clearly, if the line joining x and y is perpendicular to H_1, then $\|x - y\| = w$.

Suppose now that the line joining x and y is not perpendicular to H_1. Let H_1' and H_2' be parallel supporting hyperplanes of A that are perpendicular to the line joining x and y, as in the figure.

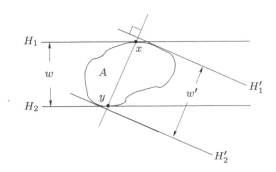

If w' is the width of the set A in the direction of the line joining x and y, then we have

$$w < \|x - y\| \leq w'.$$

However, this contradicts the maximality of w. Thus, we must have $\|x{-}y\| = w$, the maximum width of the set A.

To finish the proof, let x_0 and y_0 be any two points of A and let w_0 be the width of A in the direction of the line joining x_0 and y_0. We have

$$\|x - y\| = w \geq w_0 \geq \|x_0 - y_0\|,$$

and therefore, w is the diameter of the set A.

$\square$

Theorem 4.4.3. *If $A \subset \mathbb{R}^n$ is a compact convex set of constant width δ, then* $\mathrm{diam}(A) = \delta$.

Proof. Let a and b be two points in the set A. Also, let H_1 and H_2 be two support hyperplanes of A that are perpendicular to the segment $[a, b]$ with corresponding support points $p_1 \in H_1 \cap A$ and $p_2 \in H_2 \cap A$. Let q be the point in H_2 such that the segment $[p_1, q]$ is parallel to $[a, b]$, as in the following figure.

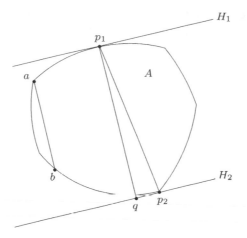

We have
$$\|a - b\| \leq \|p_1 - q\| = \delta,$$
and since a and b in A are arbitrary, then
$$\mathrm{diam}(A) = \sup\{\|a - b\| : a, b \in A\} \leq \delta.$$

Conversely, let H_1 and H_2 be two parallel supporting hyperplanes to A, and let p_1 and p_2 be the corresponding support points, as in the above figure. We have
$$\|p_1 - p_2\| \geq \|p_1 - q\| = \delta,$$
and so
$$\mathrm{diam}(A) \geq \|p_1 \quad p_2\| \geq \delta.$$
Therefore, $\mathrm{diam}(A) = \delta$.

$\square$

Note. The reasoning in the proof shows that the point q must coincide with p_2 in the above figure. In other words, if H_1 and H_2 are two parallel supporting hyperplanes of a constant width compact convex set A, and p_1, p_2 are the corresponding support points, then the segment $[p_1, p_2]$ is perpendicular to H_1 and H_2.

If A is a compact convex set of constant width δ, any chord of A that has length δ is called a **diametral chord of A**.

Theorem 4.4.4. *Let $A \subset \mathbb{R}^n$ be a nonempty compact convex subset of constant width δ. If a is any boundary point of A, then a is an endpoint of at least one diametral chord of A.*

Proof. Let H_a be a supporting hyperplane of A through a and let H_b be the supporting hyperplane of A parallel to H_a with corresponding support point b.

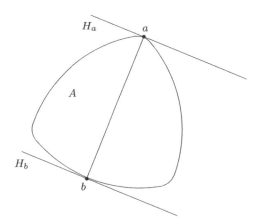

It follows from the previous theorem that $[a, b]$ is perpendicular to H_a and $\|a - b\| = \delta$; therefore, $[a, b]$ is a diametral chord of A.

$\square$

Theorem 4.4.5. *Let $A \subset \mathbb{R}^n$ be a nonempty compact convex subset of constant width δ. If $[a, b]$ is a diametral chord of A and H_a and H_b are the hyperplanes that are perpendicular to the segment $[a, b]$ and pass through a and b, respectively, then H_a and H_b are support hyperplanes of A with corresponding support points a and b.*

Proof. This follows immediately from Theorem 4.4.3, since the distance between the hyperplanes equals the maximum distance between points in the set A, that is, the diameter of the set A.

$\square$

Theorem 4.4.6. *Let $A \subset \mathbb{R}^n$ be a nonempty compact convex subset of constant width δ. If H is a supporting hyperplane of A, then $A \cap H$ consists of precisely one point.*

Proof. Suppose that H supports A at a. Let H' be a parallel supporting hyperplane that supports A at b. If there exists a point $c \neq a$ in $H \cap A$, then b and c are on parallel supporting hyperplanes H and H', but from the Pythagorean theorem, since one of $[a, b]$ or $[b, c]$ is not perpendicular to H, then either $\|a - b\| > \delta$ or $\|b - c\| > \delta$.

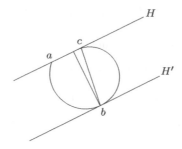

However, this contradicts the result in Theorem 4.4.3, so that $H \cap A$ consists of exactly one point.

□

Summarizing all this. We have shown that if H_1 and H_2 are parallel hyperplanes supporting a compact convex subset A of constant width δ and if x_1 and x_2 are corresponding support points for A, then $\|x_1 - x_2\| = \delta$. From this, it followed that every hyperplane that supports a set of constant width must support it at a unique point. This means that there are *no flat spots* on a set of constant width, that is, every set of constant width is ***strictly convex***.

Theorem 4.4.7. *If A and B are compact convex subsets of $\mathbb{R}^n$ and H_a and H_b are parallel supporting hyperplanes of A and B, respectively, both of which are on the same side of these hyperplanes, then the Minkowski sum $H = H_a + H_b$ is a supporting hyperplane for the compact convex subset $A + B$.*

Proof. Let u_a and u_b be the normals to the supporting hyperplanes for A and B, respectively. Since H_a and H_b are parallel, we may assume that $u_a = u_b$, so that if $u = u_a = u_b$, then

$$H_a = \{x \in \mathbb{R}^n : \langle u, x \rangle = \alpha\}$$
$$H_b = \{x \in \mathbb{R}^n : \langle u, x \rangle = \beta\},$$

so that

$$H = H_a + H_b = \{z \in \mathbb{R}^n : z = x + y, \text{ where } x \in H_a \text{ and } y \in H_b\}.$$

Thus, $z \in H$ if and only if $z = x + y$, where $\langle u, x \rangle = \alpha$ and $\langle u, y \rangle = \beta$, that is, if and only if $\langle u, z \rangle = \langle u, x + y \rangle = \alpha + \beta$, and

$$H = \{x \in \mathbb{R}^n : \langle u, x \rangle = \alpha + \beta\}$$

so that H is parallel to both H_a and H_b.

If $x_0 \in A \cap H_a$ and $y_0 \in B \cap H_b$ are support points for A and B, respectively, then

$$\langle u, x \rangle \leq \alpha \qquad \text{and} \qquad \langle u, y \rangle \leq \beta,$$

for all $x \in A$, $y \in B$, while $\langle u, x_0 \rangle = \alpha$ and $\langle u, y_0 \rangle = \beta$.

Thus, for $z = x + y \in A + B$, where $x \in A$ and $y \in B$, we have

$$\langle u, z \rangle = \langle u, x \rangle + \langle u, y \rangle \leq \alpha + \beta,$$

and if we let $z_0 = x_0 + y_0$, then $z_0 \in H$ and

$$\langle u, z_0 \rangle = \alpha + \beta.$$

Therefore, the hyperplane $H = H_a + H_b$ supports the compact convex set $A + B$ at the point $z_0 = x_0 + y_0$.

$\square$

Theorem 4.4.8. *If A and B are nonempty compact convex subsets of $\mathbb{R}^n$ and ℓ is a line in $\mathbb{R}^n$, then the width of the Minkowski sum $A + B$ in the direction ℓ is the width of A in the direction ℓ plus the width of B in the direction ℓ.*

Proof. Let u be a unit vector in the direction of the line ℓ. The widths of A, B, and $A + B$ in the direction of ℓ are

$$w(A) = \sup\{\langle u, x_1 - x_2 \rangle : x_1, x_2 \in A\},$$
$$w(B) = \sup\{\langle u, y_1 - y_2 \rangle : y_1, y_2 \in B\},$$
$$w(A + B) = \sup\{\langle u, z_1 - z_2 \rangle : z_1 - z_2 \in A + B\},$$

respectively.

Now, since A and B are nonempty compact convex sets, from Section 3.6.2, Problem 1(a), we have

$$
\begin{aligned}
w(A + B) &= \sup\{\langle u, z_1 - z_2 \rangle : z_1 - z_2 \in A + B\} \\
&= \sup\{\langle u, (x_1 + y_1) - (x_2 + y_2) \rangle : x_1, x_2 \in A, \ y_1, y_2 \in B\} \\
&= \sup\{\langle u, (x_1 - x_2) + (y_1 - y_2) \rangle : x_1, x_2 \in A, \ y_1, y_2 \in B\} \\
&= \sup\{\langle u, x_1 - x_2 \rangle : x_1, x_2 \in A\} + \sup\{\langle u, y_1 - y_2 \rangle : y_1, y_2 \in B\} \\
&= w(A) + w(B).
\end{aligned}
$$

$\square$

The following corollary shows that the class of sets of constant width is invariant under vector addition. Note that the statement in the corollary that the sets are compact can be replaced with the hypothesis that the sets are closed, since any set of constant width is bounded.

Theorem 4.4.9. *If A and B are nonempty compact convex subsets of $\mathbb{R}^n$ of constant widths δ_a and δ_b, respectively, then $A + B$ is a nonempty compact convex set of constant width $\delta = \delta_a + \delta_b$.*

Theorem 4.4.10. *If A is a nonempty compact convex subset of $\mathbb{R}^n$, then A has constant width δ if and only if*

$$A - A = \delta \cdot \overline{B}(\overline{0}, 1),$$

that is, if and only if the difference set $A - A$ is the closed ball of radius δ centered at the origin.

Proof. Let A be a nonempty compact convex set of constant width δ, since

$$\mathrm{diam}(A) = \mathrm{diam}(-A),$$

then $-A$ is also a nonempty compact convex set of constant width δ. From the preceding theorem, the set $A - A$ is also a nonempty compact convex set of constant width 2δ.

Now, if $x \in \mathrm{bdy}(A - A)$ and H is a supporting hyperplane to $A - A$ at x, since $A - A$ is symmetric with respect to the origin $\overline{0}$, then the hyperplane $H' = -H$ is parallel to H and supports $A - A$ at the point $-x$. The distance between these parallel supporting hyperplanes is 2δ and they are symmetric with respect to the origin. Since this is true for any point $x \in \mathrm{bdy}(A - A)$, then $A - A$ must be a ball of radius δ.

Conversely, if $A - A$ is a closed ball of radius δ, then the width of $A - A$ in any direction is equal to 2δ.

If δ' is the width of A in a given direction, then $-A$ has the same width δ' in this direction, and therefore, $A - A$ has width $2\delta'$ in this direction. Thus, $2\delta' = 2\delta$, so that $\delta' = \delta$. This implies that A has width δ in any direction, so that A is a compact convex subset of constant width δ.

$\square$

4.4.3 Adjunction Complete Convex Sets

A set $A \subset \mathbb{R}^n$ is said to be ***adjunction complete*** or simply ***complete****** if whenever $x \notin A$, then

$$\mathrm{diam}(A \cup \{x\}) > \mathrm{diam}(A).$$

Equivalently, A is adjunction complete if and only if it is not properly contained in any other set of the same diameter.

---

* Not to be confused with the topological notion of completeness.

Theorem 4.4.11. *If $A \subset \mathbb{R}^n$ is a compact convex set that is adjunction complete and $b \in \text{bdy}(A)$, then there exists a point $c \in A$ such that $\|b - c\| = \text{diam}(A)$.*

Proof. Since A is compact, there is a point $c \in A$ such that

$$\|b - c\| = \sup\{\|b - y\| : y \in A\}.$$

Let $\delta = \text{diam}(A)$. Since

$$\delta = \sup\{\|x - y\| : x, y \in A\},$$

if $\|b - c\| < \delta$, then $\epsilon = \delta - \|b - c\| > 0$, and since $b \in \text{bdy}(A)$, there exists a point $z \in \mathbb{R}^n \setminus A$ such that

$$\|z - b\| < \epsilon = \delta - \|b - c\|.$$

Therefore, for each $x \in A$, we have

$$\|z - x\| \leq \|z - b\| + \|b - x\| \leq \|z - b\| + \|b - c\| < \delta,$$

which implies that

$$\text{diam}(A) \leq \text{diam}(A \cup \{z\}) = \sup\{\|x - y\| : x, y \in A \cup \{z\}\} \leq \delta,$$

that is, $\text{diam}(A \cup \{z\}) = \delta$. However, this contradicts the fact that A is adjunction complete. Therefore, we must have $\|b - c\| = \delta$. $\square$

Theorem 4.4.12. *If A is a subset of $\mathbb{R}^n$ with positive diameter δ, then A is adjunction complete if and only if A is the intersection of all closed balls of radius δ whose centers belong to A.*

Proof. Let S be the intersection of all closed balls of radius δ whose centers belong to A, that is,

$$S = \bigcap \{\overline{B}(a, \delta) : a \in A\},$$

Suppose that $A = S$, then A is a nonempty compact convex subset of $\mathbb{R}^n$. If $x \in \mathbb{R}^n \setminus A$, then there exists a point $a \in A$ such that the ball $\overline{B}(a, \delta)$ does not contain x, that is, $\|x - a\| > \delta$. Therefore,

$$\text{diam}(A \cup \{x\}) \geq \|x - a\| > \delta,$$

so that A is adjunction complete.

Conversely, suppose that A is adjunction complete. Since $\text{diam}(A) = \delta$, we have

$$\text{diam}(\overline{A}) = \text{diam}(A) = \delta \qquad \text{and} \qquad \text{diam}(\text{conv}(A)) = \text{diam}(A) = \delta,$$

and since A is adjunction complete, this implies that

$$\overline{A} = A \quad \text{and} \quad \text{conv}(A) = A,$$

that is, A is a compact convex set.

If $x \in A$, since $\text{diam}(A) = \delta$, then

$$\|x - a\| \leq \delta$$

for all $a \in A$, so that $x \in S$ and $A \subset S$.

If $x \in S$, then

$$\|x - a\| \leq \delta$$

for all $a \in A$, and therefore,

$$\delta = \text{diam}(A) \leq \text{diam}(A \cup \{x\}) \leq \delta$$

and so the set $A \cup \{x\}$ has diameter δ. Since A is adjunction complete, then $x \in A$, so that $S \subset A$.

$\square$

The statement of the following lemma can be found in the monograph *Convexity* by H. G. Eggleston (without a proof), and the proof we give here can be found in the text *Convexity* by Roger Webster.

Lemma 4.4.13. *If $A \subset \mathbb{R}^n$ is a compact convex set that is adjunction complete, and $\text{diam}(A) = \delta$, then A contains every circular arc of radius δ joining two of its points.*

Proof. From the previous theorem, A is the intersection of all closed balls of radius δ with centers in A, we only have to prove the lemma for a closed ball B of radius δ.

Let B be a closed ball in $\mathbb{R}^n$ with radius δ and center x. Let a, b, and c be noncollinear points in $\mathbb{R}^n$ with $a, b \in B$ such that $\|a - c\| = \delta$ and $\|b - c\| = \delta$. We will show that the circular arc with radius δ and center c joining a and b lies entirely in B, that is, for every z in the arc C, we have $z \in B$.

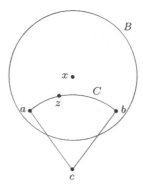

Since a and b are in B, we have

$$\|a - x\|^2 = \|a - c\|^2 + 2\langle a - c, c - x \rangle + \|c - x\|^2 \leq \delta^2$$

and

$$\|b - x\|^2 = \|b - c\|^2 + 2\langle b - c, c - x \rangle + \|c - x\|^2 \leq \delta^2,$$

so that

$$2\langle a - c, c - x \rangle + \|c - x\|^2 \leq 0 \qquad (*)$$

and

$$2\langle b - c, c - x \rangle + \|c - x\|^2 \leq 0. \qquad (**)$$

Now let z lie on the circular arc of radius δ and center c joining a and b. Since the arc lies above the chord joining a and b, we have

$$z - c = \alpha(a - c) + \beta(b - c)$$

for some $\alpha, \beta \geq 0$, and $\|z - c\| = \delta$. Since $\|a - c\| = \delta$ and $\|b - c\| = \delta$, this implies that $\alpha + \beta \geq 1$. Multiplying $(*)$ by α and $(**)$ by β and adding, we obtain

$$2\langle z - c, c - x \rangle + \|c - x\|^2 \leq 2\langle z - c, c - x \rangle + (\alpha + \beta)\|c - x\|^2 \leq 0.$$

Therefore,

$$\|z - x\|^2 = \|z - c\|^2 + 2\langle z - c, c - x \rangle + \|c - x\|^2 \leq \|z - c\|^2 = \delta^2,$$

so that $\|z - x\| \leq \delta$, that is, $z \in B$, and the circular arc of radius δ with center c joining a and b lies in B.

$\square$

The following theorem gives the relation between adjunction complete sets and sets of constant width, the proof follows that in the text *Convexity* by Roger Webster.

Theorem 4.4.14. *A compact convex set $A \subset \mathbb{R}^n$ is an adjunction complete set with diameter δ if and only if A has constant width δ.*

Proof. Suppose that A is a compact convex set of constant width δ. Since A is compact, there exist points a_1 and a_2 in A such that $\|a_1 - a_2\| = \text{diam}(A)$. The hyperplanes H_1 and H_2 through a_1 and a_2 perpendicular to the segment $[a_1, a_2]$ are parallel supporting hyperplanes to A at a_1 and a_2, respectively, and the distance between these hyperplanes is $\text{diam}(A)$. Since A has constant width δ, we have $\text{diam}(A) = \delta$.

Now suppose that $b \notin A$, and let a be the nearest point of A to b. The hyperplane H_a through a perpendicular to $[a, b]$ is a supporting hyperplane to A.

The opposite parallel supporting hyperplane H_z for A meets A in at least one point, say $z \in A$, and we may assume that z lies on the segment $[b, a]$ extended beyond a, since if not, then we would have $\mathrm{diam}(A) > \delta$.

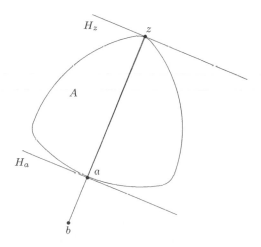

Thus, we have

$$\|b - z\| > \|a - z\| = \delta,$$

and $\mathrm{diam}(A \cup \{b\}) > \mathrm{diam}(A)$ for any $b \notin A$, so no point can be adjoined to A without increasing its diameter, that is, A is an adjunction complete set of diameter δ.

Conversely, suppose that $A \subset \mathbb{R}^n$ is an adjunction complete compact convex set with $\mathrm{diam}(A) = \delta$, and suppose that A does not have constant width. We may assume that the minimum width of A is α and that $\alpha < \delta$.

Let u be a unit vector in whose direction A has minimum width α, then by compactness of A and continuity of the chord length in the direction of u, there exists a chord $[a, b]$ in the direction of u whose length is α. If we define the hyperplanes

$$H_a = \{x \in \mathbb{R}^n : \langle u, x \rangle = \langle u, a \rangle\}$$

and

$$H_b = \{x \in \mathbb{R}^n : \langle u, x \rangle = \langle u, b \rangle\},$$

then A is contained in the slab between the hyperplanes, that is,

$$A \subseteq \{x \in \mathbb{R}^n : \langle u, a \rangle \leq \langle u, x \rangle \leq \langle u, b \rangle\}.$$

Thus, H_a and H_b are two parallel supporting hyperplanes to A, both perpendicular to u. Moreover, the distance between the hyperplanes is $\|b - a\| = \alpha < \delta$, since $b - a = \alpha u$.

Thus, the hyperplanes support A at a and b, respectively, as in the figure below.

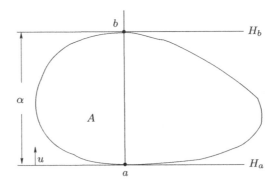

By Theorem 4.4.11, there exists a point c in A such that $\|c - b\| = \delta$, and by Theorem 4.4.12, A is the intersection of all closed balls of radius δ with centers in A. Thus, if x and y are in A, by Theorem 4.4.13, any circular arc of radius δ with endpoints x and y lies entirely in A.

Now, in the plane determined by a, b, and c, let d be the point of intersection of the two circles centered at a and c with radius δ. Let $\mathcal{C}$ be the arc of the circle centered at d with radius δ with endpoints a and c, we will show that $\mathcal{C}$ crosses the hyperplane H_a.

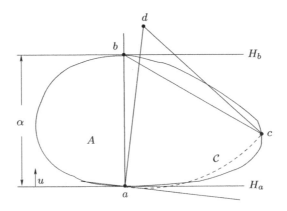

Since H_a is a supporting hyperplane, the angle between the segment $b - a$ and the segment $d - a$ is the same as the angle between the tangent to A at a and the tangent to the segment C at a. Thus, a portion of the arc C lies in the halfplane $\{x \in \mathbb{R}^n : \langle u, x \rangle < \langle u, a \rangle\}$, which is a contradiction.

$\square$

Remark. From the above, we see that every set A of constant width is a nonempty compact convex subset with nonempty interior and that every support point of A is an exposed point of the set, that is, the boundary of a set of constant width contains no line segments.

In Lemma 4.3.35, we showed that a compact convex set C had a unique circumball. Now we show that a set A of constant width has a unique inball and circumball, that they are concentric, and that the sum of their radii is equal to the constant width of A.

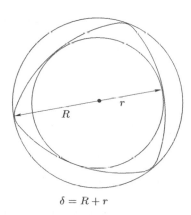

$$\delta = R + r$$

Theorem 4.4.15. *If $A \subset \mathbb{R}^n$ is a compact convex set of constant width δ and if r is the radius of an inball and R the radius of the circumball of A, then A has a unique inball, which is concentric with its circumball, and $\delta = R + r$.*

Proof. Let $\overline{B}(a, r)$ where $a \in A$ be any inball of A, we will show that

$$A \subset \overline{B}(a, \delta - r).$$

If $A \not\subset \overline{B}(a, \delta - r)$, then there exist a point $x \in A$ such that

$$\|a - x\| > \delta - r.$$

If we let

$$y = a - \frac{r(x-a)}{\|x-a\|},$$

and note that $\|y - a\| = r$, so that $y \in \overline{B}(a,r) \subset A$, that is, $y \in A$. Now,

$$x - y = (x-a)\left[1 + \frac{r}{\|x-a\|}\right]$$
$$= (x-a)\left[\frac{\|x-a\| + r}{\|x-a\|}\right],$$

and so

$$\|x - y\| = \|x - a\| + r > \delta,$$

which contradicts the fact that $\delta = \text{diam}(A)$.

Therefore, $A \subset \overline{B}(a, \delta - r)$, and since

$$R = \inf\{\rho > 0 : A \subset B, \text{ where } B \text{ is a closed ball of radius } \rho\},$$

then this implies that $R \leq \delta - r$, that is,

$$R + r \leq \delta. \qquad (*)$$

Next, let $\overline{B}(c, R)$ be the circumball of A, so that $A \subset \overline{B}(c, R)$. We will show that

$$\overline{B}(c, \delta - R) \subset A.$$

If $\overline{B}(c, \delta - R) \not\subset A$, then there exists a point $b \in \overline{B}(c, \delta - R)$ such that $b \notin A$.

Since both $\{b\}$ and A are compact convex sets, then they can be strongly separated, that is, there exists a hyperplane $H = \{x \in \mathbb{R}^n : \langle u, x \rangle = \alpha\}$, where u is a unit vector perpendicular to H, such that

$$\langle u, x \rangle < \alpha < \langle u, b \rangle$$

for all $x \in A$.

Now, if $x \in A$, then $x \in \overline{B}(c, R)$, so that

$$|\langle u, x - c \rangle| \leq \|u\| \, \|x - c\| = \|x - c\| \leq R,$$

so that

$$-R \leq \langle u, x - c \rangle \leq R,$$

that is,

$$\langle u, c \rangle - R \le \langle u, x \rangle \le \langle u, c \rangle + R$$

for all $x \in A$.

Similarly, since $b \in \overline{B}(c, \delta - R)$, then

$$|\langle u, b - c \rangle| \le \|u\| \, \|b - c\| = \|b - c\| \le \delta - R,$$

which implies that

$$\langle u, c \rangle - \delta + R \le \langle u, b \rangle \le \langle u, c \rangle + \delta - R.$$

Thus,

$$\langle u, c \rangle - R \le \langle u, x \rangle < \alpha < \langle u, b \rangle \le \langle u, c \rangle + \delta - R$$

for all $x \in A$.

For every $x \in A$, the component of x in the direction of u lies in the slab between parallel hyperplanes given by

$$\{y \in \mathbb{R}^n : \langle u, c \rangle - R \le \langle u, y \rangle < \langle u, c \rangle + \delta - R\}$$

whose width is strictly less than δ, and this says that the width of A in the direction of u is less than δ, which is a contradiction. Therefore, we must have $\overline{B}(c, \delta - R) \subset A$, and from the definition of the inradius r, this implies that $\delta - R \le r$, that is,

$$\delta \le R + r. \qquad (**)$$

Combining $(*)$ and $(**)$, we have $\delta = R + r$.

Now, if $\overline{B}(a, r)$ is an inball of A, then $\overline{B}(a, R)$ is the circumball of A, since

$$A \subset \overline{B}(a, \delta - r) = \overline{B}(a, R).$$

Thus, $a = c$ so the inball is unique and is concentric with the circumball of A.

$\square$

4.4.4 Sets of Constant Width in the Plane

Theorem 4.4.16. *If A is a compact convex subset of the plane of constant width δ, and if B is a compact convex subset of the plane with* $\mathrm{diam}(B) = \delta$ *such that $A \subset B$, then $B = A$.*

Proof. Suppose that $A \subsetneq B$, that is, there is a point $b \in B$ such that $b \notin A$. If this is the case, then there is a support line ℓ for A such that A lies in a closed halfspace H with boundary ℓ, and b does not belong to H, as in the figure below.

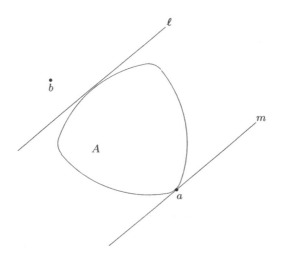

Let m be a support line for A that is parallel to ℓ, supporting A at the point a. Since A is a set of constant width δ, the distance between ℓ and m is equal to δ, so that $\|a - b\| > \delta$.

However, since $A \subset B$, both a and b are in B, and this contradicts the fact that the diameter of B is equal to δ.

$\square$

Note: A set A of constant width δ also has the property that if x_1 is a boundary point of A, then the closed ball $\overline{B}(x_1, \delta)$ contains the set A and moreover, there must be a point x_2 of the set A in the sphere $S(x_1, \delta)$. In fact, there must be parallel hyperplanes, H_1 and H_2, supporting the set A at x_1 and x_2, respectively.

4.4.5 Barbier's Theorem

We mentioned earlier in Example 4.4.1 that the perimeter of a Reuleaux triangle of constant width a is πa. We now have the machinery we need to prove the result that for any compact convex set A in $\mathbb{R}^2$ of constant width δ, the perimeter of A is $\pi \delta$. We will follow the proof given in Boltyanski and Soifer: *Geometric Études in Combinatorial Mathematics*, but first a lemma.

Lemma 4.4.17. *Let $ABCD$ be a rhombus in the plane and let ℓ and m be parallel lines at a distance δ from each other that intersect the rhombus and are perpendicular to the diagonal $[B, D]$, as in the figure below. The intersection of the rhombus and the strip between the lines ℓ and m is a hexagon whose perimeter is independent of the position of the lines ℓ and m.*

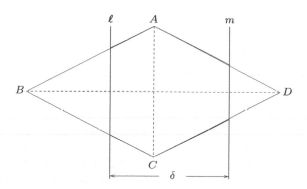

Proof. In the figure below, from symmetry, we see that the perimeter of the hexagon in question is

$$P = 2(AU + AX) + UV + XY.$$

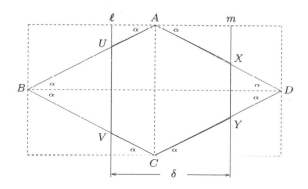

We have

$$AU \cos \alpha + AX \cos \alpha = \delta,$$

so that

$$AU + AX = \frac{\delta}{\cos \alpha}.$$

Similarly,

$$UV = AC - 2AU \sin \alpha \qquad \text{and} \qquad XY = AC - 2AX \sin \alpha,$$

so that

$$UV + XY = 2AC - 2(AU + AX) \sin \alpha = 2(AC - \delta \tan \alpha).$$

Therefore,

$$P = 2\delta \sec \alpha + 2(AC - \delta \tan \alpha) = 2AC - 2\delta(\tan \alpha - \sec \alpha),$$

which depends only on the geometry of the rhombus $ABCD$ and the distance δ between the lines ℓ and m.

□

Theorem 4.4.18. *(Barbier's Theorem)*

If Q is a convex subset of the plane with constant width δ, then the perimeter of Q is $\pi\delta$.

Proof. Let Q be a convex subset of $\mathbb{R}^2$ of constant width δ, and let K be a closed disk with the same width δ. Clearly, any squares circumscribed about Q and K are congruent and, therefore, have the same perimeter. We will use this as the basis for inductive proof that Q and K have the same perimeter.

Let $n > 2$ and assume that the 2^n-gons with equal angles circumscribed about Q and K have the same perimeter. We will show that the same is true for 2^{n+1}-gons with equal angles circumscribed about Q and K.

Consider two adjacent sides BE and BF and the opposite sides DG and DH in the 2^n-gon with equal angles circumscribed about the disk K, as in the figure below.

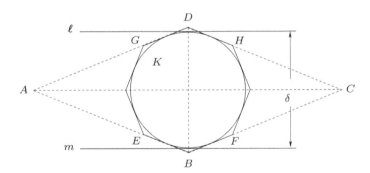

The straight lines containing these sides form a parallelogram with both altitudes equal to δ, that is, a rhombus.

Similarly, we construct the rhombus for Q, the set of constant width δ, by considering two adjacent sides $B'E'$ and $B'F'$ and the opposite sides $D'G'$ and $D'H'$ in the 2^n-gon with equal angles circumscribed about Q, as in the following figure.

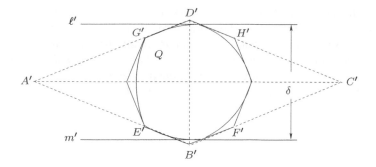

These rhombuses for K and Q are congruent, and we construct support lines ℓ and m for K that are perpendicular to the diagonal BD. Similarly, we construct support lines ℓ' and m' for Q that are perpendicular to the diagonal $B'D'$. The distance between ℓ and m is δ, and the distance between ℓ' and m' is also δ.

According to Lemma 4.4.17, the intersection of these 2^n-gons with the strips we constructed using the support lines has equal perimeters. If we carry out this construction for each pair of adjacent sides, we obtain 2^{n+1}-gons with equal angles circumscribed about K and Q, which have the same perimeter, and this completes the inductive proof.

Thus, 2^n-gons with equal angles circumscribed about K and Q have equal perimeters for each $n \geq 2$, and letting $n \to \infty$, we see that the boundaries of K and Q have the same length. Since the closed disk K of radius δ has perimeter $\pi\delta$, this implies that the perimeter of the convex set Q of constant width δ also has perimeter $\pi\delta$.

$\square$

4.4.6 Constructing Sets of Constant Width

The Reuleaux triangle in Example 4.4.1 is a set of constant width of a special type. It was constructed by drawing an equilateral triangle with vertices p, q, and r, whose edges are of unit length.

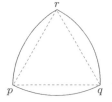

Then, with center p and radius 1, we draw arc(q, r). We draw the remaining two arcs in the same way. The three arcs form a curve, which is the boundary of a convex set of constant width, that is, a ***Reuleaux triangle***, and is just one of the whole family of constant width curves called ***Reuleaux polygons***. Other Reuleaux polygons are constructed using ***star polygons***, that is, polygons such that each edge cuts k other edges, as in the figure below.

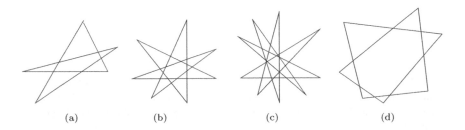

       (a)   (b)   (c)   (d)

For the construction of Reuleaux polygons, we require that no vertex lies between the arms of any of the vertex angles. In the above figure, polygons (a), (b), and (c) fulfill this condition but polygon (d) does not. In order to fulfill the required condition, the polygon must have an odd number of vertices.

We will start with ***equilateral star polygons*** and construct a Reuleaux polygon with five vertices. If p_1 through p_5 are the vertices of an equilateral star polygon whose edges are 1 unit in length as shown in the figure below, using p_1 as center and with radius 1 unit, draw the arc(p_5, p_2). Then with center p_2 and radius 1 unit, draw arc(p_1, p_3); with center p_3, draw arc(p_2, p_4); and so on, until the Reuleaux pentagon is complete.

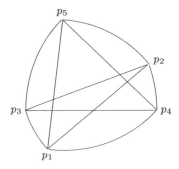

Note that unlike an equilateral triangle, an equilateral star polygon does not have to be regular; that is, it does not have to have equal angles. For example, to construct the star polygon in the above figure, one can first draw the three edges $[p_1, p_2]$, $[p_2, p_3]$, and $[p_3, p_4]$, making sure that the angles as shown in the figure below are unequal but are such that the edges $[p_1, p_2]$ and $[p_3, p_4]$ cross. The remaining vertex can be found by constructing the isosceles triangle with base $[p_1, p_4]$.

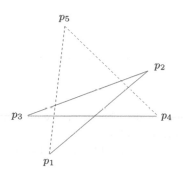

Curves of constant width drawn so far all have "sharp" points, and one may wonder if there are smooth curves of constant width other than the circle. We will show in the following example how to construct a smooth "Reuleaux triangle."

Example 4.4.19. *Construct a smooth Reuleaux triangle, that is, one without any corners.*

Solution. We begin with an equilateral triangle $p_1 p_2 p_3$ whose edges are of length ρ. With center p_1 and radius $\rho + \delta$, draw $\text{arc}(a_1, a_2)$. With center p_2 and radius δ, draw $\text{arc}(a_2, a_3)$. With center p_3 and radius $\rho + \delta$, draw $\text{arc}(a_3, a_4)$. Continue in the manner, until all six arcs have been drawn, forming a curve of constant width $\rho + 2\delta$. For this curve, there are no "sharp points" or corners, that is, at every boundary point, there is one and only one supporting line.

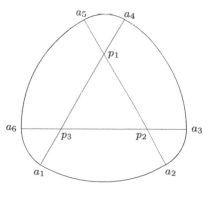

□

Example 4.4.20. *Construct a smooth Reuleaux polygon based on a triangle whose edges are all of different lengths.*

Solution. We start with a triangle ABC with edges a, b, and c where $c > a > b$, as shown in the following figure.

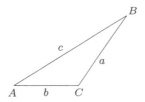

Let $\delta > 0$, we will construct a six-sided Reuleaux polygon of width $a + c - b + 2\delta$, which is smooth.

1. First, extend all sides of the triangle, and with center A, draw a circular arc of radius $c + \delta$ from $\overline{AB}$ to $\overline{AC}$
2 Next, draw with center C a circular arc of radius $c + \delta - b$ from $\overline{AC}$ to $\overline{BC}$.
3. Next, draw with center B a circular arc of radius $a + c + \delta - b$ from $\overline{BC}$ to $\overline{AB}$.
4. Next, draw with center A a circular arc of radius $a - b + \delta$ from $\overline{AB}$ to $\overline{AC}$.
5. Next, draw with center C a circular arc of radius $a + \delta$ from $\overline{AC}$ to $\overline{BC}$.
6. Finally, draw with center B a circular arc of radius δ from $\overline{BC}$ to $\overline{AB}$.

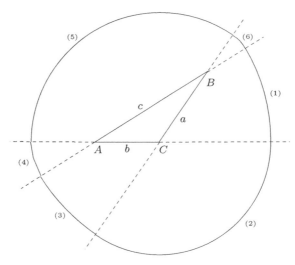

The resulting figure has constant width $w = a + c - b + 2\delta$, since this is the distance between any two parallel supporting hyperplanes.

□

Example 4.4.21. *Construct a smooth Reuleaux polygon based on a star polygon, none of whose edges are of the same length.*

Solution. We start with the star pentagon shown in the figure below, and we assume that the longest edge is BD and choose $\delta > 0$ large enough so that

$$AC + BD - AD - CE + \delta > 0,$$
$$AD - AC + \delta > 0,$$
$$AD + CE - AC - BE + \delta > 0.$$

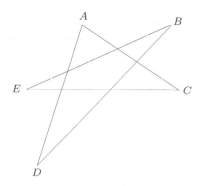

1. First, extend all sides of the star polygon, and with center D, draw a circular arc of radius $BD + \delta$ from $\overline{BD}$ to $\overline{AD}$.
2. Next, with center A, draw a circular arc of radius $BD - AD + \delta$ from $\overline{AD}$ to $\overline{AC}$.
3. Next, with center C, draw a circular arc of radius $AC + BD - AD + \delta$ from $\overline{AC}$ to $\overline{EC}$.
4. Next, with center E, draw a circular arc of radius $AC + BD - AD - CE + \delta$ from $\overline{EC}$ to $\overline{BE}$.
5. Next, with center B, draw a circular arc of radius $AC + BD + BE - AD - CE + \delta$ from $\overline{BE}$ to $\overline{BD}$.
6. Next, with center D, draw a circular arc of radius δ from $\overline{BD}$ to $\overline{AD}$.
7. Next, with center A, draw a circular arc of radius $AD + \delta$ from $\overline{AD}$ to $\overline{AC}$.
8. Next, with center C, draw a circular arc of radius $AD - AC + \delta$ from $\overline{AC}$ to $\overline{CE}$.
9. Next, with center E, draw a circular arc of radius $AD + CE - AC + \delta$ from $\overline{CE}$ to $\overline{BE}$.
10. Finally, with center B, draw a circular arc of radius $AD + CE - AC - BE + \delta$ from $\overline{BE}$ to $\overline{BD}$.

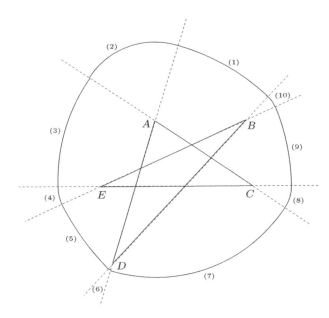

The resulting figure is a smooth Reuleaux polygon of constant width

$$w = BD + 2\delta,$$

since this is the distance between any two parallel supporting hyperplanes.

Note that if we take $\delta = 0$, the figure is no longer smooth, that is, it has pointed corners. □

The curves of constant width that we have constructed so far were composed of either n circular arcs or $2n$ circular arcs, where n was an *odd integer*. In the next example, we show that this does not have to be the case.

Example 4.4.22. *Construct a curve of constant width that is composed of four circular arcs.*

Solution. In this example, we start with an obtuse isosceles triangle with sides of length a, a, and b, where $b > a$.

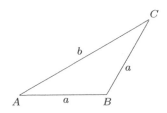

1. First, extend the sides of the triangle ABC, and with center A, draw a circular arc of radius b from $\overline{AC}$ to $\overline{AB}$.
2. Next, with center B, draw a circular arc of radius $b - a$ from $\overline{AB}$ to $\overline{BC}$.
3. Next, with center C, draw a circular arc of radius b from $\overline{BC}$ to $\overline{AC}$.
4. Finally, with center B, draw a circular arc of radius a from $\overline{AB}$ to $\overline{BC}$.

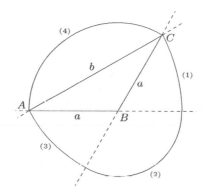

The resulting figure is a (nonsmooth) Reuleaux polygon composed of four circular arcs, and the curve is of constant width $w = b$.

$\square$

Note that by Theorem 4.4.9, we can construct a smooth Reuleaux polygon from one that is nonsmooth simply by rolling a circle on the outside of the Reuleaux polygon and adding to that polygon the area swept out by this rolling circle. The result is a compact convex body of constant width.

We should mention that there is no equivalent notion of Reuleaux polygons in three dimensions. For example, if T is a regular tetrahedron in $\mathbb{R}^3$ and R is the region common to the four balls, each of which is centered at a vertex of T and passes through the other three vertices, then R is not a set of constant width in $\mathbb{R}^3$. If we consider the six circular arcs that are the edges of R, it turns out that any two opposite ones (that is, any pair of edges without a common point) have midpoints that are slightly further apart than any two vertices of R.

Curves of constant width are used in some mechanical applications. On some fire hydrants, the top nut, which turns on the water, is shaped like a smooth regular Reuleaux polygon. An ordinary wrench cannot easily be used to turn the nut because the wrench tends to slide rather than grip. However, a wrench whose opening exactly fits the nut can easily turn it.

Note that a curve of constant width δ can be turned continuously inside a square of side δ so that it maintains contact with all four sides of the square.

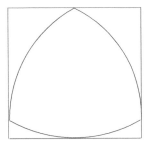

Therefore, a drill bit shaped like a Reuleaux triangle can be used to drill a hole with straight sides, where the corners of the triangle are the cutting edges.* It follows from Example 4.4.1 that any such hole cannot be a square and must have "rounded" corners.

Also, the English 50 pence coin is a regular Reuleaux 7-gon, and the Canadian dollar coin is a regular Reuleaux 11-gon. This makes them harder to counterfeit.

Most movie theaters today use digital projection. Films used to be projected onto the screen mechanically, using celluloid film and mechanical projectors. The film had to make a brief, quick movement with the shutter closed and then a momentary pause with the shutter open. The gear mechanism that provided this intermittent motion was based on the Reuleaux triangle.

Usually, manhole covers are circular in shape. The cover is slightly larger than the opening, and no matter how your turn it, you cannot drop it down the hole. Of course, if the shape of the opening were a curve of constant width, the cover would still have the same property. Apparently, some cities do have manhole covers that are sets of constant width.

Some curves of constant width occur as the cross-sectional shape of pistons in certain internal combustion engines.

We end this section by mentioning that not all curves of constant width are built from circular arcs. There are, in fact, some curves of constant width such that no part of the curve is a circular arc.

4.4.7 Borsuk's Problem

In 1933, the Polish mathematician Karol Borsuk formulated the following problem.

Borsuk's Problem: Can every subset $A \subset \mathbb{R}^n$ of diameter d be decomposed into the union of at most $n + 1$ sets, each of which is of diameter strictly less than d?

---

* Note that the axis of the drill has to move.

For ease of discussion, we call such a diameter reducing decomposition a
Borsuk decomposition. For example, in $\mathbb{R}^2$, a square whose diagonals have
length d has a Borsuk decomposition into three parallel strips.

In the plane, Borsuk's problem has a positive answer.

Theorem 4.4.23. *In $\mathbb{R}^2$, every set of diameter d has a Borsuk decomposition
into at most three parts.*

Proof. By Pál's theorem (Theorem 4.3.37), a set of diameter d can be enclosed
by a regular hexagon whose incircle has diameter d. From the figure below, it
is evident that this hexagon can be decomposed into three pentagons, each of
diameter $\sqrt{3}d/2$.

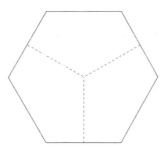

Of course, the decomposition also breaks every subset of the hexagon into
three (or fewer) pieces whose diameter is no greater than $\sqrt{3}d/2$.

$\square$

Note that the bound $n + 1$ in Borsuk's conjecture is the best possible. If A
is a regular n-dimensional simplex, or even just the set of $n + 1$ vertices of a
regular n-dimensional simplex, then no part of a Borsuk decomposition of A
can contain more than one of the simplex vertices. Thus, if $f(n)$ denotes the
smallest positive integer such that every bounded set $A \subset \mathbb{R}^n$ has a Borsuk
decomposition into $f(n)$ parts, then the example of a regular n-dimensional
simplex shows that $f(n) \geq n + 1$.

In 1932, Borsuk proved that for any $n \geq 2$, an n-dimensional ball of diameter d *cannot* be decomposed into n parts of diameter less than d.

Unknown to Borsuk, this result was obtained earlier in 1930 by two Russian mathematicians, Lazar Lusternik and Lev Schnirelmann. Borsuk's proof was obtained independently (and with a different proof).

The Borsuk problem for $\mathbb{R}^2$ was first solved by Borsuk in 1933 and led to the inaccurate name of the **Borsuk Conjecture**.

It turns out that the Borsuk conjecture is also true in $\mathbb{R}^3$. It was first proved by Julian Perkal in 1947 and then 8 years later by the British mathematician H. G. Eggleston in 1955. Eggleston's proof was rather complicated.

There are proofs that resemble that for $\mathbb{R}^2$, namely, it is possible to find a universal cover K in $\mathbb{R}^3$ with the property that K can be expressed as the union of four sets, each of which has diameter less than 1.

In fact, using an argument similar to that given in Pál's theorem in 1920, David Gale proved in 1953 that every set of diameter d can be embedded in a regular octahedron with distance d between its opposite faces.

Thus, this octahedron is a universal cover for all sets of diameter d in $\mathbb{R}^3$. In 1957, the Israeli mathematician Branko Grünbaum and the Hungarian mathematician Aladár Heppes found proofs of the Borsuk conjecture in $\mathbb{R}^3$ that were similar to the proof for $\mathbb{R}^2$ given above in Theorem 4.4.23.

Borsuk's problem was raised in 1933 and, after 60 years, was finally solved in 1993. The problem was solved by Jeff Kahn and Gil Kalai in 1993, who showed that the general answer to Borsuk's question is *no*. They showed that the Borsuk conjecture is false for $n = 1,325$ and for each $n > 2,014$.

After Andriy V. Bondarenko showed in 2013 that Borsuk's conjecture was false for all $n \geq 65$, the current best bound (as of this writing) by Thomas Jenrich in 2013, is that Borsuk's conjecture is false for all $n \geq 64$. The methods used in the proofs are very different from the proofs given here and use Euclidean representations of strongly regular graphs (see [66] and [64]).

We mentioned earlier that not much is known about universal covers in higher dimensional spaces. It is not known if the proof by universal covers can be extended even to $\mathbb{R}^4$. It should also be mentioned that there is no reason to suspect that the proof cannot be so extended, but it is still not known if the Borsuk conjecture is true for $n = 4$.

For very special types of sets in higher dimensions, Borsuk's problem is known to have an affirmative answer.

Borsuk's original 1933 paper included the Borsuk–Ulam theorem.

Theorem 4.4.24. *If* $S_n = \{x \in \mathbb{R}^n : \|x\| = 1\}$ *is the* n-*dimensional unit sphere, then every continuous map* $f : S_n \longrightarrow \mathbb{R}^n$ *maps two antipodal points of the sphere* S_n *onto the same point in* $\mathbb{R}^n$.

He used this to prove the Borsuk conjecture when $n \leq 3$ for smooth bodies A.

In 1946, Hugo Hadwiger proved the Borsuk conjecture for all n when A is a smooth convex body. We will prove this result for smooth sets of constant width in $\mathbb{R}^n$.

Theorem 4.4.25. *If* $K \subset \mathbb{R}^n$ *is a smooth set of constant width* 1, *then* K *has a Borsuk decomposition into at most* $n + 1$ *parts.*

Proof. We will give Z. A. Melzak's beautiful proof: It depends on two facts, if K is a smooth set of constant width 1 in $\mathbb{R}^n$, then

(i) Taking the closed convex hull of a subset of $\mathbb{R}^n$ does not increase its diameter (see Theorem 2.5.15 and Example 3.3.9).
(ii) The support hyperplanes for any two diametrically opposite points of K, that is, points that are separated by a distance of 1 unit, must be perpendicular to the line joining the two points (see Theorem 4.4.5).

This means, as Melzak puts it, that the north and south poles of any smooth object cannot be simultaneously seen from any finite distance.

But now the proof follows immediately: Enclose the set K in the interior of some simplex. For each of the $n + 1$ vertices of the simplex, let S_i, for $i = 1, 2, \ldots, n + 1$, be the part of the boundary of K that is visible from the ith vertex.

From what we have just said, every two points of S_i must be closer than 1 unit to each other, and since S_i is compact, from Theorem 2.6.13, it follows that each S_i has diameter strictly less than 1.

It is also evident from the fact that K is interior to the simplex and that the union of the S_i's cover the boundary of K.

Now let p be any point interior to K, and note that every point of S_i is closer to p than 1, so the set $S_i \cup p$ has diameter strictly less than 1.

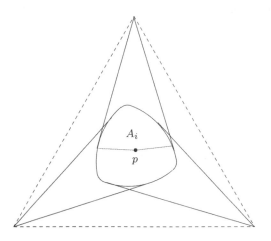

Taking

$$A_i = \operatorname{conv}(S_i \cup \{p\}),$$

for $i = 1, 2, \ldots, n + 1$, produces the desired decomposition.

$\square$

The preceding result appears to be tantalizingly close to a complete solution to the Borsuk problem for $\mathbb{R}^n$. To explain why, let us consider some reductions that can be made to the problem. We will suppose that $A \subset \mathbb{R}^n$ has diameter 1.

- The first thing to note is that the closure of A has the same diameter as A, so we may assume that A is closed.
- The second thing to note is that we may assume that A is convex, again because $\operatorname{conv}(A)$ has the same diameter as A.
- Finally, we may assume that A is adjunction complete, that is, if we take A together with any other point p not in A, then $A \cup \{p\}$ has diameter greater than 1. (If not, A may be replaced by $A \cup \{p\}$, since any Borsuk decomposition of $A \cup \{p\}$ automatically produces a Borsuk decomposition of A).

Now, A is adjunction complete with diameter 1 if and only if A is of constant width 1. In other words, the Borsuk problem reduces to the case where the set is a compact convex set of constant width 1.

Alas! Some sets of constant width are not smooth!

4.4.8 Problems

1. In $\mathbb{R}^2$, show that

   (a) the smallest square, which is a universal cover, has sides of length 1,

   (b) the smallest circle, which is a universal cover, has radius $1/\sqrt{3}$,

   (c) the smallest regular hexagon, which is a universal cover, has sides of length $1/\sqrt{3}$,

   (d) the smallest equilateral triangle, which is a universal cover, has sides of length $\sqrt{3}$.

2. Let $C \subset \mathbb{R}^2$ with $\text{diam}(C) = \delta$, show that there exists a subset A of constant width δ such that $C \subset A$.

3. In the plane, the set of constant width δ with the smallest area is the Reuleaux triangle.

4. In the plane, the set of constant width δ with the largest area is the circle.

5. Show that the minimum area A of a universal cover in $\mathbb{R}^2$ must satisfy the inequality

$$\frac{\pi}{4} \leq A \leq \frac{\sqrt{3}}{2}.$$

6. Given a subset $C \subset \mathbb{R}^2$ of constant width δ, show that a semicircle of diameter δ can be inscribed in C.

7. Show that if $C \subset \mathbb{R}^n$ has constant width $\delta > 0$, then $\text{int}(C) \neq \emptyset$.

8. Show that if $C \subset \mathbb{R}^n$ has constant width δ and is symmetric with respect to the origin, that is, $-x \in C$ whenever $x \in C$, then $C = \overline{B}(0, \delta/2)$.

9. Show that if $C \subset \mathbb{R}^n$ is a compact convex set of diameter δ, and for each $b \in \text{bdy}(C)$, there exists a point $c \in C$ such that $\|b - c\| = \delta$, then C has constant width δ.

10. Let $C \subset \mathbb{R}^n$ be a compact convex set of constant width δ.

    (a) Show that the radius R of the circumball of C satisfies the inequality

$$\frac{\delta}{2} \leq R \leq \delta \sqrt{\frac{n}{2(n+1)}}.$$

    (b) Show that both the lower and upper bounds are attained.

11. Prove that every subset $C \subset \mathbb{R}^n$ with $\text{diam}(C) = \delta > 0$ is contained in a compact convex set A of constant width δ.

12. Let $C \subset \mathbb{R}^n$ be a set of constant width δ and $x_0 \in \text{bdy}(C)$, show that there exists a closed ball B of radius δ with $x_0 \in B$ such that $C \subset B$. Thus, through every boundary point of C, there passes a *supporting ball*.

13. Let $A \subset \mathbb{R}^2$ be a set of constant width δ, and let B be the same set rotated through $180°$. Show that the Minkowski sum $A + B$ is a disk of radius δ. Use this result to give a new proof of Barbier's theorem 4.4.18.

14. If the Minkowski sum of a compact convex set $A \subset \mathbb{R}^2$ and the compact convex set B obtained by rotating A through $180°$ is a disk, then A is a compact convex set of constant width.

Bibliography

1 Aigner, M., and G. M. Ziegler, *Proofs from THE BOOK*, 2nd ed., Springer-Verlag, Berlin, 2001.

2 Alon, N., and D. J. Kleitman, Piercing convex sets, *Bulletin of the American Mathematical Society*, **27**, 252-256, 1992.

3 Apostol, T. M., *Mathematical Analysis*, 2nd ed., Addison-Wesley Publishing Company, Reading, MA, 1974.

4 Baez, J. C., Bagdasaryan, K., and P. Gibbs, The Lebesgue Universal Covering Problem, arXiv:1502.01251v3 [math.MG] 2 Mar 2015.

5 Barbier, E., Note sur le problème de l'aiguille et le jeu joint couvert, *Journal de Mathématiques Pures et Appliquées, Series 2*, **5**, 273-286, 1860.

6 Bartle, R. G., *The Elements of Real Analysis*, 2nd ed., John Wiley & Sons, Inc., New York, 1976.

7 Bellman, R., *Introduction to Matrix Analysis*, 2nd ed., McGraw-Hill Book Company, New York, 1970.

8 Benson, R. V., *Euclidean Geometry and Convexity*, McGraw-Hill Book Company, New York, 1966.

9 Berger, M., Convexity, *American Mathematical Monthly*, **97**, 650-678, 1990.

10 Berkovitz, L. D., *Convexity and Optimization in $\mathbb{R}^n$*, John Wiley & Sons, Inc., New York, 2002.

11 Birkhoff, G., Tres observaciones sobre el algebra lineal, *Revista Universidad Nacional de Tucumán, Serie A*, 5, 147-151, 1946.

12 Blashke, W., *Kreis und Kugel*, Chelsea Publishing Company, New York, 1949.

13 Boltyanski, V. G., and I. T. Gohberg, *Results and Problems in Combinatorial Geometry*, Cambridge University Press, Cambridge, 1985.

14 Boltyanski, V., and A. Soifer, *Geometric Etudes in Combinatorial Mathematics*, Center for Excellence in Mathematical Education, Colorado Springs, CO, 1991.

15 Bonnesen, T., and W. Fenchel, *Theorie der konvexen Körper*, Chelsea Publishing Company, New York, 1948.

16 Borsuk, K., Über die Zerlegung einer Euklidischen n-dimensionalen Vollkugel in n Mengen, *Verh. Internat. Math. Kongr.*, Zurich, No. 2, 192, 1932.

17 Borsuk, K., Drei Sätze über die n-dimensionale euklidische Sphäre, Fundamenta Mathematicae, **20**, 177-190, 1933.

18 Borwein, J., A proof of the equivalence of Helly's and Krasnoselski's theorems, *Canadian Mathematical Bulletin*, 20, 1, 35-37, 1977.

19 Borwein, J. M., and A. S. Lewis, *Convex Analysis and Nonlinear Optimization: Theory and Examples*, Springer Science + Business Media, Inc., New York, 2006.

20 Botts, T., Convex sets, *American Mathematical Monthly*, **49**, 527-535, 1942.

21 Boyd, S., and L. Vandenberghe, *Convex Optimization*, Cambridge University Press, Cambridge, 2004.

22 Brown, A., and C. Pearcy, *An Introduction to Analysis, Graduate Texts in Mathematics*, vol. 154, Springer-Verlag New York Inc., New York, 1995.

23 Burkill, J. C., and H. Burkill, *A Second Course in Mathematical Analysis*, Cambridge University Press, Cambridge, 1970.

24 Cadwell, J. H., *Topics in Recreational Mathematics*, Cambridge University Press, Cambridge, 1970.

25 Chakerian, G. D., and H. Groemer, Convex bodies of constant width, *Convexity and its Applications*, Birkhäuser Verlag, Basel, 1983.

26 Chakerian, G. D., and L. H. Lange, Geometric extremum problems, *Mathematics Magazine*, **44**, 57-69.

27 Cook, W. D., and R. J. Webster, Carathéodory's Theorem, *Canadian Mathematical Bulletin*, **15**, 293, 1972.

28 Coxeter, H. S. M., *Regular Polytopes*, 2nd ed., Macmillan, New York, 1963.

29 Danzer, L. W., A characterization of the circle, American Mathematical Society Symposium on Convexity, *Proceedings of Symposia in Pure Mathematics*, **7**, 99-100, 1963.

30 Danzer, L. W., Grünbaum, B., and V. Klee, Helly's theorem and its relatives, American Mathematical Society Symposium on Convexity, *Proceedings of Symposia in Pure Mathematics*, **7**, 101-180, 1963.

31 DeSantis, R., A generalization of Helly's theorem, *Proceedings of the American Mathematical Society*, **8**, 336-340, 1957.

32 Dieudonné, J., *Foundations of Modern Analysis*, Academic Press, New York, 1960.

33 Dines, L. L., On convexity, *American Mathematical Monthly*, **45**, 199-209, 1938.

34 Eggleston, H. G., Covering a three-dimensional set with sets of smaller diameter, *Journal of the London Mathematical Society*, **30**, 1, 11-24, 1955.

35 Eggleston, H. G., *Problems in Euclidean Space: Applications of Convexity*, Pergamon Press, New York, 1957.

36 Eggleston, H. G., *Convexity, Cambridge Tracts in Mathematics and Mathematical Physics*, vol. 47, Cambridge University Press, Cambridge, 1958.

37 Fejes-Tóth, L., *Lagerungen in der Ebene auf der Kugel und im Raum*, Springer-Verlag, Berlin-Göttingen-Heidelberg, 1953.

38 Fejes-Tóth, L., *Regular Figures*, Pergamon Press, London, 1964.

39 Fenchel, W., Convexity through the ages, *Convexity and its Applications*, Birkhäuser Verlag, Basel, 1983.

40 Gale, D., On inscribing n-dimensional sets in a regular n-simplex, *Proceedings of the American Mathematical Society*, **4**, 222-225, 1953.

41 Gardner, M., Curves of constant width, one of which makes it possible to drill square holes, *Scientific American*, **208**, 2, 148-156, 1963.

42 Giles, J. R., *Convex Analysis with Application in the Differentiation of Convex Functions*, Pitman Advanced Publishing Program, Pitman Publishing Inc., Marshfield, MA, 1982.

43 Giles, J. R., *Introduction to the Analysis of Metric Spaces*, Australian *Mathematical Society Lecture Series*, vol. 3, Cambridge University Press, Cambridge, 1987.

44 Goldberg, M., Circular-arc rotors in regular polygons, *American Mathematical Monthly*, **55**, 392-402, 1948.

45 Graham, R. L., The largest small hexagon, *Journal of Combinatorial Theory, Series A*, **18**, 165-170, 1975.

46 Gruber, P. M., and J. M. Willis, *Convexity and its Applications*, Birkhäuser Verlag, Basel, 1983.

47 Gruber, P. M., and J. M. Willis, *Handbook of Convex Geometry*, North-Holland, Amsterdam, 1993.

48 Grünbaum, B., A simple proof of Borsuk's conjecture in three dimensions, *Proceedings of the Cambridge Philosophical Society*, **53**, 776-778, 1957.

49 Grünbaum, B., On common transversals, *Archiv der Mathematik*, **9**, 465-469, 1958.

50 Grünbaum, B., Borsuk's problem and related questions, American Mathematical Society Symposium on Convexity, *Proceedings of Symposia in Pure Mathematics*, **7**, 271-284, 1963.

51 Grünbaum, B., *Convex Polytopes*, Interscience Publishers, A Division of John Wiley & Sons, Ltd., London, 1967.

52 Hadwiger, H., Überdeckung einer Menge durch Mengen kleineren Durchmessers, *Commentarii Mathematici Helvetici*, 18, 73-75, 1945-1946.

53 Hadwiger, H., Überdeckung einer Menge durch Mengen kleineren Durchmessers, *Commentarii Mathematici Helvetici*, 19, 72-73, 1946-1947.

54 Hadwiger, H., *Altes und neues über konvexe Körper*, Birkhäuser, Basel-Stuttgart, 1955.

55 Hadwiger, H., Debrunner, H., and V. Klee, *Combinatorial Geometry in the Plane*, Holt, Rinehart and Winston, Inc., New York, 1964.

56 Hausner, M., *A Vector Space Approach to Geometry*, Dover Publications, Inc., Mineola, NY, 1998.

57 Helly, E., Über Mengen konvexer Körper mit gemeinschaftlichen Punkten, *Jahresbericht der Deutscher Mathematiker-Vereingigung*, **32**, 175-176, 1923.

58 Hilbert, D., and S. Cohn-vossen, *Geometry and the Imagination*, Chelsea Publishing Company, New York, 1952.

59 Hille, E., *Analysis*, vols. I and II, Blaisdell Publishing Co., Waltham, MA, 1964.

60 Hoffman, K., *Analysis in Euclidean Space*, Prentice-Hall, Inc., Englewood Cliffs, NJ, 1975.

61 Holmes, R. B., *Geometric Functional Analysis and its Applications*, Graduate *Texts in Mathematics*, vol. 24, Springer-Verlag New York Inc., New York, 1975.

62 Horn, A., Some generalizations of Helly's theorem on convex sets, *Bulletin of the American Mathematical Society*, **55**, 923-929, 1949.

63 Horn, R. A., and C. R. Johnson, *Matrix Analysis*, Cambridge University Press, Cambridge, 1985.

64 Jenrich, R., and A. E. Brouwer, A 64-dimensional counterexample to Borsuk's conjecture, *The Electronic Journal of Combinatorics*, 21, 4, P4.29, 2014.

65 Jung, H. W. E., Über die kleinste Kugel, die eine räumliche Figur einschliesst, *Journal für die Reine und Angewandte Mathematik*, **123**, 241-257, 1901.

66 Kahn, J., and G. Kalai, A counterexample to Borsuk's conjecture, *Bulletin of the American Mathematical Society*, **29**(1), 1993.

67 Katchalski, M., A Helly-type theorem on the sphere, *Proceedings of the American Mathematical Society*, **66**, 1, 119-122, 1977.

68 Katchalski, M., and T. Lewis, Cutting families of convex sets, *Proceedings of the American Mathematical Society*, **79**, 3, 457-461, 1980.

69 Kelly, P. J., and M. L. Weiss, *Geometry and Convexity, A Study in Mathematical Methods*, John Wiley & Sons, Inc., New York, 1979.

70 Kirchberger, P., Über Tschebyschefsche Annäherungsmethoden, *Mathematische Annalen*, **57**, 509-540, 1903.

71 Klamkin, M. S., The circumradius-inradius inequality for a simplex, *Mathematics Magazine*, **52**, 1, 20-22, 1979.

72 Klee, V. L., A characterization of convex sets, *American Mathematical Monthly*, **56**, 247-249, 1949.

73 Klee, V. L., Convex sets in linear spaces. *Duke Mathematical Journal*, **18**, 2, 443-466, 1951.

74 Klee, V. L., On certain intersection properties of convex sets, *Canadian Journal of Mathematics*, **3**, 272-275, 1951.

75 Klee, V. L., The critical set of a convex body, *American Journal of Mathematics*, **75**, 178-188, 1953.

76 Klee, V. L., Common secants for plane convex sets, *Proceedings of the American Mathematical Society*, **5**, 639-641, 1954.

77 Klee, V. L., Convexity on Chebychev sets, *Mathematische Annalen*, **142**, 292-304, 1961.

78 Klee, V. L., What is a convex set? *American Mathematical Monthly*, **78**, 616-631, 1971.

79 Klee, V. L., Some unsolved problems in plane geometry, *Mathematics Magazine*, **52**, (3), 131-145, 1979.

80 Klee, V. L., and S. Wagon, *Old and New Unsolved Problems in Plane Geometry and Number Theory, Dolciani Mathematical Expositions*, vol. 11, The Mathematical Association of America, Washington, DC, 1991.

81 Kolmogorov, A. N., and S. V. Fomin, *Elements of the Theory of Functions and Functional Analysis*, Vols 1 and 2, Graylock Press, Albany, NY, 1961.

82 Kolmogorov, A. N., and S. V. Fomin, *Introductory Real Analysis*, Dover Publications, Inc., 1975.

83 Krasnosselsky, M. A., Sur un critère pour qu'un domain soit étoilé (Russian with French summary), *Matematicheskii Sbornik, Novaja Serija*, **19**, 61, 309-310, 1946.

84 Krein, M., and D. Milman, On the extreme points of regularly convex sets, *Studia Mathematica*, **9**, 133-138, 1940.

85 Lax, P. D., *Linear Algebra and its Applications*, John Wiley &Sons, Inc., Hoboken, NJ, 2007.

86 Lay, S. R., *Convex Sets and their Applications*, John Wiley & Sons, Inc., New York, 1982.

87 Lyusternik, L. A., *Convex Figures and Polyhedra*, Dover Publications, Inc., Mineola, NY, 1963.

88 Lyusternik, L. A., and V. J. Sobolev, *Elements of Functional Analysis*, Frederick Ungar Publishing Company, New York, 1961.

89 Marsden. J. E., *Elementary Classical Analysis*, W. H. Freeman and Company, San Francisco, CA, 1974.

90 Marti, J. T., *Konvexe Analysis*, Birkhäuser Verlag, Basel, 1977.

91 McKinney, R. L., On unions of two convex sets, *Canadian Journal of Mathematics*, **18**, 883-886, 1966.

92 Melzak, Z.A., *Invitation to Geometry*, John Wiley & Sons, Inc., New York, 1983.

93 Melzak, Z.A., *Companion to Concrete Mathematics*, vols. I and II, Dover Publications, Inc., Mineola, NY, 2007.

94 Moser, W., *Problems in Discrete Geometry*, 5th ed., Department of Mathematics, McGill University, Montréal, Québec, 1980.

95 Pál, J., Über ein elementares Variationsproblem, *Danske Vid. Selsk. Math.-Fys. Medd*, **3**(2), 1920.

96 Pál, J. F., Ein Minimumproblem für Ovale, *Mathematische Annalen*, **83**, 311-319, 1921.

97 Peressini, A. L., Sullivan, F. E., and J. J. Uhl, Jr., *The Mathematics of Nonlinear Programming*, Undergraduate Texts in Mathematics, Springer-Verlag, New York, 1988.

98 Phelps, R. R., *Lectures on Choquet's Theorem*, D. Van Nostrand Company, Inc., Princeton, NJ, 1966.

99 Rademacher, H., and I. J. Schoenberg, Helly's theorem on convex domains and Tchebycheff's approximation theorem, *Canadian Journal of Mathematics*, **2**, 245-256, 1950.

100 Rademacher, H., and O. Toeplitz, *The Enjoyment of Mathematics: Selections from Mathematics for the Amateur*, Princeton University Press, Princeton, NJ, 1970.

101 Radon, J., Mengen konvexer Körper, die einen gemeinsamen Punkt enthalten, *Mathematische Annalen*, **83**, 113-115, 1921.

102 Randolph, J. R., *Basic Real and Abstract Analysis*, Academic Press, Inc., New York, 1968.

103 Reuleaux, F., *Lehrbuch der Kinematik*, Braunschweig, Germany, 1875.

104 Roberts, A. W., and D. E. Varberg, *Convex Functions*, Academic Press, New York, 1973.

105 Rockafellar, R. T., *Convex Analysis*, Princeton University Press, Princeton, NJ, 1970.

106 Rudin, W., *Principles of Mathematical Analysis*, 3rd ed., McGraw-Hill Book Company, New York, 1976.

107 Sakuma, I., Closedness of convex hulls, *Journal of Economic Theory*, **14**, 223-227, 1977.

108 Santalo, L. A., Un teorema sôbre conjuntos de paralelipipedos de aristas paralelas, *Publ. Inst. Mat. Univ. Nac. Litoral (Rosario)*, 2, 49-60, 1940.

109 Santalo, L. A., Complemento a la Nota: un teorema sôbre conjuntos de paralelipipedos de aristas paralelas, *Publ. Inst. Mat. Univ. Nac. Litoral (Rosario)*, vol. 3, 202-210, 1942.

110 Schoenberg, I. J., *Mathematical Time Exposures*, Mathematical Association of America, Washington, DC, 1982.

111 Shimrat, M., Simple proof of a theorem of P. Kirchberger, *Pacific Journal of Mathematics*, **5**, 361-362, 1955.

112 Spivak, M., *Calculus*, 2nd ed., Publish or Perish, Inc., Houston, TX, 1980.

113 Stamey, W. L., and J. M. Marr, Union of two convex sets, *Canadian Journal of Mathematics*, **15**, 152-156, 1963.

114 Steinhaus, H., *Mathematical Snapshots*, Oxford University Press, Oxford, 1983.

115 Stromberg, K. R., *An Introduction to Classical Real Analysis*, Wadsworth, Inc., Belmont, CA, 1981.

116 Taylor, A. E., and D. C. Lay, *Introduction to Functional Analysis*, John Wiley & Sons Canada, Ltd., Rexdale, ON, 1980.

117 Valentine, F. A., *Convex Sets*, McGraw-Hill Book Company, New York, 1964.

118 Watson, D., A refinement of theorems of Kirchberger and Carathéodory, *Jounal of the Australian Mathematical Society*, **15**, 190-192, 1973.

119 Webster, R., *Convexity*, Oxford University Press, Oxford, 1994.

120 Wilansky, A., *Functional Analysis*, Blaisdell Publishing Company, New York, 1964.

121 Wilansky, A., *Topology for Analysis*, Ginn and Company, Waltham, MA, 1970.

122 Wilansky, A., *Modern Methods in Topological Vector Spaces*, McGraw-Hill, Inc., New York, 1978.

123 Yaglom, I.M., *Geometric Transformations I*, Mathematical Association of America, Washington, DC, 1962.

124 Yaglom, I.M., *Geometric Transformations II*, Mathematical Association of America, Washington, DC, 1968.

125 Yaglom, I.M., *Geometric Transformations III*, Mathematical Association of America, Washington, DC, 1973.

126 Yaglom, I.M., *Geometric Transformations IV*, Mathematical Association of America, Washington, DC, 2009.

127 Yaglom, I.M., and V. G. Boltyanski, *Convex Figures*, Holt, Rinehart and Winston, Inc., New York, 1961.

128 Yoshida, K., *Functional Analysis*, Springer-Verlag New York Inc., New York, 1968.

Index

Geometry of Convex Sets, First Edition. I. E. Leonard and J. E. Lewis.
© 2016 John Wiley & Sons, Inc. Published 2016 by John Wiley & Sons, Inc.

Printed and bound by CPI Group (UK) Ltd, Croydon, CR0 4YY

27/10/2024

14580277-0002